Sound Insulation in Buildings

Sound Insulation in Buildings

Jens Holger Rindel

CRC Press
Taylor & Francis Group
Boca Raton London New York

CRC Press is an imprint of the
Taylor & Francis Group, an **informa** business

Cover photo by Iiris Turunen-Rindel. The photo on the cover shows a concrete building from the 1960s being demolished.

CRC Press
Taylor & Francis Group
6000 Broken Sound Parkway NW, Suite 300
Boca Raton, FL 33487-2742

First issued in paperback 2020

© 2018 by Taylor & Francis Group, LLC
CRC Press is an imprint of Taylor & Francis Group, an Informa business

No claim to original U.S. Government works

ISBN 13: 978-0-367-87136-9 (pbk)
ISBN 13: 978-1-4987-0041-2 (hbk)

Visit the Taylor & Francis Web site at
http://www.taylorandfrancis.com

and the CRC Press Web site at
http://www.crcpress.com

Library of Congress Cataloging-in-Publication Data

Names: Rindel, J. H. (Jens Holger), author.
Title: Sound insulation in buildings / by Jens Holger Rindel.
Description: Boca Raton : CRC Press, [2018] | Includes bibliographical references and index.
Identifiers: LCCN 2017024121| ISBN 9781498700412 (hardback : alk. paper)
ISBN 9781498700429 (ebook : alk. paper)
Subjects: LCSH: Soundproofing. | Architectural acoustics.
Classification: LCC TH1725 .R56 2018 | DDC 693.8/34--dc23
LC record available at https://lccn.loc.gov/2017024121

Contents

6 Sound radiation from plates   155

7 Statistical energy analysis, SEA   189

10 Impact sound insulation   275

11 Flanking transmission   313

Preface

This book on sound insulation is based on my lecture notes developed over a long span of time, beginning with a note on sound insulation from 1975. After a few years of teaching basic acoustics, I wanted to establish a more advanced course on architectural acoustics, meant for those students from the building engineering faculty who wanted to study more in-depth about acoustics and might consider doing their masters in acoustics. The course included room acoustics as well as sound insulation. Hence, from 1977 to 1982, a series of lecture notes were written, covering both fields.

I am indebted to late Professor Fritz Ingerslev (1912–1994), who was my teacher and mentor in the first years of my career. Quite early in my studies at the Technical University of Denmark (DTU), I discovered his textbook on building acoustics, and there is no doubt that this book was the reason that I decided to turn my education in the direction of house building and to specialize in acoustics. Professor Ingerslev suggested the topic of my master's thesis – and later also the topic of my PhD thesis – to be related to the sound insulation of windows. I was very fortunate that the acoustical laboratories at DTU had an incredibly good library with complete collections of all relevant acoustical journals and a rich collection of textbooks and research reports. From around 1980, I was guided by another mentor, Jørgen Kristensen (1925–1992), who was head of the building acoustic section at the Danish Building Research Institute (SBI), today a part of Aalborg University, Copenhagen. He invited me to coauthor a book on building acoustics, which was published in 1989. He also suggested that I should take over his role as acoustical expert for the Danish Ministry of Housing, which in turn led to my work for the Nordic Committee for Building Regulations in 1993–1996.

My research has always been divided between building acoustics and room acoustics. From around 1985, the development of room acoustic calculation models was in focus and the software ODEON was launched in 1991. The following years were devoted to research projects, some of them in buildings acoustics, but most of them related to room acoustics, musical instruments and auralisation. In the 1990s, I also became involved with standardisation and new measurement methods.

In 2008, I left the university. I started to work as a consultant for the Norwegian company, Multiconsult and continued my work with the Danish company, ODEON. After a while, I realised that some of my old lecture notes might still be relevant in consulting and in 2014, I decided to collect my former lecture notes on sound insulation and make a book of it. Fortunately, Tony Moore, senior editor at CRC Press (imprint of Taylor and Francis), was supportive of the idea. However, it was necessary to bring everything up-to-date in accordance with the latest research results and current international standards. Concerning references, I feel it important to give credit to those who originally came up with ideas and contributed to the field, and thus, the reader will find many references to old papers and books. Science is not only about developing new knowledge but it is also important to know the history behind the theories and methods and be able to trace back to the origins, when needed.

The first five chapters of the book deal with fundamental matters, mechanical vibrations, sound reflections and basic room acoustics. Chapter 5 describes sound insulation in accordance with the generally known classical theory. This first part of the book is more or less, what I would teach in a general course on acoustics for building engineers. Chapters 6 through 15 are meant for the more advanced student, who wants a deeper understanding of the physics behind sound transmission and measurement methods. In recent years, there has been a growing focus on measurement uncertainty in building acoustics, and especially at low frequencies. Chapter 13 explains the fundamentals of measurement uncertainty, with some emphasis on the spatial variance of the sound pressure in a room. For some years, there has been a debate on the importance of the low frequencies between 50 Hz and 100 Hz in relation to sound insulation. The origin of this is the complaints from inhabitants in multi-unit houses concerning insufficient sound insulation, especially against noise from traffic and neighbours. Thus, Chapter 14 deals with annoyance and other subjective aspects of sound insulation, explaining the statistical methods used in socio-acoustic investigations, and presenting some of the findings. The final chapter, Chapter 15, provides a background for discussions and ideas on how to develop building technology, so that future houses can be built with satisfactory sound insulation.

I am very grateful to Multiconsult for supporting me in my work with this book by offering an office, where I could concentrate on writing, surrounded by my books and papers. I have enjoyed the support of all my colleagues in the acoustical section of the company. In particular, I thank the head of the acoustics section, Ingunn Milford for her enthusiasm and support. Dr Arild Brekke has provided very important comments and suggestions to improve Chapter 2 on vibrations. Special thanks go to Dr Anders Løvstad, who has read the entire manuscript and provided invaluable feedback and comments.

Oslo, August 2017
Jens Holger Rindel

About the Author

Jens Holger Rindel is currently a senior consultant at Multiconsult in Norway and senior researcher at Odeon A/S in Denmark. He holds a MSc in civil engineering (house building) and a PhD in acoustics, both from the Technical University of Denmark. Until 2008, he was professor in acoustics at the Technical University of Denmark, Copenhagen, Denmark. He is also past president of the Nordic Acoustical Association and past chairman of the Technical Committee for Building and Room Acoustics of the European Association of Acoustical Societies. He is a fellow of the Institute of Acoustics and of the Acoustical Society of America. He has been guest professor at universities in Sydney, Auckland and Tokyo and worked as guest researcher at research institutions in Braunschweig, Germany and Oslo, Norway.

Introduction

The science of architectural acoustics is divided into room acoustics and building acoustics. The origin of room acoustics as a scientific discipline is closely connected to the pioneering work done by Wallace C. Sabine (1868–1919) around 1896–1900. A theoretical model and a calculation method for reverberation time were developed together with the first measurement method. Carefully tuned organ pipes were used as sound sources and the human ear as measuring instrument.

The beginning of building acoustics was quite different. Although there was increasing awareness of the importance of sound insulation in the beginning of the twentieth century, the theoretical models and calculation methods available then were very simple and insufficient. The importance of mass for the sound insulation of massive single constructions was known since the work of Berger (1911). The need for completely separate wall constructions for high sound insulation was also a known principle, which was used in special cases, such as theatres and concert halls. Sabine (1915) described three cases where he had tried to design for good sound insulation but, to his frustration, the results did not meet his expectations. With a touch of bitterness, he says "It is always easier to explain why a method does not work than to know in advance whether it will or not. It is especially easy to explain why it does not work when not under the immediate necessity of correcting it or of supplying a better." This was in direct response to the unsuccessful attempts at insulation for the New England Conservatory of Music.

Airborne sound insulation was measured with organ pipes as an extension of the method used for reverberation time. For laboratory measurements, a test wall was built between a large reverberant room and a smaller receiving room with high acoustic damping. The operator used a stopwatch to measure the decay time from when the source was stopped until the time that the sound was inaudible. When measured in the receiving room, the sound intensity was lower and thus, the audible decay time was much shorter than in the source room, and the transmission coefficient of the test wall could be calculated from the difference in decay times. The reverberation method for sound insulation is described in detail by Paul E. Sabine

(1930), who succeeded Wallace C. Sabine as head of the acoustical test laboratories at Riverbank (Kopec, 1997). In 1930, the organ pipes had been replaced by loudspeakers and 17 test tones were used from 128 Hz to 4096 Hz.

Barkhausen (1926) developed an instrument for measuring the loudness level in phon. The operator listened to the noise with one ear and adjusted the level of a 1000 Hz test sound heard in the other ear, to get equal loudness. This made it possible to do sound insulation measurements with a stationary noise source, and several mechanical devices were put to use in the following years. Dr H. Reier from the technical university in Stuttgart used a very noisy rattle machine with a frequency range from 100 Hz to 2000 Hz. He was asked to do measurements in a very complicated case in Copenhagen, where new broadcasting studios had been built on top of a theatre using separated constructions. The measurement report by Reier (1931) is reproduced in the book by Oelsner (1935). The measured result was a very high airborne sound insulation of 83 phons. The unit was phon since the sound insulation was measured as the difference between the loudness levels on either side of the construction.

Vern O. Knudsen (1930) describes three measurement methods for airborne sound insulation, which he applied in the newly built acoustic test facilities at the University of California. One of them is Sabine's reverberation method with loudspeakers as sound sources. The second method is the insertion loss, i.e., a measurement with and without the test panel. The third method uses an electric attenuator to adjust the level of the loudspeaker in the source room until the sound is inaudible. The reported results of the three methods are in surprisingly good agreement (within about 1 dB) with each other. A few years later, a new test method was developed by Meyer and Keidel (1935) using a logarithmic attenuator and a voltage meter to draw a curve of the sound pressure level as a function of frequency. For the first time, it was possible to measure sound insulation without using the human ear as a measuring instrument.

Another issue that would need attention was the impact sound from floors, especially footfall noise. An early method of measuring impact sound used the dropping of a steel ball from different heights above the floor; the height of fall that made the sound just audible in the receiving room was taken as a measure of the impact sound. Later, the tapping machine, as it is known today, was developed and the impact sound level of the tapping noise was measured as a loudness level in phon. The comprehensive book by Schoch (1937) presents the state of the art in building acoustics at the time. Although the measuring methods for airborne and impact sound insulation had been developed, the physical and technical foundation for understanding sound insulation problems was still very poor. It was restricted to the mass law and the advantage of separated structures.

A scientific approach to the physics of sound insulation began in 1942, when Lothar Cremer (1905–1990) explained the importance of bending waves and introduced the concept of coincidence in sound transmission in his pioneering article (Cremer, 1942). Albert London (1950) was the first to present a theoretical model for sound transmission through double walls. Today, there is no doubt that bending waves, the coincidence effect and the critical frequency are key points in understanding sound insulation. However, the knowledge of Cremer's findings spread very slowly and did not reach the majority of those working with building acoustics until several decades later. Looking into the textbooks by Brüel (1946), Ingerslev (1949) and Knudsen and Harris (1950), the coincidence effect is not even mentioned. The book by Ingerslev was in use at the Technical University of Denmark until 1971, and not until then, would the teaching in building acoustics include models of sound transmission that were more advanced than the mass law. The book by Knudsen and Harris was reprinted in 1978, because it was still in use in education in many countries.

From around 1930, building technology changed dramatically. The introduction of concrete slabs with thickness of only 80 mm to 100 mm caused severe noise problems in houses. It became very clear that dimensions determined solely from the point of view of structural engineering were not sufficient from an acoustical point of view. Several countries introduced minimum requirements for sound insulation between dwellings in the 1940s or 1950s, and these first requirements were typically in terms of specific acceptable constructions, e.g. a 250 mm full-brick wall with plaster on both sides. The industrialised building technology, which started in the 1960s, and the extensive use of reinforced concrete led to buildings that were uniform and relatively cheap, but not particularly satisfactory for the inhabitants. These houses became synonymous with noise problems and still today, concrete as a building material is often believed by the layman to cause poor sound insulation. The introduction of building codes with minimum requirements for sound insulation had some negative effects; building apartment houses was a matter of money investment, and thus, it was important to keep the building costs low. The result was that any sound insulation better than the minimum requirement was considered a waste of money. This led to houses with a sound insulation that was just acceptable in those days, but is clearly insufficient today, about 50 years later. Since it is always difficult and expensive to improve the sound insulation in an existing house, these houses from the 1960s are often pulled down and replaced by completely new buildings.

New building technologies, new powerful sound sources in homes (hi-fi loudspeakers, home cinemas with subwoofers, etc.) and increased awareness of noise in the society have all contributed to the development of better sound insulation design. An important part of this development is the acousticians' understanding of the physical principles in sound insulation. Another important part of the development is the knowledge about how individuals react to noise from neighbours, and how the sound insulation of a building is related to the degree of satisfaction that can be expected among the inhabitants. In recent years, this has led to raised minimum requirements for sound insulation in many countries. The idea of a sound classification system for houses, with specified sound classes better than the minimum requirements, has aroused interest, and an international standard for sound classification of houses is under preparation (ISO/DIS 19488, 2017). Projects with experimental houses that have shown new directions for building technology with better sound insulation are very important. Much more can be done along this line in the future, and it is important to spread the information to the public that good sound insulation in houses is possible. A major problem may be to have a general acceptance among architects and in the building industry that sound insulation is a design parameter that must be taken seriously.

REFERENCES

H. Barkhausen (1926). Ein neuer Schallmesser für die Praxis (A new sound measuring device for practical application, in German). *Zeitschrift für technische Physik 7*, 599–601.

R. Berger (1911). *Über die Schalldurchlässigkeit.* (On the Sound Transmission, in German). Dissertation, Munich, Germany.

P.V. Brüel (1946). *Lydisolation og rumkustik* (Sound insulation and room acoustics, in Danish). Jul. Gjellerups Forlag, Copenhagen.

L. Cremer (1942): Theorie der Schalldämmung dünner Wände bei schrägem Einfall. (Theory of sound insulation of thin walls at oblique incidence, in German). *Akustische Zeitschrift 7*, 81–104.

F. Ingerslev (1949). *Akustik. Lærebog i bygningsakustik for ingeniører* (Acoustics. Textbook on building acoustics for engineers, in Danish). Teknisk Forlag, A-S Dansk Ingeniørforenings Forlag, Copenhagen.

ISO/DIS 19488 (2017). *Acoustics—Acoustic classification of dwellings.* (Under development.) International Organization for Standardization, Geneva, Switzerland.

V.O. Knudsen (1930). Measurement and calculation of sound-insulation. *Journal of the Acoustical Society of America 2*, 129–140.

V.O. Knudsen and C.M. Harris (1950). *Acoustical Designing in Architecture.* John Wiley & Sons, Inc., New York. Reprint 1978 by the Acoustical Society of America, New York.

J.W. Kopec (1997). *The Sabines at Riverbank. Their Role in the Science of Architectural Acoustics.* Acoustical Society of America, New York.

A. London (1950). Transmission of reverberant sound through double walls. *Journal of the Acoustical Society of America 22*, 270–279.

E. Meyer and L. Keidel (1935). Röhrenvoltmeter mit logarithmischer Anzeige und seine Anwendungen in der Akustik, (Vacuum-tube voltmeter with logarithmic display and its applications in acoustics, in German). *Elektrische Nachrichtentechnik* (ENT) 12(2), 37–46.

W. Oelsner (1935). *Det teknisk-videnskabelige Grundlag for Rumakustik og Lydisolation* (The technical-scientific foundation of room acoustics and sound insulation, in Danish). Hjorths Tryk, Copenhagen.

P.E. Sabine (1930). The transmission of sound by walls. *Journal of the Acoustical Society of America* 1, 181–201.

W.C. Sabine (1915). The insulation of sound. *The Brickbuilder* 24(2). Paper No. 10 in *Collected Papers on Acoustics*. Harvard University Press, 1922. Published 1964 by Dover Publications Inc., New York.

A. Schoch (1937). *Die physikalischen und technischen Grundlagen der Schalldämmung im Bauwesen* (The physical and technical fundamentals of sound insulation in buildings, in German). S. Hirzel Verlag, Leipzig.

Chapter 1

Basic concepts in acoustics

This chapter summarizes the most important basic concepts used in building acoustics. The definitions and symbols are in accordance with International Organization for Standardization (ISO) technical report (TR) 25417.

1.1 INTRODUCTION

The sound insulation of walls and floors in a building is often of major concern for the people using the building. It is of special interest in residential buildings, but also in, e.g. offices, hospitals and school buildings. Sound insulation in buildings has been under research and standardization for a long time – many contributions were made in the 1950s. However, since that time, the behaviour and habits of people, as well as their expectations, have changed. Noise sources have become louder and cover a broader frequency spectrum, and it is possible that the demands of the tenants have increased. At the same time, there are a gradual development and change in building elements and methods. Thus, knowledge of physical and other aspects of sound insulation in buildings is essential for an acoustician. Moreover, basic knowledge of sound insulation is also important in other branches of acoustics, e.g. in the development of vehicles, ships and aeroplanes.

Sound insulation is divided into airborne sound insulation, where the noise source is in the air inside or outside the building, and structure-borne sound insulation, where the noise source is located at a building structure. The most important case of structure-borne sound insulation is the impact noise sound insulation, which includes noise generated by people walking. Chapter 5 gives a basic overview of the principles of sound insulation. Chapters 8 and 9 deal more thoroughly with airborne sound insulation, and Chapter 10 addresses the impact noise.

Airborne noise sources in a building are typically people talking, sound from stereo equipment, televisions and musical instruments. The development of hi-fi stereo equipment means that these sources are much louder and cover more of the low frequencies than was the case when the building regulations for sound insulations were developed. It is thus not given that a regulation originally intended to avoid noise discomfort still does so. It is therefore most often recommended to use a broader frequency spectrum (50 Hz to 5000 Hz) than is used in the original measurement description (100 Hz to 3150 Hz), and sometimes it is also important to perform a subjective evaluation of the noise situation in more experimental buildings. Examples of external airborne noise sources are traffic and industrial noise, where the windows of a building are of major interest.

1.2   THE DECIBEL SCALE

Sound is an audible vibration, either in the air or in liquids or solid materials. The human ear is sensible to the pressure variations in a sound wave, and since the audible dynamic range is immense, the logarithmic decibel scale is used. The sound pressure level L_p is

$$L_p = 10 \cdot \lg \frac{p^2}{p_{ref}^2} \quad (\text{dB}) \tag{1.1}$$

where $p_{ref} = 20$ µPa is the reference sound pressure. The sound pressure is the root-mean-square (RMS) value of the time-varying sound pressure. For harmonic vibrations, $p^2 = 1/2 \ |p|^2$, where $|p|$ is the amplitude of the sound pressure. In rough terms, the range of audible sound is 0 dB to 120 dB.

A sound source is characterized by its emitted sound power, which can also be expressed in decibels. If P is the sound power in watts, the sound power level is

$$L_W = 10 \cdot \lg \frac{P}{P_{ref}} \quad (\text{dB}) \tag{1.2}$$

Table 1.1 Definition of levels in decibels

Quantity	Symbol	Reference	Level (dB)
Sound power	P	1 pW	$10\lg\left(\dfrac{\text{Quantity}}{\text{Reference}}\right)$
Sound intensity	I	1 pW/m²	
Sound pressure (in air)	p	20 µPa	$10\lg\left(\dfrac{\text{RMS quantity}}{\text{Reference}}\right)^2$
Force	F	1 µN	
Acceleration	a	1 µm/s²	$= 20\lg\left(\dfrac{\text{RMS quantity}}{\text{Reference}}\right)$
Velocity	v	1 nm/s	
Displacement	x	1 pm	

Note the use of prefixes: $\mu = 10^{-6}$, $n = 10^{-9}$, $p = 10^{-12}$.

where $P_{ref} = 10^{-12}$ W is the reference sound power. The relation between sound power level and sound pressure level is addressed in Chapter 4.

The decibel scale is also used for many other acoustic parameters, e.g. vibratory acceleration, vibratory velocity and sound intensity (see Table 1.1). The reference values are defined in International Organization for Standardization (ISO) 1683.

1.3 FREQUENCY AND RELATED CONCEPTS

1.3.1 Harmonic vibrations

If the velocity of particle motions is small compared to the speed of sound propagation in the medium, the vibrations can be considered *harmonic vibrations*. Concepts such as sound pressure, velocity, acceleration, and force can be described mathematically by either real or complex numbers as

$$y = |y| \cos(\omega t + \varphi) \tag{1.3}$$

or in the complex notation as

$$y = |y| e^{j(\omega t + \varphi)} = |y| \left(\cos(\omega t + \varphi) + j \sin(\omega t + \varphi) \right) \tag{1.4}$$

In the latter case, y is a complex number that can be considered in two different ways: either having a size (the modulus |y|) and an angle $(\omega t + \varphi)$ in the complex plane or having a real part (the cosine term) and an imaginary part (the sine term). Whatever notation is used, the physical vibration that can be observed is just the cosine part. However, the complex notation has some mathematical advantages.

The imaginary unit $j = \sqrt{-1}$ may also be written as $j = 1 e^{j(\pi/2)}$, i.e. as the unit vector turned perpendicular to the real axis.

In the aforementioned expressions, t is the time, φ is a phase angle and ω is the *angular frequency*:

$$\omega = 2\pi f = \frac{2\pi}{T_1} \quad \text{(rad/s)} \tag{1.5}$$

where f is the *frequency* (Hz) and T_1 is the *period duration* (s).

A wave propagating in the direction of the x-axis can be described mathematically by

$$y = |y| e^{j(\omega t - kx)} = |y| e^{j\omega t} e^{-jkx} \tag{1.6}$$

The factor $e^{j\omega t}$ is called the *time factor,* and it is often omitted in theoretical derivations if it does not directly affect the result or calculations. Sometimes, the time factor $e^{-i\omega t}$ is used instead, which is equally fine; but it is very important to be consistent and use the same time factor throughout any derivations or calculations.

The place factor e^{-jkx} describes the propagation in the positive direction of the x-axis, and it contains the *angular wave number k,* which is

$$k = \frac{2\pi}{\lambda} = \frac{\omega}{c} \quad (\text{m}^{-1}) \tag{1.7}$$

where λ is the wavelength (m) and c is the speed of sound (m/s).

1.3.2 Speed of sound

The speed of sound is the distance travelled by sound per unit time. It is found as the distance travelled during one period (one wavelength λ) multiplied by the number of periods per second (the frequency f):

$$c = \lambda f \quad (\text{m/s}) \tag{1.8}$$

In air and other gases or fluids, the speed of sound is a function of the *bulk modulus K_s* (Pa) and the *density ρ* (kg/m³) of the media it travels in

$$c = \sqrt{\frac{K_s}{\rho}} \quad (\text{m/s}) \tag{1.9}$$

In indoor air with a temperature of ~ 20 °C, the density is $\rho \approx 1.205\,\text{kg/m}^3$ and the speed of sound is $c \approx 344\,\text{m/s}$.

1.3.3 Frequency bands

Acoustic measurements are usually made in frequency bands. The *bandwidth Δf* is the difference between the upper and lower limiting frequencies:

$$\Delta f = f_2 - f_1 \quad (\text{Hz}) \tag{1.10}$$

The *centre frequency* is the geometrical mean of the limiting frequencies:

$$f_{\text{center}} = \sqrt{f_1 f_2} \quad (\text{Hz}) \tag{1.11}$$

The *relative bandwidth* is the ratio between the bandwidth and the centre frequency:

$$\frac{\Delta f}{f_{center}} = \sqrt{\frac{f_2}{f_1}} - \sqrt{\frac{f_1}{f_2}} \tag{1.12}$$

In building acoustics, the most often used frequency bands are *one-third octave bands*:

$$\frac{f_2}{f_1} = 10^{1/10} \cong \sqrt[3]{2} \quad \Delta f \cong 0.23 f_{center} \tag{1.13}$$

Some times *octave bands* are used in stead:

$$\frac{f_2}{f_1} = 10^{3/10} \cong 2 \quad \Delta f \cong 0.71 f_{center} \tag{1.14}$$

In music, the octave is defined as a frequency interval where the upper frequency is exactly two times the lower frequency. (The name comes from the Latin *Octavus*, meaning 'eight'.) Playing an ascending scale on the white keys of a keyboard, the octave is step number 8. Using both black and white keys, the octave is divided into 12 semitones. Thus, the one-third octave is the interval of four semitones, which is called a *major third* in music. (In German literature "Terz" is often used for one-third octave.) Using just intonation, the major third has the frequency interval$=5/4=1.25$. In the modern equal temperament, the semitones all have equal interval$=2^{1/12}$, and the major third is $2^{1/3} \approx 1.259921$. For comparison, the exact interval of the one-third octave is $10^{1/10} \approx 1.258925$.

When applied to filters for acoustic measurements, ISO 266 defines the exact and nominal center frequencies to be periodic within a decade, i.e. the one-third octave bands are, strictly speaking, one-tenth decade bands (see Table 1.2). If n denotes the band number, the exact center frequency is calculated from the following equation:

$$f_{center} = 10^{n/10} \tag{1.15}$$

The standard range for the one-third octave center frequencies is 100 Hz to 3150 Hz, although an extended frequency range of 50 Hz to 5000 Hz is often preferred. Sometimes, even lower frequencies are used for impact sound, and for vibrations, the standard range is 1 Hz to 80 Hz.

Table 1.2 Exact and nominal centre frequencies in hertz of the one-third octave bands often used in building acoustics

Band no.	Exact	Nominal	Band no.	Exact	Nominal
14	25.12	25	26	398.11	400
15	31.62	31.5	27	501.19	500
16	39.81	40	28	630.96	630
17	50.12	50	29	794.33	800
18	63.10	63	30	1000.00	1000
19	79.43	80	31	1258.93	1250
20	100.00	100	32	1584.89	1600
21	125.89	125	33	1995.26	2000
22	158.49	160	34	2511.89	2500
23	199.53	200	35	3162.28	3150
24	251.19	250	36	3981.07	4000
25	316.23	315	37	5011.87	5000

1.4   LOUDNESS LEVEL AND A-WEIGHTED SOUND PRESSURE LEVEL

The sensation of sound is very frequency dependent. The subjective loudness of a low-frequency sound (e.g. below 200 Hz) is much lower than a sound with the same physical sound pressure level in the frequency range of 1 kHz to 5 kHz. This is displayed in Figure 1.1 as equal-loudness contours, i.e. curves showing combinations of sound pressure level and frequency that give the same subjective loudness. The *loudness level* of a sound is defined as the sound pressure level of a 1000 Hz tone that has the same loudness as the sound. The unit of the loudness level is *phon*.

In building acoustics, it is usual that sound is measured in one-third octave bands because transmission properties are very frequency dependent. However, it is practical to express the sound level as a single-number value, and the commonly used value is the A-weighted sound pressure level. A-weighting applies a frequency-dependent attenuation, which is in the A-weighting filter in a sound level meter (IEC 61672-1). The A-weighted sound pressure level can also be calculated from measured sound pressure levels in one-third octave bands $L_{p,i}$ using the following equation:

$$L_A = 10 \lg \sum_i 10^{0,1(L_{p,i} + L_{corr,i})} \quad \text{(dB)} \qquad (1.16)$$

where $L_{corr,i}$ is the correction at frequency band i (Table 1.3). Actually, the A-weighting filter is a rough approximation of the 40 phon curve.

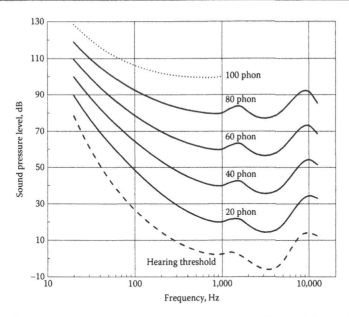

Figure 1.1 Equal loudness contours and hearing threshold. (Adapted from ISO 226, *Acoustics – Normal Equal-Loudness-Level Contours*, International Organization for Standardization, Geneva, Switzerland, 2003.)

Table 1.3 A-weighting corrections in one-third octave bands

Frequency (Hz)	A-weighting (dB)	Frequency (Hz)	A-weighting (dB)
10	−70.4	500	−3.2
12.5	−63.4	630	−1.9
16	−56.7	800	−0.8
20	−50.5	1000	0
25	−44.7	1250	0.6
31.5	−39.4	1600	1
40	−34.6	2000	1.2
50	−30.2	2500	1.3
63	−26.2	3150	1.2
80	−22.5	4000	1
100	−19.1	5000	0.5
125	−16.1	6300	−0.1
160	−13.4	8000	−1.1
200	−10.9	10000	−2.5
250	−8.6	12500	−4.3
315	−6.6	16000	−6.6
400	−4.8	20000	−9.3

Source: IEC 61672-1, Electro acoustics – Sound level meters – part 1: specifications. International Electrotechnical Commission, Geneva, Switzerland, 2002.

1.5 PROPAGATION OF SOUND IN PLATES

In order to understand the physics behind sound insulation, it is necessary to have some basic knowledge about structure-borne sound, especially the wave phenomena in plates. A single wall is basically a plate.

Solid materials differ from fluids (such as air and water) in that they can resist shear forces. This implies that apart from the longitudinal wave (that is present in fluid), there will also be shear waves as well as combinations of these two basic wave types (Figure 1.2). Combined waves are typical for plates, rods and beams. The bending wave is the most important combined wave type.

The two basic wave types are generally uncoupled in extended materials and therefore independent of each other. At a boundary, there will however be coupling and transformation from one wave type to the other. This is the reason for the combined waves in plates, rods and beams.

The basic wave types are both controlled by the wave equation; in the case of a plane wave propagating in the x-direction, it will be the ordinary wave equation, which can be written as follows:

$$\frac{\partial^2 u}{\partial x^2} = \frac{1}{c^2}\frac{\partial^2 u}{\partial t^2} \tag{1.17}$$

where u is the particle velocity. In the longitudinal case, this velocity will be in the x-direction (particle movement in the same direction as the propagation), whereas in the transversal case, it will be in the y and z directions (more or less perpendicular to the direction of propagation). Note, however, that in the case of general non-plane wave motion, the differential equations will be a little bit more complicated, especially in the transverse case. As the waves are described by the ordinary wave equation, they will

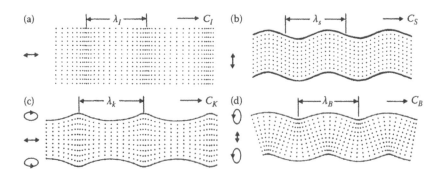

Figure 1.2 Important types of waves. (a) Longitudinal; (b) quasilongitudinal; (c) shear; (d) bending. (Reproduced from Kristensen, J., Rindel, J.H., *Bygningsakustik – Teori og praksis* (Building acoustics – Theory and practice, in Danish), SBI-Anvisning 166, Danish Building Research Institute, Hørsholm, Denmark, 1989. With permission.)

behave in the same way as the acoustic wave in air; only the magnitude of the wave speed c will differ. Also, note that c will be constant with respect to the frequency. However, a combined wave such as the bending wave cannot be described by the ordinary wave equation, and the wave speed will be frequency dependent. This is called *dispersion*.

1.5.1 Longitudinal waves

The *bulk modulus* of a solid is the ratio between a small pressure increase and the resulting decrease in volume or increase in density. It is slightly different under adiabatic or isothermal conditions, which depends on the thermal conductivity of the material. In solid structures, the bulk modulus and thus the speed of longitudinal waves depend on the actual shape of the structure. The simplest case is that of a bar whose sides are free to move as the compression travels through the bar. Hence, the bulk modulus of the bar equals the *Young's modulus E*, and the speed of the compression wave is

$$c_{L,\text{bar}} = \sqrt{\frac{E}{\rho_m}} \quad \text{(m/s)} \tag{1.18}$$

where ρ_m is the density of the solid material. In a plate, the *Poisson ratio* μ tends to increase the bulk modulus, and the speed of a longitudinal wave travelling along the plate is

$$c_{L,\text{plate}} = \sqrt{\frac{E}{\rho_m(1-\mu^2)}} \quad \text{(m/s)} \tag{1.19}$$

In an extended solid medium, i.e. without nearby surfaces, the speed is

$$c_{L,\text{extended}} = \sqrt{\frac{E(1-\mu)}{\rho_m(1+\mu)(1-2\mu)}} \quad \text{(m/s)} \tag{1.20}$$

Most solid materials have Poisson ratios between 0.1 and 0.5, the latter being the theoretical upper limit. Some examples are given in Table 1.4.

1.5.2 Bending and shear waves in plates

In plates and beams, there exists wave motion in the form of bending – that is, the same type of motion as when you bend a stick. The motion is perpendicular to both the direction of propagation and the surface of the structure (Figures 1.2 and 1.3). The bending wave is very important in acoustics in general as it can radiate sound very efficiently to the

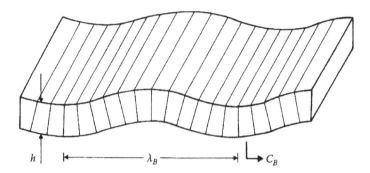

Figure 1.3 A bending wave in a plate with thickness *h*.

surrounding air. The reason for this is partly due to the transverse out-of-plane motion and the low impedance associated with this motion, and partly due to the frequency dependency of the wave speed, as will be discussed later.

A plate with thickness *h* can be characterized by the mass per unit area,

$$m = \rho_m h \quad (\text{kg/m}^2) \tag{1.21}$$

the bending stiffness per unit width

$$B = \frac{Eh^3}{12(1-\mu^2)} \quad (\text{Nm}) \tag{1.22}$$

and the shear modulus

$$G = \frac{E}{2(1+\mu)} \quad (\text{N/m}^2) \tag{1.23}$$

The pure bending wave, typical for thin plates and long wavelengths, is controlled by the bending wave equation. In the case of a plane wave propagating in the *x*-direction, the bending wave equation is

$$B\frac{\partial^4 v}{\partial x^4} + m\frac{\partial^2 v}{\partial t^2} = 0 \tag{1.24}$$

where *v* is the vibration velocity perpendicular to the plate. It should be noted that compared to the ordinary wave equation, the bending wave equation is of the fourth order in the spatial derivative and of the second order in the time derivative. This implies that the wave speed is frequency dependent and given by

$$c_B = \sqrt{\omega} \sqrt[4]{\frac{B}{m}} = \sqrt{2\pi f} \sqrt[4]{\frac{Eh^2}{12\rho_m(1-\mu^2)}} \quad \text{(m/s)} \tag{1.25}$$

The bending wave speed is thus proportional to the square root of the frequency. This is in good agreement with reality at low frequencies (and thin plates). However, it is unrealistic to have a wave speed that approaches infinity in the limit of high frequencies. The bending wave equation and wave speed are low-frequency approximations.

Another transversal wave is the shear wave, in which the sections of a plate are shifted in parallel, perpendicular to the direction of propagation (Figure 1.2). The speed of a shear wave is

$$c_S = \sqrt{\frac{G}{\rho_m}} = \sqrt{\frac{E}{2\rho_m(1+\mu)}} \quad \text{(m/s)} \tag{1.26}$$

In contrast to bending waves, shear waves depend neither on the frequency nor on the thickness of the plate. If the thickness of the plate exceeds approximately one-sixth of the wavelength, the resistance of the bending wave exceeds the resistance of the shear wave. Thus, with increasing frequency, there is a gradual transition from bending to shear, which can be approximated by the following expression for the effective transverse wave speed (Rindel 1994):

$$c_{B,\text{eff}} \cong \left(\frac{1}{c_B^3} + \frac{1}{c_S^3} \right)^{-1/3} \tag{1.27}$$

The frequency dependency of the effective transverse wave speed compared to the bending wave speed is shown in Figure 1.4. When the speed of propagation depends on the frequency, it is called dispersion, and the curve in Figure 1.4 is a dispersion curve.

The effect of dispersion of bending waves can be observed in wintertime when skating on the ice on a lake. An impact on the ice some distance away can be heard as a characteristic chirp sound, starting at high frequencies and quickly descending to low frequencies. The explanation of this phenomenon is that the vibrations propagate as bending waves along the plate of ice, and the high frequencies travel faster than the low frequencies.

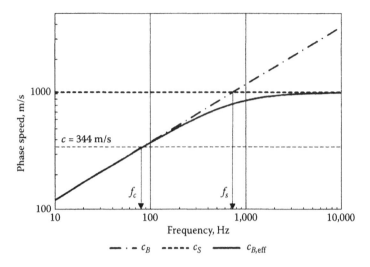

Figure 1.4 The effective speed of transverse waves in a plate showing the transition from bending to shear. In this example, the speed of shear waves is three times c, the critical frequency is 80 Hz, and the crossover frequency is 720 Hz.

1.5.3 Critical frequency

The acoustic wave speed in air is independent of the frequency ($c \approx 344$ m/s at 20°C), whereas the bending wave speed and the effective transverse wave speed are frequency dependent. This has the consequence that there exists a frequency where the two wave types have the same speed and wavelength, and therefore couple easily. A good coupling means that the plate will easily radiate sound at and above this frequency. Actually, it is tempting to adopt the *Mach number M* known from fluid dynamics and aircraft technology. This is the ratio of speed of propagation to the speed of sound in air. Thus, the Mach number of a structural wave is $M = c_{B,\mathrm{eff}}/c$.

The *critical frequency* f_c is defined as the frequency at which $c_B = c = 344$ m/s (Mach 1). Thus, from Equation 1.25, it follows that

$$f_c = \frac{c^2}{2\pi}\sqrt{\frac{m}{B}} = \frac{c^2}{\pi h}\sqrt{\frac{3\rho_m(1-\mu^2)}{E}} = \frac{K_c}{h} \tag{1.28}$$

where K_c is material constant. It should be noted that for a given material, the critical frequency is inversely proportional to the plate thickness, h.

The *crossover frequency* from bending waves to shear waves, f_s is defined as the frequency at which $c_B = c_S$:

$$f_s = \frac{c_s^2}{2\pi}\sqrt{\frac{m}{B}} = \frac{1}{2\pi h}\sqrt{\frac{3E(1-\mu)}{\rho_m(1+\mu)}} \tag{1.29}$$

Table 1.4 Examples of typical parameters for common building materials

Material	Density, ρ_m (10^3 kg/m^3)	Young's modulus, E (10^9 N/m^2)	Poisson ratio, μ	Loss factor, internal η (10^{-3})	K_c (Hz·m)	Shear wave speed, c_s (m/s)
Concrete	2.30	26	0.20	10	19	2150
Clinker concrete	1.70	4.6	0.10	15	39	1100
Light clinker concrete	1.30	3.8	0.10	15	38	1150
Porous concrete	0.60	0.9	0.10	15	53	820
Brick	1.90	24	0.10	10	18	2400[a]
Light brick	1.60	14	0.10	10	22	2000[a]
Gypsum board	0.84	3	0.30	15	33	1150
Glass	2.50	60	0.24	1	13	3000
Steel	7.80	210	0.33	0.1	12	3200
Aluminium	2.70	72	0.33	0.1	12	3150
Lead	11.30	17	0.45	100	48	720
Wooden chipboard	0.70	4.6	0.20	20	25	1650
Wood, pine, parallel to fibres	0.50	9.8	0.10	10	15	2980
Wood, pine, perpendicular to fibres	0.50	0.2	0.10	10	103	420
Laminated wood, parallel to fibres	0.60	6.2	0.10	28	20	2150
Laminated wood, perpendicular to fibres	0.60	2.3	0.10	28	33	1300

Source: Cremer, L., M. Heckl, *Structure-Borne Sound, Structural Vibrations and Sound Radiation at Audio Frequencies*, 2nd edition, Springer, Berlin and Heidelberg, 1988; Fasold, W., E. Sonntag, *Bauakustik. Bauphysikalische Entwurfslehre*, (Building acoustics. Designing in building physics, in German). Band 4, 3rd edition, VEB Verlag, Berlin, 1978; Bodlund, K., Luftlydisolering. En sammanstälning av tillämplig teori. (Airborne sound insulation. A collection of applicable theory, in Swedish). Rapport R60:1980, Statens råd för Byggnadsforskning, Stockholm, 1980; Rindel, J.H., Appl. Acoust., 41, 97–111, 1994.

[a] Brick and light brick walls are not homogeneous, and thus, the actual shear wave speed is 2–3 times lower than that calculated from the equations.

Thus, the ratio between f_s and f_c is

$$\frac{f_s}{f_c} = \frac{c_S^2}{c^2} = \frac{E}{2c^2 \rho_m (1+\mu)} \tag{1.30}$$

i.e. only dependent on the material properties. Examples for some building materials are given in Table 1.4.

1.6   AVERAGING

Four different kinds of averaging are applied in acoustics – arithmetic, geometric, harmonic and energetic.

1.6.1   Arithmetic average

$$a = \frac{1}{2}(x+y) \tag{1.31a}$$

$$a = \frac{1}{n}\sum_{i=1}^{n}x_i \tag{1.31b}$$

Example: The average reverberation time in a room is the arithmetic average of the reverberation times in n positions.

1.6.2   Geometric average

$$a = \sqrt{x\,y} \tag{1.32}$$

Example: The centre frequency of an octave band is the geometric average of the upper and lower limiting frequency.

1.6.3   Harmonic average

$$a = \frac{2\,x\,y}{x+y} \tag{1.33}$$

The harmonic average is, in fact, the arithmetic average of the reciprocal values, $1/a=(1/x+1/y)/2$.
   Example: The characteristic distance from a reflecting surface is the harmonic average of the distance to the source and the distance to the receiver.

1.6.4   Energetic average

$$L_a = 10\lg\left[\frac{1}{n}\sum_{i=1}^{n}10^{0.1\cdot L_i}\right]\ (\text{dB}) \tag{1.34}$$

Example: L_a is the sound pressure level averaged over n positions in a room, and L_i is the sound pressure level at position i.

1.7   USEFUL MATHEMATICAL FUNCTIONS AND FORMULAE

1.7.1   Trigonometrical formulae

$$2\cos(x)\sin(x) = \sin(2x) \tag{1.35}$$

$$\cos^2(x) - \sin^2(x) = \cos(2x) \tag{1.36}$$

$$2\cos^2(x) = 1 + \cos(2x) \tag{1.37}$$

$$2\sin^2(x) = 1 - \cos(2x) \tag{1.38}$$

1.7.2   Euler's formulae

$$\cos(x) = \frac{e^{jx} + e^{-jx}}{2} \tag{1.39}$$

$$\sin(x) = \frac{e^{jx} - e^{-jx}}{2j} \tag{1.40}$$

$$e^{jx} = \cos(x) + j\sin(x) \tag{1.41}$$

1.7.3   The cardinal sine function

$$\mathrm{sinc}(x) = \frac{\sin(x)}{x} \tag{1.42}$$

Note that sinc (0)=1. This function is also called the 0th order spherical Bessel function of the first kind, $j_0(x)$. The function is shown graphically in Figure 1.5.

1.7.4   Logarithmic functions

The common logarithm, base 10:

$$y = 10^x \iff x = \log_{10}(y) = \lg(y) \tag{1.43}$$

$$\lg(xy) = \lg(x) + \lg(y) \tag{1.44}$$

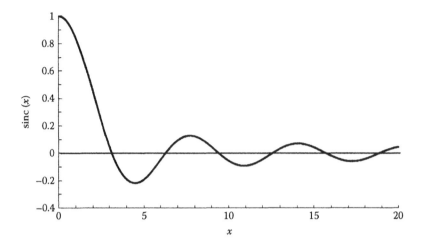

Figure 1.5 The sinc function.

$$\lg\left(x^{n}\right) = n \lg(x) \tag{1.45}$$

The natural logarithm, base e:

$$y = e^{x} \Leftrightarrow x = \ln(y) \tag{1.46}$$

Conversion between common and natural logarithms:

$$\ln(x) = \frac{\lg(x)}{\lg(e)} \tag{1.47}$$

$$\lg(x) = \frac{\ln(x)}{\ln(10)} \tag{1.48}$$

1.7.5   Basic equations for complex numbers

$$\text{Complex number}: z = \text{Re}\{z\} + j\,\text{Im}\{z\} = |z|e^{j\theta} \tag{1.49}$$

$$\text{Real part}: \text{Re}\{z\} = |z|\cos\theta \tag{1.50}$$

$$\text{Imaginary part}: \text{Im}\{z\} = |z|\sin\theta \tag{1.51}$$

$$\text{Modulus}: |z| = \sqrt{\text{Re}\{z\}^2 + \text{Im}\{z\}^2} \tag{1.52}$$

$$\text{Argument}: \theta = \arctan\left(\frac{\text{Im}\{z\}}{\text{Re}\{z\}}\right) \tag{1.53}$$

$$\text{Complex conjugate}: z^* = \text{Re}\{z\} - j\,\text{Im}\{z\} = |z|\,e^{-j\theta} \tag{1.54}$$

$$\text{Imaginary unit}: j = e^{j\pi/2}, \quad j^2 = e^{j\pi} = -1 \tag{1.55}$$

$$\text{Multiplication}: z_1 z_2 = |z_1||z_2|\,e^{j(\theta_1+\theta_2)} \tag{1.56}$$

$$\text{Division}: \frac{z_1}{z_2} = \frac{|z_1|}{|z_2|}\,e^{j(\theta_1-\theta_2)} \tag{1.57}$$

REFERENCES

K. Bodlund (1980). *Luftlydisolering. En sammanstälning av tillämplig teori*, (Airborne sound insulation. A collection of applicable theory, in Swedish). Rapport R60:1980, Statens råd för Byggnadsforskning, Stockholm, Sweden.

L. Cremer and M. Heckl (1988). *Structure-Borne Sound, Structural Vibrations and Sound Radiation at Audio Frequencies*. Translated and revised by E.E. Ungar, 2nd edition. Springer-Verlag, Berlin.

W. Fasold and E. Sonntag (1978). *Bauakustik. Bauphysikalische Entwurfslehre* (Building acoustics. Designing in building physics, in German), Band 4, 3rd edition. VEB Verlag, Berlin.

IEC 61672-1 (2002). *Electro acoustics – Sound level meters – Part 1: specifications*. International Electrotechnical Commission, Geneva, Switzerland.

ISO 1683 (2008). *Acoustics – Preferred reference values for acoustical and vibratory levels*. International Organization for Standardization, Geneva, Switzerland.

ISO 226 (2003). *Acoustics – Normal equal-loudness-level contours*. International Organization for Standardization, Geneva, Switzerland.

ISO 266 (2003). *Acoustics – Preferred frequencies*. International Organization for Standardization, Geneva, Switzerland.

ISO/TR 25417 (2007). *Acoustics – Definition of basic quantities and terms*. Technical report. International Organization for Standardization, Geneva, Switzerland.

J. Kristensen and J.H. Rindel (1989), *Bygningsakustik – Teori og praksis* (Building acoustics – Theory and practice, in Danish). SBI-Anvisning 166. Danish Building Research Institute, Hørsholm, Denmark.

J.H. Rindel (1994). Dispersion and absorption of structure-borne sound in acoustically thick plates. *Applied Acoustics* 41, 97–111.

Chapter 2

Mechanical vibrations

2.1 A SIMPLE MECHANICAL SYSTEM

In this chapter, we shall look at one-dimensional vibrations in a simple mechanical system consisting of a mass, a spring and a damping element. Since the vibrations can be fully described in one dimension, this is also called a system with one degree of freedom.

First, the use of complex notation is shown for harmonic vibrations as a basis for later applications in acoustic vibrations and sound fields. Next, the vibrations in a resonant mechanical system are dealt with, as a basis for numerous applications in resonant acoustic systems. Finally, the theory for vibration isolation is explained as this has fundamental importance for noise control of machines and equipment in practice. The description follows the terminology and concepts in ISO 2041.

Figure 2.1 shows a mechanical system consisting of a mass m placed on an elastic layer on top of a rigid, unmovable surface. The elastic layer has the stiffness k symbolized as a spring and the resistance r symbolized as a dashpot. Some materials have slightly different stiffness for static load and for a dynamic excitation, and thus, we will distinguish between those cases. The acceleration due to gravity means that the system has a static excitation, which leads to a compression of the elastic layer. The static displacement is by definition:

$$x_s = \frac{mg}{k_s} \tag{2.1}$$

where $g=9.81$ ms^{-2} is the acceleration due to gravity and k_s is the static stiffness (N/m).

In the case of a dynamic excitation of the system by the force F, we have the equation of motion from Newton's second law:

$$m\frac{\mathrm{d}^2 x}{\mathrm{d}t^2} + r\frac{\mathrm{d}x}{\mathrm{d}t} + k_d x = F \tag{2.2}$$

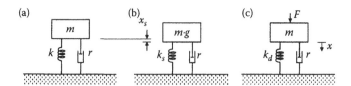

Figure 2.1 (a) Simple mechanical system. (b) Static deflection due to mass load and gravity. (c) Dynamic excitation by an external force.

where x is the displacement from the rest position, t is the time and k_d is the dynamic stiffness (N/m).

As long as the external force is active, the system is said to display forced vibrations. However, when the force stops, the system can still exhibit vibrations for a shorter or longer time after the force is stopped. These are resonant or free vibrations.

Let us assume that the external force is *harmonic*, i.e. it varies with time as a sinusoidal function:

$$F = F_0 \cos(\omega t + \varphi) \tag{2.3}$$

where F_0 is the amplitude of the force, ω is the angular frequency and φ is the phase angle. Using the complex notation, the harmonic force can be written as

$$F = F_0 \cos(\omega t + \varphi) + jF_0 \sin(\omega t + \varphi) = F_0 e^{j(\omega t + \varphi)} = \hat{F} e^{j\omega t} \tag{2.4}$$

The force is expressed here as a complex phasor. $\hat{F}$ is the complex amplitude, and the magnitude of the phasor is the amplitude of the force, $|F| = F_0$. The advantage of the complex notation appears when the force is expressed as a phasor, and we want to do multiplications, divisions, or other mathematical operations with harmonic signals. However, only the real part has a physical reality, i.e. we can neglect the imaginary part when the force is seen in relation to physical conditions.

As a consequence of the complex notation, the harmonic varying signal can be displayed in the complex plane, where the real and imaginary parts are marked on two orthogonal axes. At a certain time, the signal is displayed as a vector, starting from origin, and with a length equal to the amplitude (the modulus). The direction of the vector is determined by the argument, which is the angle relative to the real axis. The force in Equation 2.4 can be displayed as shown in Figure 2.2.

The harmonically varying signal is represented in the complex plane as a rotating vector, the phasor $\hat{F}$. In the scientific literature, it is common not to include the time factor $e^{j\omega t}$, as it is understood.

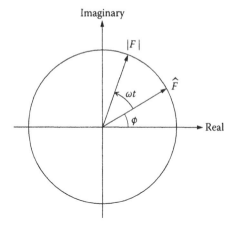

Figure 2.2 A harmonically varying force displayed as a phasor in the complex plane.

2.2 MECHANICAL IMPEDANCE AND MOBILITY

Again, we consider the simple mechanical system in Figure 2.1. The system is excited by the harmonic force F:

$$F = |F| e^{j(\omega t + \varphi)} \qquad (2.5)$$

The vibrations exhibited by the system can be characterized by the deflection x, the velocity v, or the acceleration a. As long as the force is acting, the system exhibits forced vibrations, implying that the vibrations have the same angular frequency ω as the acting force. Thus, the velocity can be written as

$$v = |v| e^{j\omega t} \qquad (2.6)$$

where we have chosen to set the phase angle of the velocity to zero.

Mechanical impedance Z_m of a system is defined as the complex ratio between a harmonic force and the resulting velocity:

$$Z_m = \frac{F}{v} = \frac{|F|}{|v|} e^{j\varphi} \qquad (2.7)$$

The unit is N s/m. The phase angle between force and velocity is φ.

The force and the velocity can be determined either in the same point or in two different points. Therefore, we distinguish between the *driving point impedance* and the *transfer impedance*. The latter is defined by

$$Z_{m,12} = \frac{F_1}{v_2} \qquad (2.8)$$

where F_1 is a harmonic force that excites the system in one point, and v_2 is the resulting velocity in another point of the system, both taken as complex quantities.

The *mobility* Y_m of a mechanical system is the inverse of mechanical impedance, i.e.

$$Y_m = \frac{v}{F} = \frac{|v|}{|F|} e^{-j\varphi} \tag{2.9}$$

The mobility is sometimes called *mechanical admittance*.

For a mass m, we assume a harmonic vibration (Equation 2.6) and get the following relation from Newton's second law:

$$F = m\frac{dv}{dt} = j\omega m v \tag{2.10}$$

Considering a spring with the dynamic stiffness k_d, we get

$$F = k_d x = k_d \int v \, dt = \frac{k_d v}{j\omega} \tag{2.11}$$

For a dashpot with resistance r, we get by definition

$$F = rv \tag{2.12}$$

Table 2.1 shows the point impedance and mobility for the three basic mechanical elements and for an extended system, namely, a large homogeneous plate.

An important system for applications in building acoustics is a large and thin, homogeneous plate. This may be excited in a point by a harmonic force in the normal direction perpendicular to the surface of the plate. Bending waves are generated propagating radially in the plate outgoing from the excitation point. The driving point impedance of the plate is found to be surprisingly simple (Cremer and Heckl, 1967):

$$Z_0 = \frac{F_0}{v_0} = 8\sqrt{m''B} = \frac{4c^2 m''}{\pi f_c} \tag{2.13}$$

where m'' is the mass per unit area and B is the bending stiffness per unit width. The critical frequency f_c (Equation 1.28) is inserted in the last term. It is noted that the point impedance is independent of frequency and real, i.e. force and velocity are in phase. It can be shown that the impedance (Equation 2.13) is also a good approximation for the average velocity due to point excitation of a finite plate.

Table 2.1 Examples of mechanical point impedance Z_m and mobility Y_m

System		Mechanical impedance Z_m	Mobility Y_m
	Mass: m (kg)	$j\omega m$	$\dfrac{1}{j\omega m}$
	Spring: k_d (N/m)	$\dfrac{k_d}{j\omega}$	$\dfrac{j\omega}{k_d}$
	Dashpot: r (Ns/m)	r	$\dfrac{1}{r}$
	Plate Mass per unit area: m'' (kg/m²) Bending stiffness: B (Nm)	$8\sqrt{m''B}$	$\dfrac{1}{8\sqrt{m''B}}$

2.3 FREE VIBRATIONS

Free vibrations of the mechanical system are described through the solution of the equation of movement (Equation 2.2) when the force is $F=0$:

$$m\frac{\mathrm{d}^2x}{\mathrm{d}t^2} + r\frac{\mathrm{d}x}{\mathrm{d}t} + k_d x = 0 \qquad (2.14)$$

The solution of this homogeneous differential equation is well known from mathematical physics. Three categories of solutions appear:

$r < mk_d$: The system is weakly damped and can exhibit free harmonic vibrations.

$r = mk_d$: The system is critically damped.

$r > mk_d$: The system is overdamped and a deflection slowly returns to the steady-state position.

Only the first case is relevant for acoustical applications and is discussed in the following.

The equation of movement (Equation 2.14) can be rewritten in the case of a weakly damped system as

$$\frac{\mathrm{d}^2x}{\mathrm{d}t^2} + 2\delta\frac{\mathrm{d}x}{\mathrm{d}t} + \omega_0^2 x = 0 \qquad (2.15)$$

where δ is the *damping coefficient*:

$$\delta = \frac{r}{2m} \qquad (2.16)$$

and ω_0 is the angular frequency of the equivalent *undamped* system (i.e. for $r=0$):

$$\omega_0 = \sqrt{\frac{k_d}{m}} \qquad (2.17)$$

A solution to Equation 2.15 is

$$x = x_0 e^{-\delta t} \cos(\omega_r t - \theta) \qquad (2.18)$$

where x_0 and the phase angle θ depend on the initial conditions and the angular frequency of the *damped* vibration is

$$\omega_r = \sqrt{\omega_0^2 - \delta^2} \qquad (2.19)$$

The time function of the vibration is shown in Figure 2.3. The phase angle is set to zero. The amplitude of the vibration is $x_0 e^{-\delta t}$, and thus, it decreases exponentially with time. It can be noted from Equation 2.19 that the angular frequency of the system, ω_r, is lower than the angular frequency of the undamped system ω_0.

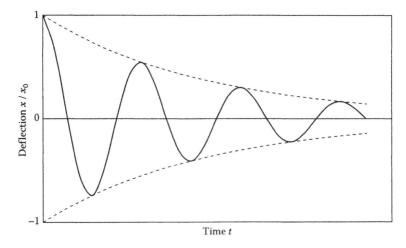

Figure 2.3 Time function of the free vibration of a weakly damped system with one degree of freedom.

The system is said to exhibit a natural mode of vibration, and the associated frequency is called the *natural frequency* $f_r = \omega_r/2\pi$.

The undamped natural frequency $f_0 = \omega_0/2\pi$ is often referred to as the resonance frequency. However, this is only correct if a velocity resonance is understood (see more about this in Section 2.5). The undamped natural frequency is

$$f_0 = \frac{\omega_0}{2\pi} = \frac{1}{2\pi}\sqrt{\frac{k_d}{m}} \quad \text{(Hz)} \tag{2.20}$$

The *period duration* of the damped natural vibration is the reciprocal of the natural frequency:

$$T_1 = \frac{1}{f_r} = \frac{2\pi}{\omega_r} = \frac{2\pi}{\sqrt{\omega_0^2 - \delta^2}} \quad \text{(s)} \tag{2.21}$$

From Equation 2.18, it can be seen that the ratio between deflections x at time t and $t + T_1$ is $e^{\delta T_1}$. The natural logarithm of this ratio is called the *logarithmic decrement* Λ:

$$\Lambda = \ln\frac{x(t)}{x(t + T_1)} = \delta T_1 \tag{2.22}$$

The damped vibrations can also be characterized by the *time constant* τ defined as the reciprocal of the damping coefficient:

$$\tau = \frac{1}{\delta} \quad \text{(s)} \tag{2.23}$$

The energy of the vibration is related to the velocity v, which is easily found when Equation 2.18 is written in the complex notation:

$$x = x_0 e^{-\delta t} e^{j\omega_r t} e^{-j\theta} \tag{2.24}$$

$$v = \frac{dx}{dt} = (j\omega_r - \delta)x_0 e^{-\delta t} e^{j\omega_r t} e^{-j\theta} \tag{2.25}$$

$$|v| = \sqrt{\omega_r^2 + \delta^2}\, x_0 e^{-\delta t} = \omega_0 x_0 e^{-\delta t} \tag{2.26}$$

The energy as a function of time is

$$E = \frac{1}{2}m|v|^2 = \frac{1}{2}m\omega_0^2 x_0^2 e^{-2\delta t} = E_0 e^{-2\delta t} \tag{2.27}$$

where E_0 is the initial energy at $t=0$.

The *loss factor* η is the most used descriptor of the damping of vibrating systems in building acoustics. It is defined as the relative energy loss per radian:

$$\eta = \frac{-dE/dt}{\omega_0 E} = \frac{2\delta}{\omega_0} = \frac{r}{\sqrt{mk_d}} = \frac{r}{\omega_0 m} \qquad (2.28)$$

The *reverberation time* T is primarily used in room acoustics, but it may also be used in relation to the decay of free vibrations. By definition, T is the time for the energy to decrease to 10^{-6} of the initial energy, which corresponds to a decay of 60 dB. For $t=T$, we get from Equation 2.27

$$\frac{E}{E_0} = e^{-2\delta T} = 10^{-6} = e^{-6\ln(10)} \qquad (2.29)$$

and the reverberation time is

$$T = \frac{6\ln(10)}{2\delta} \approx \frac{6.9}{\delta} \approx \frac{2.2}{f_0\eta} \qquad (2.30)$$

The latter relation is very useful for experimental determination of the loss factor at the resonance frequency – this can be done by measuring the reverberation time of the system.

2.4 FORCED VIBRATIONS

We now consider the situation where the mechanical system is excited by a harmonic force $F = \hat{F}e^{j\omega t}$, which means that there are forced vibrations with the same angular frequency as the force, i.e. the deflection is $x = \hat{x}\,e^{j\omega t}$. By insertion in the equation of movement (Equation 2.2), and since

$$\frac{dx}{dt} = j\omega x \quad \text{and} \quad \frac{d^2x}{dt^2} = -\omega^2 x \qquad (2.31)$$

we get

$$\hat{x} = \frac{\hat{F}}{-\omega^2 m + j\omega r + k_d} = \frac{\hat{F}}{j\omega\left[r + j(\omega m - k_d/\omega)\right]} \qquad (2.32)$$

The velocity of the vibration is

$$v = \frac{dx}{dt} = j\omega\hat{x}e^{j\omega t} = \frac{F}{r + j(\omega m - k_d/\omega)} \qquad (2.33)$$

From the definition of mechanical impedance Z_m (Equation 2.7), we get

$$Z_m = \frac{F}{v} = r + j(\omega m - k_d / \omega) \tag{2.34}$$

It may be noted that this formula is analogous to Ohm's law for a simple electrical circuit.

The modulus and argument of the mechanical impedance are

$$|Z_m| = \sqrt{r^2 + (\omega m - k_d / \omega)^2} \tag{2.35}$$

$$\theta = \arctan\left(\frac{\omega m - k_d / \omega}{r}\right) \tag{2.36}$$

where $Z_m = |Z_m| e^{j\theta}$.

Applying the mechanical impedance of the system, we get the following formulas for the force, velocity and deflection in complex notation:

$$F = |F| e^{j\omega t} \tag{2.37}$$

$$v = \frac{F}{Z_m} = \frac{|F|}{|Z_m|} e^{j(\omega t - \theta)} \tag{2.38}$$

$$x = \frac{F}{j\omega Z_m} = \frac{|F|}{\omega |Z_m|} e^{j(\omega t - \theta - \pi/2)} \tag{2.39}$$

Here we have chosen the force as a reference for the phase angle. Figure 2.4 shows these three parameters as phasors in the complex plane. In the physical world, the parameters appear as projections on the real axis with harmonic variation (the exponential functions are replaced by cosine functions).

The displacement x can be found from Equation 2.39 by inserting Equation 2.34 and using Equations 2.17 and 2.28:

$$x = \frac{F}{j\omega\left(r + j\frac{m}{\omega}(\omega^2 - \omega_0^2)\right)} = \frac{F / k_d}{j\eta \cdot \frac{\omega}{\omega_0} + \left(1 - \frac{\omega^2}{\omega_0^2}\right)} \tag{2.40}$$

This leads us to the following frequency dependency of the displacement amplitude relative to the amplitude of the force:

$$\frac{|x|}{|F|} = \frac{1 / k_d}{\sqrt{\left(\eta \cdot \frac{\omega}{\omega_0}\right)^2 + \left(1 - \frac{\omega^2}{\omega_0^2}\right)^2}} \tag{2.41}$$

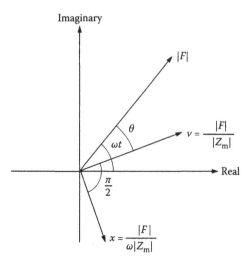

Figure 2.4 Display of force, velocity and displacement as vectors in the complex plane. The vectors rotate with the angular frequency ω, but the relative positions remain constant.

The phase angle between the force and deflection is

$$\theta + \frac{\pi}{2} = \arctan\left(\eta \cdot \left(\frac{\omega_0}{\omega} - \frac{\omega}{\omega_0}\right)^{-1}\right) \tag{2.42}$$

Figure 2.5 shows the amplitude and phase of the displacement as functions of frequency for different values of the loss factor. (Note that the relative angular frequency and the relative frequency are identical, i.e. $\omega/\omega_0 = f/f_0$.)

At low frequencies ($\omega \ll \omega_0$), the deflection amplitude is almost independent of the frequency and the phase angle is close to zero, which means that the deflection follows the force without delay. The mechanical impedance is dominated by the stiffness term $|Z_m| \approx k_d/\omega$.

At the resonance frequency ($\omega = \omega_0$), the deflection amplitude increases dramatically for small loss factors, and more moderately for large loss factors. The phase angle is $\pi/2$ and the deflection is one-fourth period behind the force, while the velocity is in phase with the force. The mechanical impedance is dominated by the damping term $|Z_m| \approx r$.

At high frequencies ($\omega \gg \omega_0$), the deflection amplitude decreases rapidly with frequency and the phase angle is close to π, which means that deflection and the force are almost in opposite phase. The velocity is behind the force and the acceleration is in phase with the force. The mechanical impedance is dominated by the mass term $|Z_m| \approx \omega m$.

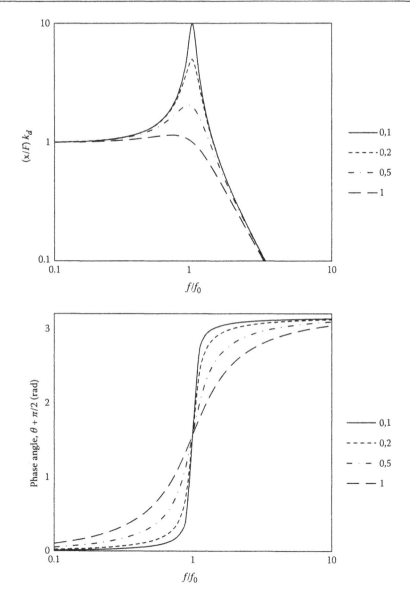

Figure 2.5 Amplitude of deflection (upper graph) and phase angle between deflection and force (lower graph) for different values of the loss factor.

The described behaviour of the system is typical for a resonant system. When the system is excited with a frequency close to the natural frequency of the system, the amplitude of the deflection is significantly larger than at other frequencies.

2.5 RESONANCE

Resonance in a system is defined as a state in which any change in excitation frequency results in a decreased response. Thus, the resonance frequency depends on what kind of response is observed – deflection, velocity or acceleration.

2.5.1 Resonance frequency

If we consider the *deflection* of the system, it follows from Equation 2.40 that the maximum deflection occurs when the denominator

$$\left(\eta \cdot \frac{\omega}{\omega_0}\right)^2 + \left(1 - \frac{\omega^2}{\omega_0^2}\right)^2$$

is minimum, which happens for the angular resonance frequency:

$$\omega_{res,x} = \omega_0 \sqrt{1 - \frac{1}{2}\eta^2} = \sqrt{\omega_0^2 - 2\delta^2} \tag{2.43}$$

The maximum deflection at resonance is

$$\left|x\right|_{max} = \frac{|F|}{k_d \eta \sqrt{1 - \frac{1}{4}\eta^2}} = \frac{|F|}{r\sqrt{\omega_0^2 - \delta^2}} \tag{2.44}$$

However, if we instead chose to consider the *velocity* of the system, insertion of Equations 2.17 and 2.28 in Equation 2.33 yields

$$v = \frac{F}{\omega_0 m\eta + jm / \omega(\omega^2 - \omega_0^2)} = \frac{F / \sqrt{mk_d}}{\eta + j(\omega / \omega_0 - \omega_0 / \omega)} \tag{2.45}$$

From this, it is obvious that the angular resonance frequency for velocity resonance is

$$\omega_{res,v} = \omega_0 \tag{2.46}$$

and the corresponding maximum velocity is

$$\left|v\right|_{max} = \frac{|F|}{\eta \sqrt{mk_d}} = \frac{|F|}{r} \tag{2.47}$$

Table 2.2 Natural frequencies and resonance frequencies of a simple mechanical system with one degree of freedom

	Natural frequency	
	Undamped	Damped
Angular frequency (rad/s)	$\omega_0 = \sqrt{\dfrac{k_d}{m}}$	$\omega_r = \sqrt{\omega_0^2 - \delta^2}$
Frequency (Hz)	$f_0 = \dfrac{1}{2\pi}\sqrt{\dfrac{k_d}{m}}$	$f_r = f_0\sqrt{1 - \dfrac{\eta^2}{4}}$

	Resonance frequency		
	Deflection	Velocity	Acceleration
Angular frequency (rad/s)	$\sqrt{\omega_0^2 - 2\delta^2}$	ω_0	$\dfrac{\omega_0^2}{\sqrt{\omega_0^2 - 2\delta^2}}$
Frequency (Hz)	$f_0\sqrt{1 - \dfrac{\eta^2}{2}}$	f_0	$\dfrac{f_0}{\sqrt{1 - \dfrac{\eta^2}{2}}}$

Finally, if we chose to consider the *acceleration* of the system, we find in a similar way that the maximum acceleration occurs at the angular resonance frequency for acceleration:

$$\omega_{res,a} = \frac{\omega_0}{\sqrt{1 - \dfrac{1}{2}\eta^2}} = \frac{\omega_0}{\sqrt{\omega_0^2 - 2\delta^2}} \tag{2.48}$$

An overview of the various resonance frequencies and natural frequencies is found in Table 2.2.

It is particularly interesting to note that the resonance frequency for velocity is identical to the natural frequency of the undamped system, $\omega_{res,v} = \omega_0$. This is a very convenient coincidence.

2.5.2 Bandwidth of a resonant system

The mechanical energy of a vibrating system is $(\frac{1}{2})mv^2$ and thus proportional to the velocity squared. The ratio between this and the force squared is from Equation 2.45:

$$\frac{|v|^2}{|F|^2} = \frac{1/(mk_d)}{\eta^2 + (\omega/\omega_0 - \omega_0/\omega)^2} = \frac{1/(mk_d)}{\eta^2 + (f/f_0 - f_0/f)^2} \tag{2.49}$$

In this section, the frequency f in Hz is used instead of the angular frequency. Equation 2.49 is illustrated as a function of the relative frequency in Figure 2.6. The maximum is at the velocity resonance, i.e. at the frequency $f_{\text{res},v}=f_0$.

Resonance covers a certain frequency range and in order to characterize this, we introduce the *half-power bandwidth* B_r defined as the frequency interval within which the energy is at least half the maximum energy at resonance. If f_1 and f_2 are the lower and upper limiting frequencies, respectively, it can be seen from Equation 2.49 that they can be found as solutions to

$$\eta^2 = (f / f_0 - f_0 / f)^2$$

$$\Rightarrow f = \pm \frac{1}{2}\eta f_0 \pm \sqrt{\frac{1}{4}\eta^2 f_0^2 + f_0^2}$$

Since we accept only positive solutions, we get

$$f_1 = f_m - \frac{1}{2}\eta f_0$$

$$f_2 = f_m + \frac{1}{2}\eta f_0$$

$\qquad\qquad\qquad\qquad\qquad\qquad\qquad\qquad$ (2.50)

where the mean frequency f_m is

$$f_m = \frac{1}{2}(f_1 + f_2) = f_0\sqrt{1 + \frac{1}{4}\eta^2}$$

$\qquad\qquad\qquad\qquad\qquad\qquad$ (2.51)

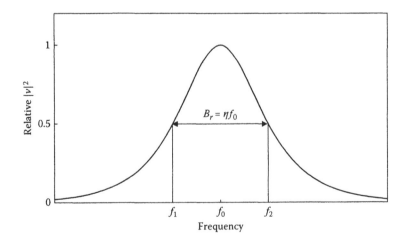

Figure 2.6 The frequency course of energy in a resonant system.

and the half-power bandwidth is

$$B_r = f_2 - f_1 = \eta f_0 \tag{2.52}$$

However, instead of the arithmetic average frequency, a frequency band is usually described by the centre frequency f_c, which is calculated as the geometrical average of the limiting frequencies:

$$f_c = \sqrt{f_1 f_2} = \sqrt{f_m^2 - \frac{1}{4}\eta^2 f_0^2} = f_0 \tag{2.53}$$

Hence, the centre frequency is identical to the resonance frequency for velocity, which again is the same as the natural frequency of the undamped system.

By means of a frequency generator, we can find the resonance frequency and the bandwidth. Thus, it is possible to apply Equation 2.52 for the measurement of the loss factor in a weakly damped system.

2.6 VIBRATION INSULATION

An important practical application of the mechanical vibration model is the vibration isolation of machines and other equipment from the building construction as a means of noise and vibration control.

2.6.1 The power input

When the force F excites a mechanical system, the power input P is defined as the time-average of the power multiplied by the velocity at the point of excitation, measured in the same direction as the force:

$$P = \overline{v \cdot F} = \frac{1}{T_1} \int_0^{T_1} v \cdot F \, dt \tag{2.54}$$

where $T_1 = 1/f = 2\pi/\omega$ here is the period duration of the forced vibration. The velocity and the force are assumed to be harmonic functions and, thus, the effective RMS (root-mean-square) values and the amplitudes are related as

$$v_{\text{eff}}^2 = \frac{1}{2}|v|^2 \text{ and } F_{\text{eff}}^2 = \frac{1}{2}|F|^2.$$

Inserting the real part of v and F from Equations 2.37 and 2.38 yields

$$P = \frac{1}{T_1} \int_0^{T_1} |v| \cdot \cos(\omega t - \theta) \cdot |F| \cdot \cos(\omega t) dt$$

$$= |v| \cdot |F| \frac{1}{T_1} \int_0^{T_1} \frac{1}{2} [\cos(2\omega t - \theta) + \cos\theta] dt \qquad (2.55)$$

$$= \frac{1}{2} |v| \cdot |F| \cos\theta$$

Note the importance of the phase angle θ.

The same derivation can be made by using complex notation:

$$P = \frac{1}{2} |v| \cdot |F| \cdot \mathrm{Re}\{e^{j\theta}\}$$

$$= \frac{1}{2} \cdot \mathrm{Re}\{|v| e^{-j\omega t} e^{j\theta} \cdot |F| \cdot e^{j\omega t}\} \qquad (2.56)$$

$$= \frac{1}{2} \cdot \mathrm{Re}\{v * \cdot F\}$$

where $v*$ is the complex conjugate of v (see Equation 1.54).The power input can be related to the velocity at the excitation point by means of the mechanical point impedance Z_m (Equation 2.7):

$$P = \frac{1}{2} \cdot \mathrm{Re}\left\{\frac{v}{v} v * \cdot F\right\} = \frac{1}{2} |v|^2 \cdot \mathrm{Re}\left\{\frac{F}{v}\right\} = v_{\mathrm{eff}}^2 \cdot \mathrm{Re}\{Z_m\} \qquad (2.57)$$

In a similar way, the power input can be related to the excitation force and the mobility Y_m (Equation 2.9):

$$P = \frac{1}{2} |F|^2 \cdot \mathrm{Re}\left\{\frac{v}{F}\right\} = F_{\mathrm{eff}}^2 \cdot \mathrm{Re}\{Y_m\} \qquad (2.58)$$

It may be noted that the following relation exists:

$$\mathrm{Re}\{Y_m\} = \mathrm{Re}\left\{\frac{1}{Z_m}\right\} = \frac{\mathrm{Re}\{Z_m\}}{|Z_m|^2} \qquad (2.59)$$

2.6.2 Force transmitted to the foundation

Figure 2.7 shows a machine that excites the mechanical vibration system with a harmonic force F:

$$F = |F| e^{j\omega t} = Z_m v = (r + j(\omega m - k_d/\omega)) v \qquad (2.60)$$

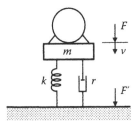

Figure 2.7 Principle of a machinery supported on vibration isolators.

where ω is the angular frequency of the excitation force, Z_m is the mechanical impedance (Equation 2.34) and v is the velocity of the mechanical vibration system. The vibrating mass m may be that of the machine, but the mass is often increased by attaching the machine rigidly to a heavy plate. A heavy mass is an advantage for setting up good vibration isolation.

The force F' that is transferred to the supporting building structure is determined by the excitation of the spring and the dashpot:

$$F' = r\frac{dx}{dt} + k_d x = \left(r - jk_d/\omega\right)v \qquad (2.61)$$

From this, we define the *transmissibility* H of the system as the ratio between the transferred force and the excitation force:

$$H = \frac{F'}{F} = \frac{r - jk_d/\omega}{r + j(\omega m - k_d/\omega)} \qquad (2.62)$$

Introducing the natural frequency ω_0 (Equation 2.17) and the loss factor η (Equation 2.28), yields

$$H = \frac{\eta - j\omega_0/\omega}{\eta + j(\omega/\omega_0 - \omega_0/\omega)} \qquad (2.63)$$

Figure 2.8 displays the numerical value of the transmissibility, i.e. the ratio between the amplitude of the forces expressed in dB, as a function of the relative frequency for various values of the loss factor. Note that $\omega/\omega_0 = f/f_0$. The transmissibility in dB is calculated from

$$20\lg|H| = 10\lg\frac{|F'|^2}{|F|^2} = 10\lg\frac{\eta^2 + (\omega_0/\omega)^2}{\eta^2 + (\omega/\omega_0 - \omega_0/\omega)^2} \qquad (2.64)$$

At low frequencies, i.e. for $\omega \ll \omega_0$, the vibration isolator is stiff and the force is transmitted to the floor without attenuation. If the excitation frequency

ω is near ω_0, the transmitted force is stronger than the excitation force. The amplification is due to resonance, and if the loss factor is small, the amplification can grow very high. Thus, the frequency range near the resonance frequency should always be avoided.

Only when the frequency $\omega > \sqrt{2}\omega_0$, does the attenuation start. It is vital that vibration isolators are designed to give a resonance frequency well below the lowest excitation frequency, and a factor of three may be a practical minimum ($\omega \approx 3\omega_0$) to obtain an efficient isolation.

If the loss factor is small and $\eta \ll \omega_0/\omega \ll 1$, the approximate transmissibility is

$$|H|^2 \cong \left(1-(\omega/\omega_0)^2\right)^{-2} \cong \left(\omega_0/\omega\right)^4 \qquad (2.65)$$

This means a slope of -12 dB per octave. However, at very high frequencies, $\omega_0/\omega \ll \eta$, the slope changes to -6 dB per octave, which follows from the approximation:

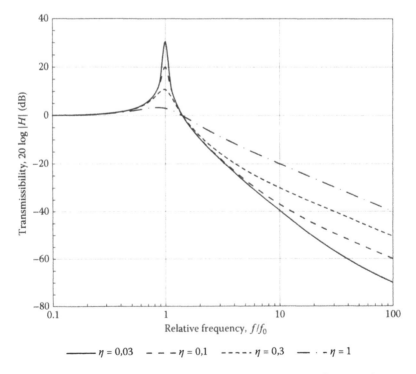

Figure 2.8 The numerical value of the transmissibility with viscous damping, shown as a function of the relative frequency for four different values of the loss factor.

$$|H|^2 \cong \frac{\eta^2}{\eta^2 + (\omega/\omega_0)^2} \cong \left(\frac{\eta \omega_0}{\omega}\right)^2 \tag{2.66}$$

In practice, attenuation at high frequencies is limited due to wave transmission through the isolator, and high-frequency resonances in the isolator may occur.

2.6.3 Vibration isolators with hysteretic damping

The viscous damping (dashpot type) assumed in the previous section is usual in cars and other places, but not in buildings. Common vibration isolators made of rubber have the losses inherent in the spring, and the dynamic stiffness is complex: $k_d(1+j\eta)$. This type of damping is called hysteretic damping, and the losses are proportional to the deflection, not the velocity. Inserting the complex stiffness and $r=0$ in Equation 2.62 yields the transmissibility of a system with hysteretic damping:

$$H = \frac{-jk_d(1+j\eta)/\omega}{j(\omega m - k_d(1+j\eta)/\omega)} = \frac{\eta - j}{\eta + j\left((\omega/\omega_0)^2 - 1\right)} \tag{2.67}$$

This leads to the transmissibility in dB:

$$20\lg|H| = 10\lg\frac{\eta^2 + 1}{\eta^2 + \left((\omega/\omega_0)^2 - 1\right)^2} \tag{2.68}$$

This is displayed in Figure 2.9. The main difference between this and viscous damping is that the damping above the resonance frequency is practically independent of the loss factor. The slope is $-12\,dB$ per octave.

2.6.4 The Q-factor

When a resonant system is excited with a frequency near the resonance frequency f_0, the system will cause an amplification instead of an attenuation. The Q-*factor* is defined for a lightly damped system as a measure of the sharpness of the resonance peak as

$$Q = \frac{f_0}{B_r} = \frac{1}{\eta} \tag{2.69}$$

Hence, the Q-factor is just the reciprocal of the loss factor.

For vibration isolators, the resonance peak in the transmissibility is found from Equation 2.64 or 2.68 by setting $\omega = \omega_0$:

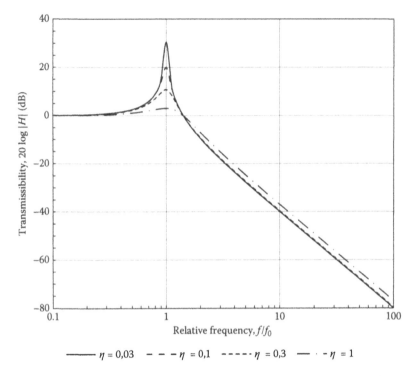

Figure 2.9 The numerical value of the transmissibility with hysteretic damping, shown as a function of the relative frequency for four different values of the loss factor.

$$20\lg|H| = 10\lg\left(1+\frac{1}{\eta^2}\right) = 10\lg\left(1+Q^2\right) \approx 20\lg Q \qquad (2.70)$$

2.6.5 Insertion loss

The efficiency of vibration isolators is not solely determined by the transmissibility of the vibrating system. The supporting floor may also influence the result. Hence, a better measure for the efficiency of vibration isolators is the *insertion loss*, defined by

$$\Delta L = 10\lg\frac{P}{P'} \quad (\text{dB}) \qquad (2.71)$$

where P and P' refer to the power transferred to the floor without or with the vibration isolator, respectively (Figure 2.10).

The point impedance of the floor is called Z_m and the velocity generated in the floor is v or v' without or with the vibration isolator, respectively. From Equation 2.57, we get

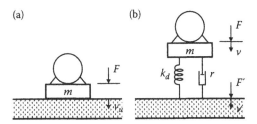

Figure 2.10 Vibrating machinery on a supporting floor (a) without vibration isolators, (b) with vibration isolators.

$$\Delta L = 10 \lg \frac{v_{\text{eff}}^2 \cdot \text{Re}\{Z_m\}}{v_{\text{eff}}'^2 \cdot \text{Re}\{Z_m\}} = 10 \lg \frac{|v|^2}{|v'|^2} \quad \text{(dB)} \tag{2.72}$$

The insertion loss is simply the difference between the velocity levels measured on the floor without and with the vibration isolator.

The force F acts against the point impedance, which is $Z_m + j\omega m$ without the vibration isolator, where m is the total mass of the machine. With reference to Figure 2.10, we have

$$\frac{F}{v} = Z_m + j\omega m \ , \quad \frac{F'}{v'} = Z_m \tag{2.73}$$

Insertion in Equation 2.72 yields

$$\Delta L = 10 \lg \frac{|F|^2 \cdot |Z_m|^2}{|F'|^2 \cdot |Z_m + j\omega m|^2} = -20 \lg |H| - 20 \lg \left| 1 + \frac{j\omega m}{Z_m} \right| \quad \text{(dB)} \tag{2.74}$$

where H is the transmissibility. If the point impedance of the floor is very large, $|Z_m| \gg \omega m$, the last term in Equation 2.74 is negligible. Thus, the insertion loss is, with a good approximation, described by the graphs shown previously in Figure 2.9. If the floor is a homogeneous plate, e.g. a concrete slab, the point impedance can be taken from Table 2.1, and we get

$$\Delta L = -20 \lg |H| - 10 \lg \left(1 + \frac{(\omega m)^2}{64 m'' B} \right) \quad \text{(dB)} \tag{2.75}$$

where m'' is the mass per unit area (kg/m²) and B is the bending stiffness (Nm) of the floor. Obviously, the efficiency of the vibration isolators depends on whether the mass and stiffness of the floor are sufficient to provide a reasonable withhold against the isolators. In order to insulate against

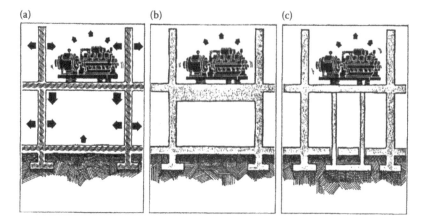

Figure 2.11 Vibration isolation of a machinery: (a) Insufficient resistance from the floor, (b) increased mass of floor, (c) increased stiffness of floor. Drawing by Bernt Forsblad. (Reproduced with permission from Arbetarskyddsnämnden, *Buller bekämpning. Principper och tillämpning.* (Noise abatement. Principles and applications, in Swedish), Arbetarskyddsfonden, Sweden, 1977.)

large forces, it may be necessary to take special steps in order to increase the mass or the stiffness of the supporting floor as illustrated in Figure 2.11.

2.6.6 Insulation against structure-borne vibrations

Vibration insulation is not only applied in cases of vibrating equipment and machines. It is equally important in cases of sensitive equipment that must be protected against vibrations from the environment. Examples include delicate equipment in laboratories or hospitals, and rooms that need very low background noise levels like concert halls and recording studios.

In principle, the vibrating system in such situations is not different from the one discussed in the previous sections, except that the excitation of the system is caused by vibrations in the supporting floor (Figure 2.12). The velocity v_e causes a vertical force that reacts on the spring and the dashpot of the system, and this, in turn, causes the mass of the equipment m to vibrate with the velocity v. Assuming harmonic vibrations, i.e. $d^2x/dt^2 = j\omega v$ and $x = v/j\omega$, application of Newton's second law (Equation 2.2) leads to

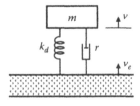

Figure 2.12 Principle of a mass that is isolated from vibrations in the floor.

$$r(v_e - v) + k_d \frac{v_e - v}{j\omega} = j\omega m v \tag{2.76}$$

leading to

$$\frac{v}{v_e} = \frac{r - jk_d / \omega}{r + j(\omega m - k_d / \omega)} = H \tag{2.77}$$

Comparing this with Equation 2.62, we realize that the transmissibility H of the system describes the vibration insulation in this case too. The only difference being that H is now a ratio of velocities, instead of a ratio of forces.

An example is the large anechoic chamber at the Technical University of Denmark, DTU. For such a facility, it is crucial that the insulation against noise and vibrations from the environment is as good as possible. The building is a box-in-box system, where the inner box is made from 40 cm thick concrete. The total mass is 1200 tons supported by 24 vibration isolators made of rubber. Six vibration isolators are placed near each corner (Figure 2.13).

To check the efficiency, vertical vibrations were measured simultaneously below and above the vibration isolators. The accelerometers were mounted on small steel plates, which can be seen in Figure 2.13. The source of vibrations was pile driving for the foundation of a neighbour building. The measured difference between the velocity levels is shown in Figure 2.14 as a function of the frequency. The resonance of the system appears as an amplification around 8 Hz; but at frequencies above 20 Hz, the attenuation is better than 10 dB.

Figure 2.13 Vibration isolators of rubber under the large anechoic chamber at DTU. (Photo by J. H. Rindel.)

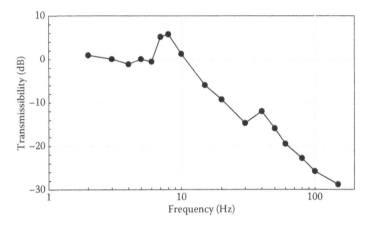

Figure 2.14 Measured attenuation of vibrations in the vertical direction obtained by the vibration isolators under the large anechoic chamber at DTU.

2.6.7 Design of vibration isolators

For the practical design of vibration isolators, the first step is to identify the lowest excitation frequency f that should be attenuated. If the source is an engine with rotating parts, e.g. an electric motor, and the number of rotations per minute is n (min⁻¹), the lowest excitation frequency is

$$f = \frac{n}{60} \quad \text{(Hz)} \tag{2.78}$$

The resonance frequency of the system can then be decided. If there are no detailed requirements, we may apply the simple criterion:

$$f_0 \cong \frac{f}{3} \quad \text{(Hz)} \tag{2.79}$$

From Figure 2.9 it can be seen that at frequency $3f_0$, the attenuation is around 15 dB to 18 dB.

We shall now recapitulate Equation 2.1 for the static load and Equation 2.17 for the dynamic vibration. Combining these equations leads to

$$f_0 = \frac{1}{2\pi} \sqrt{\frac{k_d g}{k_s x_s}} \quad \text{(Hz)} \tag{2.80}$$

where
  g is the gravity acceleration (9.81 m/s²)
  k_d is the dynamic stiffness (N/m)

k_s is the static stiffness (N/m)

x_s is the static deflection (m)

The ratio of dynamic stiffness and static stiffness depends on the actual construction and material of the vibration isolator. A typical value of $k_d/k_s \approx 1.3$ is applicable for many rubber-based isolators. Insertion of this and the value of g yield the simple relationship between resonance frequency and static deflection:

$$f_0 \cong \sqrt{\frac{1}{3x_s}} \quad (\text{Hz}) \tag{2.81}$$

Applying the design criterion (Equation 2.79) leads to the very simple design guide:

$$x_s \cong \frac{1}{3f_0^2} \cong \frac{3}{f^2} \quad (\text{m}) \tag{2.82}$$

Figure 2.15 is a graphical representation of Equation 2.82. With this graph, it is easy to find the necessary static deflection from the known frequency of

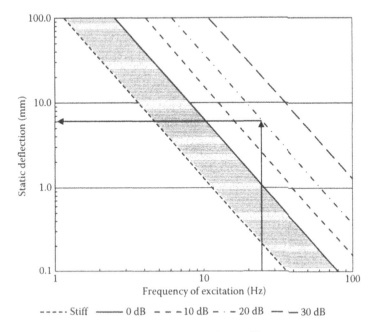

Figure 2.15 Diagram for the design of vibration isolators. The necessary static deflection is determined from the excitation frequency and the attenuation in dB. The shaded area is the resonance region and must be avoided. Example: excitation at 24 Hz, 20 dB attenuation requires 6 mm static deflection.

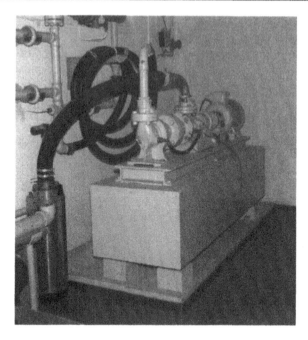

Figure 2.16 Water pump and electric motor mounted on a concrete block, which is supported by four vibration isolators (not visible on the photo). (Photo by J. H. Rindel.)

excitation and the wanted amount of attenuation. The resonance region as clearly marked and should always be avoided. The example shown in Figure 2.15 is for a water pump driven by an electric motor with 1450 turns per minute; that gives $f \approx 24\,\text{Hz}$, and 20 dB attenuation would then require $x_s = 6\,\text{mm}$.

A photo of this water pump is seen in Figure 2.16. In this case, vibrations cannot be tolerated in the building, so the vibration isolation had to be extremely good. Several practical measures were taken to obtain this. First, the vibrating equipment was mounted on a big, heavy concrete block; this means that the velocity created by the equipment is reduced to a minimum, even before the vibration isolators. Figure 2.16 also shows that all tubes and connections to and from the equipment are made flexible – this is vital to avoid any rigid connections between the vibrating equipment and building.

2.7 HUMAN RESPONSE TO VIBRATIONS

The human body can sense vibrations in the frequency range from 1 Hz to 80 Hz. The sensitivity depends on the direction of vibration relative to the body. Hence, in a standing position, vertical vibrations are sensed particularly well for accelerations in the range from 4 Hz to 8 Hz, while in a lying position, the body is most sensitive in the frequency range from 1 Hz to 2 Hz. Thus, it is common practice to apply a combined sensitivity curve (Figure 2.17).

The base curves represent curves of equal human response with respect to annoyance or complaints. The base curves can be considered the approximate lower threshold for vibrations, i.e. no adverse sensations or complaints are expected below these curves.

Vibrations are measured either with an accelerometer or with a geophone. While the former responds to acceleration and is commonly used in acoustics, the latter converts velocity into voltage and is commonly used in seismology and geo-techniques.

It is convenient to apply a frequency weighting filter for measurements (Figure 2.18). These filters approximately mirror the base curves for human sensitivity. This is analogous to using the A-weighting filter for measuring sound pressure levels.

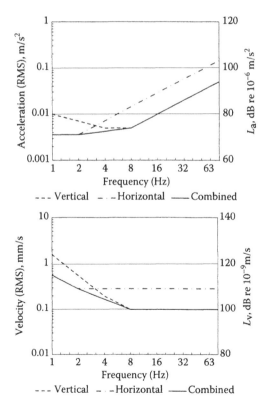

Figure 2.17 Base curves for building vibrations: (top): acceleration; (bottom): velocity. (Adapted from ISO 2631-2, *Evaluation of human exposure to whole-body vibration – Part 2: Continuous and shock-induced vibration in buildings (1 to 80 Hz)*, International Organization for Standardization, Geneva, Switzerland, 1989.)

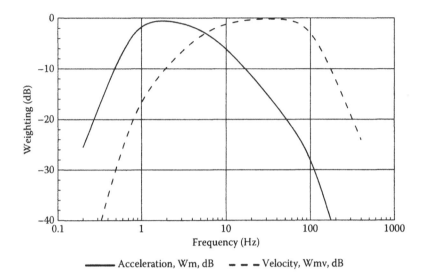

Figure 2.18 Frequency weighting applied to vibration measurements, acceleration and velocity. (Adapted from ISO 2631-2, *Evaluation of human exposure to whole-body vibration – Part 2: Continuous and shock-induced vibration in buildings (1 to 80 Hz)*. International Organization for Standardization, Geneva, Switzerland, 2003; NS 8176, *Vibration and shock – Measurement of vibration in buildings from land based transport and guidance to evaluation of effects on human beings*. (In Norwegian, 2nd edition, Standard Norge, Lysaker, Norway, 2005.)

An interesting feature of the frequency weighting filters is that there is a simple relationship between the weighted acceleration and the weighted velocity:

$$v_w \cong k \cdot a_w \quad \text{(m/s)} \tag{2.83}$$

where $k = 0.028\,\text{s}$ is a constant. Similarly, for the frequency weighted vibration levels, the relationship is

$$L_{v,w} \cong L_{a,w} + 29\,\text{dB} \tag{2.84}$$

Without the weighting filters, the relationship is frequency dependent, $a = j\omega v$.

2.7.1 Examples of evaluation criteria

The threshold for whole-body sensation of vibrations are, according to the curves in Figure 2.17, approximately a weighted acceleration level of 71 dB re 1 μm/s^2 (RMS acceleration 3.6 mm/s^2). Using the velocity level, the threshold is 100 dB re 1 nm/s (RMS velocity 0.1 mm/s). This may be used as an evaluation criterion for critical areas, e.g. hospital operation theatres or precision laboratories. Other examples are shown in Table 2.3.

Table 2.3 Examples of evaluation criteria in terms of weighted acceleration level
or weighted velocity level

Place	$L_{a,w}$ dB re 1 µm/s²		$L_{v,w}$ dB re 1 nm/s	
	Continuous	Transient	Continuous	Transient
Critical working areas	71	71	100	100
Residential, day	74–83	107–110	103–112	136–139
Residential, night	74	74–97	103	103–126
Office	83	107–113	112	136–142
Workshop	89	107–113	118	136–142

Source: ISO 2631-2, *Evaluation of human exposure to whole-body vibration – Part 2: Continuous and shock-induced vibration in buildings (1 to 80 Hz)*. International Organization for Standardization, Geneva, Switzerland, 1989.

REFERENCES

Arbetarskyddsnämnden (1977). *Buller bekämpning. Principper och tillämpning* (Noise abatement. Principles and applications, in Swedish). Arbetarskyddsfonden, Stockholm, Sweden.

L. Cremer, and M. Heckl (1967). *Körperschall.* (*Structure-borne sound*, translated by E.E. Ungar, 1973). Springer, Berlin, Germany.

ISO 2041 (2009). *Mechanical vibration, shock and condition monitoring – Vocabulary*. International Organization for Standardization, Geneva, Switzerland.

ISO 2631-2 (1989). *Evaluation of human exposure to whole-body vibration – Part 2: Continuous and shock-induced vibration in buildings (1 to 80 Hz)*. International Organization for Standardization, Geneva, Switzerland.

ISO 2631-2 (2003). *Mechanical vibration and shock – Evaluation of human exposure to whole-body vibration – Part 2: Vibration in buildings (1 to 80 Hz)*. International Organization for Standardization, Geneva, Switzerland.

NS 8176 (2005). *Vibration and shock – Measurement of vibration in buildings from land based transport and guidance to evaluation of effects on human beings.* (In Norwegian), 2nd edition. Standard Norge, Lysaker, Norway.

NT ACOU 082 (1991). *Buildings – Vibration and shock, evaluation of annoyance.* Nordtest, Espoo, Finland. ISSN 0283-7145.

Chapter 3

The sound field in front of a wall

3.1 NORMAL INCIDENCE OF A PLANE WAVE

We will consider the sound field in front of a wall, which is assumed to be perfectly plane, smooth and infinitely large. However, the results will also be valid approximations for walls encountered in practice as long as (1) their dimensions are much larger than the wavelength and (2) any irregularities in the surface are much smaller than the wavelength of the sound.

First, the case of normal incidence is considered, i.e. the wall is perpendicular to the direction of propagation of the incoming sound wave, which we assume to be a single harmonic sound with angular frequency ω. We place the x-axis in the direction of propagation, so that $x=0$ at the surface (Figure 3.1).

The sound pressure p_i and the particle velocity u_i in the incoming sound wave can be written as

$$p_i = |p_i| e^{j(\omega t - kx)} \tag{3.1a}$$

$$u_i = \frac{|p_i|}{\rho c} e^{j(\omega t - kx)} \tag{3.1b}$$

For a harmonic sound wave, the particle velocity component along a given axis direction x can be obtained through partial differentiation of the sound pressure:

$$u_x = -\frac{\partial \Phi}{\partial x} = -\frac{1}{j\omega\rho} \cdot \frac{\partial p}{\partial x} \tag{3.2}$$

with $\Phi = |\Phi| e^{j\omega t}$ being the velocity potential, and the sound pressure being

$$p = \rho \frac{\partial \Phi}{\partial t} = j\omega\rho \cdot \Phi \tag{3.3}$$

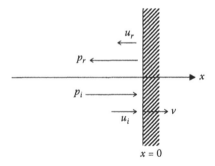

Figure 3.1 Reflection of a plane wave at normal incidence.

Normally, the reflected sound wave will have lower amplitude and a different phase compared to the incoming sound wave. Both can be considered by introducing a complex reflection factor, R. Consequently, the sound pressure and particle velocity of the reflected wave can be written as

$$p_r = R \cdot |p_i| e^{j(\omega t + kx)} \tag{3.4a}$$

$$u_r = -R \cdot \frac{|p_i|}{\rho c} e^{j(\omega t + kx)} \tag{3.4b}$$

Notice that the change in direction of propagation implies a shift in the sign of kx. Also, the sign of the particle velocity has changed compared to that of the incoming sound wave in Equation 3.1b. The sound pressure and particle velocity in the combined sound field in front of the wall are found by adding the expressions in Equations 3.1 and 3.4:

$$p = |p_i| \cdot \left(e^{-jkx} + R \cdot e^{jkx} \right) e^{j\omega t} \tag{3.5a}$$

$$u = \frac{|p_i|}{\rho c} \cdot \left(e^{-jkx} - R \cdot e^{jkx} \right) e^{j\omega t} \tag{3.5b}$$

The particle velocity at the wall surface is perpendicular to the wall and equal to the normal component of the wall velocity: v. Setting $x=0$ in Equation 3.5b yields

$$v = u_{(x=0)} = \frac{|p_i|}{\rho c} \cdot (1 - R) e^{j\omega t} \tag{3.6}$$

If the wall is sufficiently hard and heavy, $v=0$ resulting in $R=1$, and the wall is said to be totally reflecting. In front of such a wall, we can find p and u from Equation 3.5 as

$$p = 2|p_i| \cdot \cos(kx) e^{j\omega t} \tag{3.7a}$$

$$u = -2j \frac{|p_i|}{\rho c} \cdot \sin(kx) e^{j\omega t} \tag{3.7b}$$

Notice that none of these expressions contain a term of the form e^{jkx}, i.e. the combined wave is not propagating; we have a 'standing wave' in front of the wall.

The energy density in the sound field is distributed between a potential energy density and a kinetic energy density given by

$$w = w_{\text{pot}} + w_{\text{kin}} \tag{3.8}$$

where

$$w_{\text{pot}} = \frac{1}{2} \cdot \frac{\tilde{p}^2}{\rho c^2} = \frac{1}{4} \cdot \frac{|p|^2}{\rho c^2} \tag{3.9a}$$

$$w_{\text{kin}} = \frac{1}{2} \cdot \rho \tilde{u}^2 = \frac{1}{4} \cdot \rho |u|^2 \tag{3.9b}$$

These energy densities are proportional to the amplitudes of the pressure squared and the velocity squared, respectively. For the incident sound field, which is a plane propagating wave, we have

$$w_i = \frac{1}{4} \cdot \frac{|p_i|^2}{\rho c^2} + \frac{1}{4} \cdot \rho \frac{|p_i|^2}{(\rho c)^2} = \frac{1}{2} \cdot \frac{|p_i|^2}{\rho c^2} \tag{3.10}$$

In the standing wave in front of the wall, we have from Equation 3.7:

$$|p|^2 = 4|p_i|^2 \cdot \cos^2(kx) = 2|p_i|^2 \cdot (1 + \cos(2kx)) \tag{3.11a}$$

$$|u|^2 = 4 \frac{|p_i|^2}{(\rho c)^2} \cdot \sin^2(kx) = 2 \frac{|p_i|^2}{(\rho c)^2} \cdot (1 - \cos(2kx)) \tag{3.11b}$$

From this it can be seen that the potential energy has a maximum and the kinetic energy has a minimum (zero) when $2kx=n2\pi$, i.e. for $x=n\lambda/2$ where

$n=0, 1, 2, 3$, etc. The kinetic energy is maximum and the potential energy is minimum (zero) when $2kx=\pi+n2\pi$, i.e. for $x=\lambda/4+n\lambda/2$. These are the characteristic nodal points and maxima in a standing wave in front of a wall.

The total energy density is found by inserting Equation 3.11 in Equations 3.9 and 3.8:

$$w = \frac{1}{4}\cdot\frac{|p_i|^2}{\rho c^2}2\big(1+\cos(2kx)\big)+\frac{1}{4}\cdot\rho\frac{|p_i|^2}{(\rho c)^2}2\big(1-\cos(2kx)\big)=\frac{|p_i|^2}{\rho c^2} \quad (3.12)$$

It appears that the total energy density is independent of the distance from the wall. By comparison with Equation 3.10, it can also be seen that the energy density is simply doubled compared to that of the incoming plane wave alone.

3.2 OBLIQUE INCIDENCE OF A PLANE WAVE

We now assume that the angle between the direction of propagation of the plane sound wave and the normal of the wall equals θ. We also assume the direction of propagation to be in the xy plane as shown in Figure 3.2. This corresponds to a transformation of the coordinates in which x in Equation 3.1 is substituted by $x\cos\theta+y\sin\theta$:

$$p_i = |p_i|e^{-jk(x\cos\theta+y\sin\theta)}e^{j\omega t} \tag{3.13a}$$

$$u_{ix} = \frac{|p_i|}{\rho c}\cos\theta\cdot e^{-jk(x\cos\theta+y\sin\theta)}e^{j\omega t} \tag{3.13b}$$

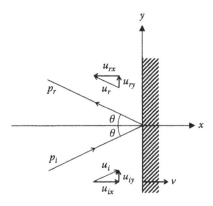

Figure 3.2 Reflection of a plane wave at oblique incidence.

$$u_{iy} = \frac{|p_i|}{\rho c} \sin\theta \cdot e^{-jk(x\cos\theta + y\sin\theta)} e^{j\omega t} \tag{3.13c}$$

in which the components of the particle velocity in the two axis directions have been found using Equation 3.2.

Likewise for the reflected wave, and remembering the change in sign for the x term, we get

$$p_r = R \cdot |p_i| e^{-jk(-x\cos\theta + y\sin\theta)} e^{j\omega t} \tag{3.14a}$$

$$u_{rx} = -R \cdot \frac{|p_i|}{\rho c} \cos\theta \cdot e^{-jk(-x\cos\theta + y\sin\theta)} e^{j\omega t} \tag{3.14b}$$

$$u_{ry} = R \cdot \frac{|p_i|}{\rho c} \sin\theta \cdot e^{-jk(-x\cos\theta + y\sin\theta)} e^{j\omega t} \tag{3.14c}$$

Through addition of Equations 3.13 and 3.14, we obtain for the total sound field in front of a hard wall with reflection factor $R = 1$:

$$p = 2 \cdot |p_i| \cos(kx\cos\theta) e^{-jky\sin\theta} e^{j\omega t} \tag{3.15a}$$

$$u_x = -2j \cdot \frac{|p_i|}{\rho c} \cos\theta \cdot \sin(kx\cos\theta) e^{-jky\sin\theta} e^{j\omega t} \tag{3.15b}$$

$$u_y = 2 \cdot \frac{|p_i|}{\rho c} \sin\theta \cdot \cos(kx\cos\theta) e^{-jky\sin\theta} e^{j\omega t} \tag{3.15c}$$

By comparing with Equation 3.7, we see that the sound field corresponds to a standing wave with the distance $\lambda/(2\cos\theta)$ between the nodal points and between maxima; but at the same time, this field propagates parallel with the wall with the speed given by

$$c_y = \frac{c}{\sin\theta} \tag{3.16}$$

Notice that the speed of the phase is $c_y > c$, i.e. supersonic. The 'trace' of the sound field seems to move with the phase speed along the y-axis.

By squaring the amplitudes in Equation 3.15, we obtain

$$|p|^2 = 2 \cdot |p_i|^2 \left(1 + \cos(2kx\cos\theta)\right) \tag{3.17a}$$

$$|u_x|^2 = 2 \cdot \frac{|p_i|^2}{(\rho c)^2} \cos^2 \theta \cdot \left(1 - \cos\left(2kx\cos\theta\right)\right)$$

(3.17b)

$$|u_y|^2 = 2 \cdot \frac{|p_i|^2}{(\rho c)^2} \sin^2 \theta \cdot \left(1 + \cos\left(2kx\cos\theta\right)\right)$$

(3.17c)

Thus, we find for the resulting particle velocity:

$$|u|^2 = |u_x|^2 + |u_y|^2 = 2 \cdot \frac{|p_i|^2}{(\rho c)^2} \cdot \left(1 - \cos\left(2\theta\right)\cos\left(2kx\cos\theta\right)\right)$$

(3.18)

Here we have made use of the mathematical relation (Equation 1.36).

While the potential energy equals zero in certain nodal points (see Equation 3.17a), this is not the case for the kinetic energy when $\theta \neq 0$. The reason is that only the normal component of the particle velocity, u_x is zero at the wall surface. The tangential component u_y reaches a maximum value at the wall and is proportional to p:

$$|u_y|^2 = \frac{|p|^2}{(\rho c)^2} \sin^2 \theta$$

(3.19)

The total energy density can be found by entering Equations 3.17a and 3.18 into Equations 3.8 and 3.9 and making use of the mathematical relation (Equation 1.38):

$$w = \frac{|p_i|^2}{\rho c^2} \cdot \left(1 + \sin^2 \theta \cdot \cos\left(2kx\cos\theta\right)\right)$$

(3.20)

It is worth mentioning that the total energy density is not constant in all parts of the sound field in the case of oblique incidence. The maxima occur, e.g. at the wall surface.

3.3 RANDOM INCIDENCE IN A DIFFUSE SOUND FIELD

In this section, we shall assume a hard wall only with reflection factor $R=1$. If the intensity of the incoming wave is the same from all angles of incidence (or the probability of all angles of incidence is the same) and if all incoming sound waves are uncorrelated, then it is possible to calculate the potential and kinetic energy densities in the diffuse field by integrating the expressions valid for a single angle of incidence over the solid angle 2π. From Equation 3.17a, we get for the pressure squared:

$$|p|^2 = \frac{1}{2\pi} \int_0^{2\pi} d\phi \int_0^{\pi/2} 2 \cdot |p_i|^2 \left(1 + \cos\left(2kx \cos\theta\right)\right) \cdot \sin\theta \, d\theta$$

$$= 2 \cdot |p_i|^2 \left(1 + \frac{\sin(2kx)}{2kx}\right) = 2 \cdot |p_i|^2 \left(1 + \mathrm{sinc}\left(2kx\right)\right) \tag{3.21}$$

Here and in the following, the sinc function (Equation 1.42) is applied for simplicity. Notice that in large distances from the wall and very close to the wall, we have, respectively:

$$|p|^2_\infty = 2 \cdot |p_i|^2 \quad \text{for } x \to \infty \tag{3.22a}$$

$$|p|^2_{x=0} = 4 \cdot |p_i|^2 = 2 \cdot |p|^2_\infty \quad \text{for } x \to 0 \tag{3.22b}$$

Likewise, from Equation 3.18, we get for the squared particle velocity:

$$|u|^2 = 2 \cdot \frac{|p_i|^2}{\rho c^2} \left(1 - \frac{\sin(2kx)}{2kx} + \frac{\dfrac{\sin\left(2kx\right)}{2kx} - \cos\left(2kx\right)}{(kx)^2}\right)$$

$$= 2 \cdot \frac{|p_i|^2}{\rho c^2} \left(1 - \mathrm{sinc}\left(2kx\right) + \frac{\mathrm{sinc}\left(2kx\right) - \cos\left(2kx\right)}{(kx)^2}\right) \tag{3.23}$$

In a large distance from the wall, Equation 3.23 yields

$$|u|^2_\infty = 2 \cdot \frac{|p_i|^2}{(\rho c)^2} \quad \text{for } x \to \infty \tag{3.24a}$$

Close to the wall, Equation 3.23 cannot be used, but we find the limit by setting $x=0$ in Equation 3.18 and repeating the integration:

$$|u|^2_{x=0} = 2 \cdot \frac{|p_i|^2}{\rho c^2} \int_0^{\pi/2} \left(1 - \cos\left(2\theta\right)\right) \cdot \sin\theta \, d\theta$$

$$= |u|^2_\infty \int_0^{\pi/2} 2 \cdot \sin^3\theta \, d\theta$$

$$= \tfrac{4}{3} |u|^2_\infty \quad \text{for } x \to 0 \tag{3.24b}$$

The total energy density is found by inserting Equations 3.21 and 3.23 in Equation 3.8:

$$w = \frac{|p_i|^2}{\rho c^2}\left(1 + \frac{\mathrm{sinc}(2kx) - \cos(2kx)}{2(kx)^2}\right) \tag{3.25}$$

Far from the wall, Equation 3.25 yields, as expected, recalling Equation 3.12:

$$w_\infty \to \frac{|p_i|^2}{\rho c^2} = 2w_i \quad \text{for } x \to \infty \tag{3.26a}$$

Very close to the wall, we find by applying Equations 3.22b and 3.24b:

$$w_{x=0} \to \frac{|p_i|^2}{\rho c^2}\left(1 + \frac{2}{3}\right) = \frac{5}{3}w_\infty \quad \text{for } x \to 0 \tag{3.26b}$$

Hence, we see that in a diffuse sound field, the total energy density at the walls (with $R=1$) is 10 log $(5/3)=2.2$ dB higher than the asymptotic value far from the walls. Likewise, from Equations 3.22 and 3.24, we see that the sound pressure is 3 dB higher and the particle velocity is 1.2 dB higher than the respective asymptotic values far from the walls. In Table 3.1, the results for the sound field at the surface of a hard wall relative to the asymptotic values far from the wall are summarized.

The main results so far are illustrated in Figure 3.3. For normal, oblique and random sound incidence on a hard wall, the patterns of potential, kinetic and total energy densities are shown – all relative to the asymptotic values far from the wall (two times the corresponding values in the incoming wave). In the case of random incidence, notice how the amplitudes of the fluctuations decrease quickly with the distance from the wall. Already at the distance of $\lambda/4$, the sound pressure has reached its asymptotic value, and only minor deviations from this value occur further away from the wall.

Table 3.1 Levels of sound pressure, particle velocity and total energy density at the surface of a hard wall, relative to the asymptotic values far from the wall

	Sound pressure (dB)	Particle velocity	Total energy density
Normal incidence	3	$\to -\infty$ dB	0 dB
Oblique incidence	3	10 lg $(2 \sin^2\theta)$	10 lg $(1+\sin^2\theta)$
Random incidence	3	1.2 dB	2.2 dB

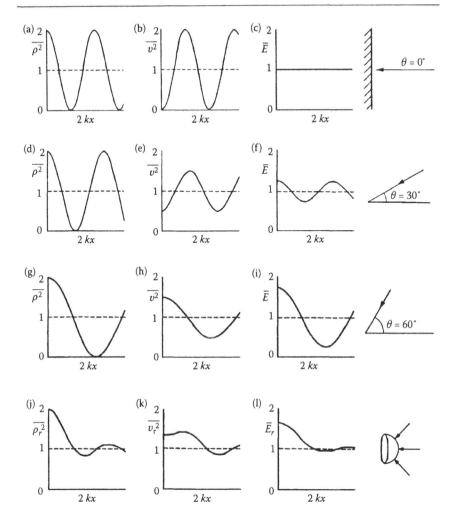

Figure 3.3 Sound field energy components in front of a totally reflecting wall. Four rows from above: plane incident sound waves with angles of incidence 0°, 30°, 60° and diffuse field random incidence. In each case, the three columns show potential, kinetic and total energy densities relative to the values far from the wall. (Reproduced from Waterhouse, R.V., J. Acoust. Soc. Am., 27, 247–258, 1955, Figure 1. With the permission of the Acoustical Society of America.)

For physical reasons, one should expect the total energy density to be equally distributed between potential and kinetic energy in the total sound field, i.e.

$$\int_0^\infty w_{\text{pot}} \, dx = \int_0^\infty w_{\text{kin}} \, dx \tag{3.27}$$

As we see immediately from Equations 3.9, 3.22a and 3.24a, the asymptotic values are equal:

$$w_{\text{pot},\infty} = w_{\text{kin},\infty} = 2 \cdot \frac{|p_i|^2}{\rho c^2} \quad \text{for } x \to \infty \tag{3.28}$$

Consequently, we only need to calculate any addition to this value due to the increase in potential energy density close to the wall:

$$\int_0^\infty \frac{w_{\text{pot}} - w_{\text{pot},\infty}}{w_{\text{pot},\infty}} \, dx = \int_0^\infty \left(\frac{|p|^2}{|p|_\infty^2} - 1 \right) dx = \int_0^\infty \text{sinc}(2kx) \, dx = \frac{1}{2k} \cdot \frac{\pi}{2} = \frac{\lambda}{8} \tag{3.29}$$

As shown by Waterhouse (1955), the same result is obtained for the kinetic energy density and, thus, for the total energy density, too. It is remarkable that acoustic energy is slightly increased near the walls in a diffuse sound field. The additional energy corresponds to a sound field with constant energy density also occupying the volume extending $\lambda/8$ beyond the physical walls. In a large room (dimensions $\gg$ wavelength), the total energy is approximately

$$E_{\text{tot}} = \langle w \rangle \cdot \left(V + S \frac{\lambda}{8} \right) \tag{3.30}$$

where $\langle w \rangle$ is the energy density averaged over the room volume avoiding regions very close to the walls, V is the room volume and S the total area of the inner wall surfaces.

This implies that to find the total energy in a room based on a measurement of the average energy density in the central part of the room, the measurement result should be corrected with the amount:

$$C_W = 10 \lg \left(1 + \frac{\lambda S}{8V} \right) (\text{dB}) \tag{3.31}$$

This correction is sometimes called the 'Waterhouse correction' because it was first derived by Waterhouse (1955) in a pioneering paper that also treats the sound field at the edge between two walls and in wall corners. While negligible at high frequencies, the correction amounts to a few dB below 500 Hz, depending on the volume. Let us summarize the assumptions on which the Waterhouse correction is based:

- Totally reflecting walls. This may be approximately right at low frequencies in a room with heavy walls, but questionable with lightweight drywalls.
- Random incidence on the surfaces, i.e. a diffuse sound field. This assumption is not fulfilled at very low frequencies in small rooms

where axial modes dominate the sound field. Axial modes correspond to the case of normal incidence, where there is no correction.

- Only the energy condition of a large surface is considered. The energy conditions near edges and corners are not included.

3.4 THE SOUND FIELD AT EDGES AND CORNERS IN A ROOM

In the previous sections, we have derived expressions for the behaviour of the sound field in front of a totally reflecting wall. In the same way, expressions can be derived for the sound field in front of room edges and corners as performed by Waterhouse (1955). In the following, only a qualitative description and presentation of the main results will be given and only in the case of diffuse sound field incidence.

As can be seen from Figure 3.4, at a certain point of observation, the sound field can be regarded as the combination of the undisturbed, incident sound field at this point and at one or more images of this point. A single wall only provides one image point and hereby, the sound field consists of two contributions. A right-angled edge results in four contributions of which one is a second-order image, and a right-angled corner results in eight contributions totally.

The properties of the resulting sound field depend on the correlation between the individual contributions to the sound field. If the point of observation is sufficiently far away from the reflecting surfaces (in a diffuse field at least $\lambda/4$), we can regard the different contributions as mutually uncorrelated, and the resulting sound field can, therefore, be regarded as the energy sum of the contributions. With the pressure of the incoming sound wave denoted as p_i and the resulting sound field far from the walls (the far field) denoted as p_∞, we have

$$\frac{|p_\infty|^2}{|p_i|^2} = \begin{cases} 2 & \text{at a wall} \\ 4 & \text{at an edge} \\ 8 & \text{in a corner} \end{cases} \tag{3.32}$$

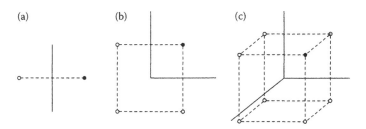

Figure 3.4 Illustration of (a) a point of observation (black spot) and its images in a wall; (b) a right-angled edge; and (c) a right-angled corner.

However, if the point of observation is moved close to the walls so that the distance between the point of observation and its images becomes very small compared to the wavelength, all the contributions to the sound field become correlated and in phase. The resulting sound field can, therefore, be described by the pressure sum of the contributions, and we get

$$\frac{|p|_{x=0}}{|p_i|} = \begin{cases} 2 & \text{at a wall} \\ 4 & \text{at an edge} \\ 8 & \text{in a corner} \end{cases} \tag{3.33}$$

In a diffuse sound field (random incidence), the squared sound pressure in front of a totally reflecting wall relative to the asymptotic value, from Equation 3.21, is

$$\frac{|p|^2}{|p|_\infty^2} = 1 + \text{sinc}(2kx) \tag{3.34}$$

Similarly, near a rectangular edge lying along the x-axis, the result is (Waterhouse 1955)

$$\frac{|p|^2}{|p|_\infty^2} = 1 + \text{sinc}(2ky) + \text{sinc}(2kz) + \text{sinc}(2kd_x) \tag{3.35}$$

where $d_x = \sqrt{(y^2 + z^2)}$. Near a rectangular corner with the x-, y- and z-axis along the three edges forming the corner, the result becomes

$$\frac{|p|^2}{|p|_\infty^2} = 1 + \text{sinc}(2kx) + \text{sinc}(2ky) + \text{sinc}(2kz) + \text{sinc}(2kd_x) \tag{3.36}$$

$$+ \text{sinc}(2kd_y) + \text{sinc}(2kd_z) + \text{sinc}(2kd)$$

where $d_y = \sqrt{(x^2 + z^2)}$, $d_z = \sqrt{(x^2 + y^2)}$ and $d = \sqrt{(x^2 + y^2 + z^2)}$.

The relative sound pressure level (Figure 3.5) is

$$\Delta L = 10 \lg \frac{|p|^2}{|p|_\infty^2} \, (\text{dB}) \tag{3.37}$$

as function of the distance d, which in this case means either the distance perpendicular from a wall, or the distance from a 2D edge along the axis of symmetry ($y=z$ in Equation 3.35), or the distance from a 3D corner along the axis of symmetry ($x=y=z$ in Equation 3.36). It is obvious that the fluctuations of ΔL near the reflecting surfaces are exaggerated when going from

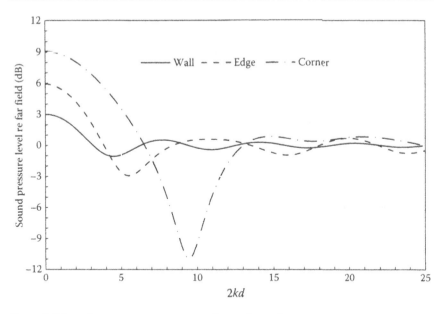

Figure 3.5 Interference pattern near a wall, an edge and a corner.

Table 3.2 Some characteristics of the sound pressure level near walls, edges and corners assuming hard reflective surfaces and a diffuse sound field

	Relative sound pressure level ΔL at $d=0$ (dB)	Distance d to first $\Delta L=0$ dB	Distance d to first minimum	Relative sound pressure level ΔL at first minimum (dB)
Wall	3	0.25λ	0.35λ	−1
Edge (2D)	6	0.32λ	0.45λ	−3
Corner (3D)	9	0.53λ	0.78λ	−11

Note: ΔL is sound pressure level relative to the asymptotic values in the far field.

a single wall to an edge and further to a corner. In the latter case, we have variations from +9 dB to −11 dB and the distance from the corner must be $d > \lambda$ in order to reach small fluctuations within ca. ±1 dB. More details from the curves are listed in Table 3.2.

3.5 THE INFLUENCE OF BANDWIDTH ON REFLECTION OF NOISE

In the previous section, we have treated the sound field in front of a wall for a single frequency. If the sound field contains white or pink noise in a certain frequency band $\Delta f = f_2 - f_1$, we can carry out a simple energy summation of the mutually uncorrelated frequency components.

For noise incident from one direction with angle of incidence θ relative to the normal of the totally reflecting wall, Equation 3.17a gives us the squared sound pressure amplitude. Using $k=2\pi f/c$ and the distance from the wall $x=d$, we get

$$\frac{|p|^2}{2\cdot|p_i|^2} = \frac{1}{f_2-f_1}\int_{f_1}^{f_2}\left(1+\cos\left(\frac{4\pi}{c}fd\cos\theta\right)\right)\mathrm{d}f$$

$$= \frac{1}{f_2-f_1}\left(f_2-f_1+\frac{\sin\left(\dfrac{4\pi}{c}f_2d\cos\theta\right)-\sin\left(\dfrac{4\pi}{c}f_1d\cos\theta\right)}{\dfrac{4\pi}{c}f_2d\cos\theta}\right)$$

$$= \left(1+\frac{2\cdot\cos\left(\dfrac{4\pi(f_2+f_1)}{2c}d\cos\theta\right)\cdot\sin\left(\dfrac{4\pi(f_2-f_1)}{2c}d\cos\theta\right)}{\dfrac{4\pi}{c}(f_2-f_1)d\cos\theta}\right)$$

$$= 1+\cos(2X)\cdot\mathrm{sinc}(Y) \tag{3.38}$$

where

$$X = \frac{2\pi}{c}f_md\cos\theta \quad \left(f_m = \tfrac{1}{2}(f_2+f_1)\right) \tag{3.39a}$$

$$Y = \frac{2\pi}{c}\Delta fd\cos\theta \quad (\Delta f = f_2-f_1) \tag{3.39b}$$

This result was first derived and validated by Rindel (1978).

It appears that the interference pattern in front of the wall depends on two parameters – X, which is controlled by the arithmetic mean frequency of the frequency band, and Y, which is controlled by the bandwidth. In addition, the distance from the wall and the angle of incidence are variables in both parameters. The difference between the centre frequency, which is the geometric mean, and the arithmetic mean frequency of the frequency band is small for octave bands ($f_m/f_c=1.06$) and negligible for one-third octave bands ($f_m/f_c=1.0066$).

Unless the bandwidth is zero, Equation 3.38 has the asymptotic value of unity. This value corresponds to a simple energy summation of incident and reflected noise without any interference. Consequently, this is a natural normalization factor for the interference function in Equation 3.38. Figure 3.6 shows four examples of what the interference pattern looks like for different bandwidths of noise. The abscissa is $X=(2\pi/c)f_md\cos\theta$. As can be seen, a standing wave occurs near the wall – like in the case of a pure tone of frequency f_m; but the larger the bandwidth, the faster the

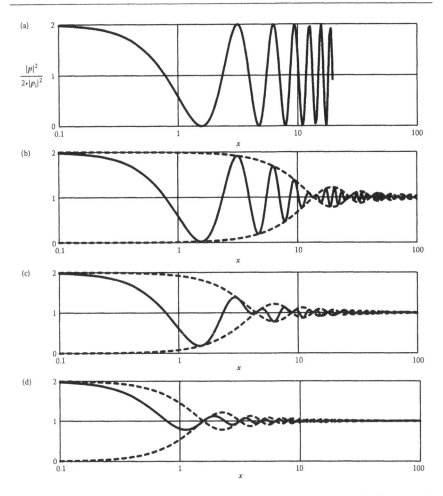

Figure 3.6 Examples of interference pattern corresponding to Equation 3.38 for four different values of the noise bandwidth. Valid for totally reflecting walls and a single angle of incidence. (a) pure tone; (b) one-third octave bandwidth; (c) octave bandwidth; and (d) wide band. The dashed lines are the sinc (Y) function.

interference pattern converges towards the asymptotic value. The bandwidth determines an envelope function that limits the fluctuations. For wide band noise, $\Delta f = 2f_m$, noticeable interference occurs only very close to the wall, see Figure 3.6d.

The interference in the sound pressure level close to the wall can be expressed as a correction relative to the sound pressure level when incident and reflected sound are added on a simple energy basis:

$$\Delta L_{int} = 10\lg\left(\frac{|p|^2}{2\cdot|p_i|^2}\right) = 10\lg\left(1 + \cos(2X)\cdot\text{sinc}(Y)\right)(\text{dB}) \qquad (3.40)$$

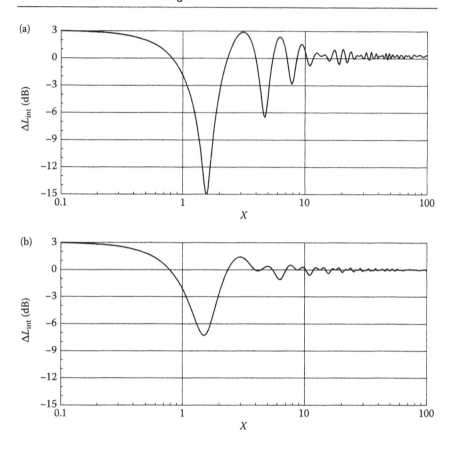

Figure 3.7 Interference correction in front of a large reflecting wall as a function of the parameter X. (a) one-third octave bands and (b) octave bands.

In Figure 3.7, tha interference correction is shown for one-third octave bands and octave bands as a function of frequency, distance and angle of incidence; the abscissa is $X=(2\pi/c)f_m d \cos \theta$. Significant negative interference is noted around $X=1.6$, which is at the distance $d=\lambda/(4 \cos \theta)$.

When measuring the sound pressure level in front of a large reflecting surface, e.g. a building façade, there are two possibilities for the choice of microphone position; either very close to the surface where the interference causes $\Delta L_{int}=3$ dB, or in a sufficiently long distance from the surface where the interference is negligible, $\Delta L_{int} \approx 0$ dB. The former requires the distance to be $d < 0.075\lambda/\cos \theta$ in order to ensure $\Delta L_{int} > 2$ dB. The most critical here are the high frequencies and normal incidence of sound. Hence, the distance should be $d < 8$ mm at 3150 Hz. This may just be obtained with a ½ inch microphone mounted with axis parallel to the wall, whereas a microphone with 1 inch diaphragm is too big. Alternatively, the microphone may be mounted into a hole in the wall with the diaphragm flush with the wall.

The other option is to measure in distance $d > 1.7\lambda/\cos\theta$ in order to ensure $|L_{int}| < 1\,\text{dB}$. In this case, the most critical are low frequencies and oblique incidence of sound (see examples in Table 3.3). Hence, at 100 Hz one-third octave, the minimum distance should be around 6 m, and even more if the angle of incidence exceeds 15°. This means that in practice it is not possible to avoid measurement errors due to interference.

When measuring façade sound insulation in accordance with ISO 16283-3, there are two possibilities for the sound source. Either it can be a loudspeaker with an angle of incidence 45°, or the source can be traffic noise emitted from a line, preferably covering a range of angles within ±60°. For such a line source, the time-averaged sound field can be assumed to contain a constant sound intensity from all directions of incidence within the angular range, under which the road is seen from the measurement position. The resulting interference pattern is calculated by numerical integration of Equation 3.40 and the result is shown in Figure 3.8. It is to be noted that

Table 3.3 Minimum distance (in m) from a reflecting wall in order to have $|\Delta L_{int}|$ < 1 dB at 100 Hz one-third octave band or 125 Hz octave band for oblique incidence at various angles or for a line source covering the range from −60° to +60°

Angle of incidence (°)	One-third octave100 Hz	Octave125 Hz
0	5.8	2.7
15	6.0	2.8
30	6.7	3.1
45	8.2	3.8
60	11.6	5.4
75	22.3	10.5
Line (±60°)	2.2	1.2

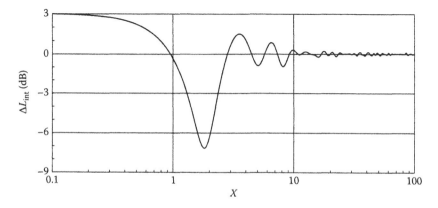

Figure 3.8 Interference correction for one-third octave bands, averaging angles of incidence with in the range from −60° to +60°. The parameter is $X = (2\pi/c)f_m d$.

the interference pattern is less pronounced compared to the case of a fixed angle of incidence.

The standard ISO 16283-3 prescribes measurements to be made either flush with the façade or at a distance of 2.0 m±0.2 m. The theoretical measurement errors in the lower one-third octave bands are shown in Table 3.4 for the distance 2 m. Below 400 Hz, the interference causes errors around ±2 dB for the loudspeaker measurements, whereas the error is around ±1 dB for the traffic noise measurements.

Measurements in front of building façades put great demands on the choice of microphone position; but if the bandwidth can be extended, e.g. to octave instead of one-third octave, the interference problems are substantially reduced. By measurements of A-weighted traffic noise levels, the bandwidth is so large that interference is no longer a problem. However, in

Table 3.4 Theoretical measurement errors due to interference for one-third octave bands with the microphone 2 m in front of the façade

Frequency (Hz)	45° ΔL_{int} (dB)	Traffic ΔL_{int} (dB)
100	1.7	−1.3
125	2.7	1.2
160	−2.3	0.8
200	−2.2	−0.9
250	2.0	0.8
315	−2.0	−1.0
400	−0.6	0.1
500	0.1	0.1

Note: Either loudspeaker measurements using angle of incidence 45° or measurements with traffic noise covering the range from −60° to +60°.

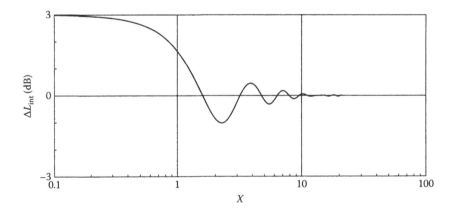

Figure 3.9 Interference correction for one-third octave bands and random incidence (3D diffuse field). The parameter is $X=(2\pi/c)f_m d$.

case the noise contains substantial pure tones, the problems occur again. Sometimes, the problem can be solved by carrying out a spatial average in an area in front of the façade, as is the normal procedure in closed rooms.

For random incidence or 3D diffuse sound field, the effect of the bandwidth can be found through numerical integration of Equation 3.21 over the frequency range of relevance. For the one-third octave band, the result is shown in Figure 3.9. The first minimum occurs at a distance $d=0.35\lambda$ from the wall, i.e. at a little longer distance than in the case of normal incidence (0.25λ).

REFERENCES

ISO 16283-3 (2016). *Acoustics – Field measurements of sound insulation in buildings and of building elements – Part 3: Façade sound insulation.* International Organization for Standardization, Geneva, Switzerland.

J.H. Rindel (1978). Lydfeltet foran en stor, reflekterende plan (The sound field in front of a large reflecting surface, in Danish). *Proceedings of NAM-78*, 16–18 August 1978, Odense, Denmark, pp. 101–104.

R.V. Waterhouse (1955). Interference patterns in reverberant sound fields. *Journal of the Acoustical Society of America* 27, 247–258.

Chapter 4

Introduction to room acoustics

This chapter summarizes the basic concepts in room acoustics as a basis for sound insulation theory and building acoustic measurement methods.

4.1 SOUND WAVES IN ROOMS

4.1.1 Standing waves in a rectangular room

A rectangular room has the dimensions l_x, l_y and l_z. The wave equation can then be written as (e.g. Kuttruff 1973, p. 51)

$$\frac{\partial^2 p}{\partial x^2} + \frac{\partial^2 p}{\partial y^2} + \frac{\partial^2 p}{\partial z^2} + k^2 p = 0 \qquad (4.1)$$

where p is the sound pressure and $k=\omega/c$ is the angular wave number, ω is the angular frequency and c is the speed of sound in air. The equation can be solved by separation of the variables, and it is assumed that the solution can be written in the form:

$$p = X(x) \cdot Y(y) \cdot Z(z) \cdot e^{j\omega t}$$

Insertion of this expression in Equation 4.1 and division by p gives

$$\frac{1}{X} \frac{\partial^2 X}{\partial x^2} + \frac{1}{Y} \frac{\partial^2 Y}{\partial y^2} + \frac{1}{Z} \frac{\partial^2 Z}{\partial z^2} + k^2 = 0$$

This can be separated in various directions, and for the x-direction, it yields

$$\frac{1}{X} \frac{\partial^2 X}{\partial x^2} + k_x^2 = 0$$

Similar equations hold for the y and z directions. The angular wave number k is divided into three along the directions as

$$k^2 = k_x^2 + k_y^2 + k_z^2 \tag{4.2}$$

The general solution to the aforementioned one-dimensional equation is

$$X(x) = C_x \cos(k_x x + \varphi_x)$$

in which the constants C_x and φ_x are determined from the boundary conditions.

The room surfaces are now assumed to be rigid, i.e. the normal component of the particle velocity is zero at the boundaries

$$u_x = -\frac{1}{j\omega\rho}\frac{\partial p}{\partial x} = 0 \quad \text{for } x = 0 \text{ and } x = l_x$$

This means that $\varphi_x = 0$ and

$$k_x = \frac{\pi}{l_x} \cdot n_x, \quad \text{where } n_x = 0, 1, 2, 3, \dots \tag{4.3}$$

Two similar boundary conditions hold for the y and z directions. With these conditions, the solution to Equation 4.1 is

$$p = p_0 \cdot \cos\left(\pi n_x \frac{x}{l_x}\right) \cdot \cos\left(\pi n_y \frac{y}{l_y}\right) \cdot \cos\left(\pi n_z \frac{z}{l_z}\right) \tag{4.4}$$

The time factor $e^{j\omega t}$ is omitted. The amplitude of the sound pressure does not move with time, so the waves that are solutions to Equation 4.4 are called *standing waves*. They are also called the *modes* of the room, and each of them is related to a certain *natural frequency* (or eigenfrequency) given by

$$f_n = \frac{\omega_n}{2\pi} = \frac{ck}{2\pi} = \frac{c}{2\pi}\sqrt{k_x^2 + k_y^2 + k_z^2}$$

$$f_n = \frac{c}{2}\sqrt{\left(\frac{n_x}{l_x}\right)^2 + \left(\frac{n_y}{l_y}\right)^2 + \left(\frac{n_z}{l_z}\right)^2} \tag{4.5}$$

Modes can be divided into three groups:

- *Axial modes* are one-dimensional, i.e. the amplitude of the sound pressure varies in one direction, only – one of n_x, n_y, n_z is > 0.

- *Tangential modes* are two-dimensional, i.e. the amplitude of the sound pressure varies in two directions – two of n_x, n_y, n_z are > 0.
- *Oblique modes* are three-dimensional, i.e. the amplitude of the sound pressure varies in three directions – all three of n_x, n_y, n_z are > 0.

Some examples are shown in Figure 4.1. It is observed that the set of numbers (n_x, n_y, n_z) indicate the number of *nodes* (places with $p=0$) along each coordinate axis. Although it is usual to display the room modes as the distribution of p^2, it may actually be more relevant to look at the distribution of the sound pressure level in dB; see the last example, mode (2,2,0), in Figure 4.1.

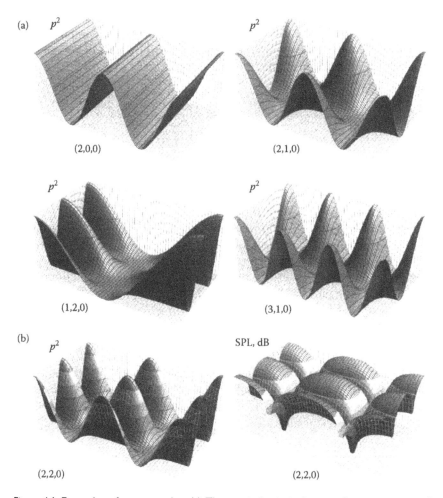

Figure 4.1 Examples of room modes. (a) The vertical axis is the sound pressure squared, and the horizontal axes are along the length and width of the room. Mode (2,0,0) is an axial mode and the others are tangential modes and (b) Mode (2,2,0) is shown both as the sound pressure squared and as the sound pressure level in dB.

4.1.2 Modal reverberation times

A rectangular room with dimensions lx, ly, lz is considered. Using Euler's formula (Equation 1.39) for the solution (Equation 4.4) to the wave equation, the sound pressure in a room mode can be written as

$$p = \frac{p_0}{8} \sum \exp\left[j\pi \left(\pm\frac{n_x}{l_x} x \pm \frac{n_y}{l_y} y \pm \frac{n_z}{l_z} z \right) \right] \tag{4.6}$$

As before, the time factor $e^{j\omega t}$ is omitted, and the summation is taken over all eight possible combinations of the + and −signs. Equation 4.6 shows that the total sound pressure can be interpreted as the interference between eight plane waves travelling in different directions. The wave number of the propagation along the x-axis is $k_x = \pi n_x/l_x$. The direction of propagation expressed as an angle relative to the x-axis is

$$\cos\varphi_x = \frac{\dfrac{n_x}{l_x}}{\sqrt{\left(\dfrac{n_x}{l_x}\right)^2 + \left(\dfrac{n_y}{l_y}\right)^2 + \left(\dfrac{n_z}{l_z}\right)^2}} = \pm\frac{n_x \cdot c}{2 \cdot l_x \cdot f_n} \tag{4.7}$$

It is analogous for the y and z directions.

Two examples of tangential modes (A and B) in a rectangular room are visualized in Figure 4.2. The nodal lines (actually vertical planes) represent the standing wave pattern, whereas the corresponding wave fronts represent the plane waves travelling in four different directions. The normal distance between parallel wave fronts is the wave length $\lambda = c/f$. Figures 4.2c of modes A and B show the four possible directions of propagation that are perpendicular to the wave fronts. The direction of propagation and the directional cosines in Equation 4.7 are displayed in Figure 4.3 in the case of an oblique mode.

The sound absorption coefficients of the six surfaces of the room are denoted by α_{x1}, α_{x2}, α_{y1}, α_{y2}, α_{z1} and α_{z2}. The reverberation time T_n of a room mode with modal number (n_x, n_y, n_z) can be derived – see Rindel (2016) for the details – and the result is

$$T_n = \frac{55.3 \cdot f_n}{-c^2} \left[\begin{array}{l} \dfrac{n_x}{l_x^2} \cdot \ln\left((1-\alpha_{x1})(1-\alpha_{x2})\right) + \dfrac{n_y}{l_y^2} \cdot \ln\left((1-\alpha_{y1})(1-\alpha_{y2})\right) \\[2mm] + \dfrac{n_z}{l_z^2} \cdot \ln\left((1-\alpha_{z1})(1-\alpha_{z2})\right) \end{array} \right]^{-1} \tag{4.8}$$

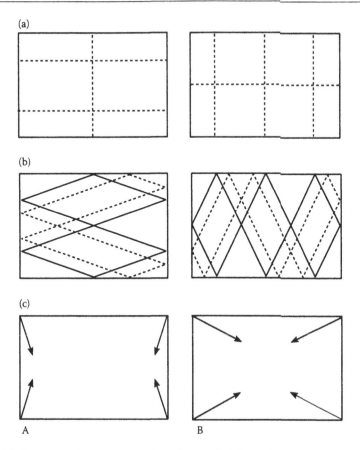

Figure 4.2 Two examples of tangential modes. A: (1,2,0) mode, left side and B: (3,1,0) mode, right side. (a) nodal lines; (b) wave fronts in two positions with a short time delay; and (c) directions of propagation, perpendicular to the wave fronts.

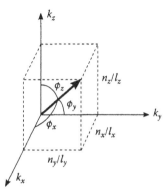

Figure 4.3 Direction of propagation and directional cosines of an oblique mode.

If we assume all absorption coefficients to be equal to the mean absorption coefficient α_m, we get the modal reverberation time:

$$T_n = \frac{55.3 \cdot f_n}{-c^2} \left[2 \cdot \left(\frac{n_x}{l_x^2} + \frac{n_y}{l_y^2} + \frac{n_z}{l_z^2} \right) \cdot \ln(1 - \alpha_m) \right]^{-1} \tag{4.9}$$

This shows that the axial modes have longer reverberation times than tangential modes, and oblique modes have shorter reverberation times.

4.1.3 Transfer function in a room

The transfer function is the frequency response from a source position to a receiver position in a room. When a sound source submits a tone at a certain frequency, a large number of room modes will be excited and the sound pressure at a point in the room can be described as the addition of the contributions from all modes with phase (e.g. Kuttruf 1973, eq. (III.30)):

$$p(\omega) = \sum_n \frac{A_n}{\omega^2 - \omega_n^2 - j2\delta_n\omega_n} \tag{4.10}$$

Here A_n is the amplitude of mode n, ω is the angular frequency emitted by the source, ω_n is the angular frequency of room mode n and δ_n is the damping constant.

An example of a calculated transfer function is shown in Figure 4.4. It fluctuates very much with frequency, and the maxima can be identified as the natural frequencies of the room. The example in Figure 4.4 has the same room dimensions as was used for the calculations in Table 4.1. The calculations are made as explained in Rindel (2015). While the first five modes are clearly separated, there are several examples of coinciding or overlapping modes that cannot be separated: modes (1,0,1) and (2,1,0) around 54 Hz; modes (1,1,1) and (0,2,0) around 62 Hz; and the three modes (2,0,1), (1,2,0) and (3,0,0) around 67 Hz.

4.1.4 Modal density

A closer inspection of Equation 4.5 shows that the natural frequencies of a rectangular room may be interpreted in a geometrical way. A three-dimensional frequency space is shown in Figure 4.5. The natural frequencies of the one-dimensional modes are marked on each of the axes, representing the axial modes of the length, the width and the height, respectively. Interestingly, the points in the grid represent the oblique modes, the distance to each point from the origin is the natural

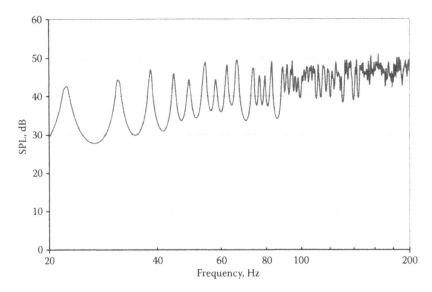

Figure 4.4 Transfer function in a rectangular room with dimensions 7.7 m, 5.5 m and 3.5 m. At low frequencies, it is possible to identify the modes by their modal numbers (compare with Table 4.1).

Table 4.1 Calculated natural frequencies at low frequencies using Equation 4.5 in a rectangular room with dimensions 7.7 m, 5.5 m and 3.5 m

n_x	n_y	n_z	f_n (Hz)
1	0	0	22.3
0	1	0	31.2
1	1	0	38.4
2	0	0	44.6
0	0	1	49.0
1	0	1	53.9
2	1	0	54.4
0	1	1	58.1
1	1	1	62.3
0	2	0	62.4
2	0	1	66.3
1	2	0	66.3
3	0	0	66.9
2	1	1	73.3
3	1	0	73.8
2	2	0	76.7
0	2	1	79.4

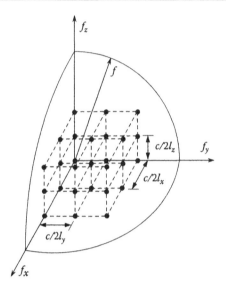

Figure 4.5 Frequency grid in which each grid point represents a room mode.

frequency of that mode. Hence, the number of oblique modes below a certain frequency f is equal to the number of grid points inside the sphere with radius f.

The volume is one-eighth of the sphere with radius f, i.e. $(4\pi f^3/3)/8 = \pi f^3/6$. Each mode occupies a volume $c^3/(8l_x l_y l_z) = c^3/(8V)$. Hence, the number of modes below f is approximately

$$N = \frac{\pi f^3}{6} \cdot \frac{8V}{c^3} = \frac{4\pi V}{3} \cdot \left(\frac{f}{c}\right)^3 \tag{4.11}$$

The axial modes are marked along each of the three axes and the tangential modes are found in each of the three planes between two axes. If these modes are also taken into account, the number of modes with natural frequencies below the frequency f, the mode count, is (Pierce 1989, p. 293)

$$N = \frac{4\pi V}{3}\left(\frac{f}{c}\right)^3 + \frac{\pi S}{4}\left(\frac{f}{c}\right)^2 + \frac{L}{8}\frac{f}{c} + \frac{1}{8} \tag{4.12}$$

where V is the volume of the room, S is the total area of the surfaces and $L = 4\,(l_x + l_y + l_z)$ is the total length of all edges. At high frequencies, the oblique modes dominate and the first term in Equation 4.12 is a good approximation for any room, not only for rectangular rooms.

The modal density is the average number of modes per hertz. From Equation 4.12, we get

$$n(f) = \frac{\mathrm{d}N}{\mathrm{d}f} = \frac{4\pi V}{c^3} f^2 + \frac{\pi S}{2c^2} f + \frac{L}{8c} \tag{4.13}$$

In Figure 4.6, this is compared to the actual modal density in a room. At very low frequencies, the estimate can give fractional values between 0 and 1, whereas the actual number of modes within a frequency band must be a whole number. In the example room, there is one mode in the 20 Hz and the 31.5 Hz bands, but none in the 25 Hz band.

For high frequencies (e.g. above 200 Hz), it may sometimes be sufficient to use only the first term in Equation 4.12 for the modal density:

$$n(f) \cong \frac{4\pi V}{c^3} f^2 \tag{4.14}$$

A more detailed analysis of the modal grid points in the frequency domain (Rodríguez Molares 2010, Table 7.1) leads to the modal densities of axial, tangential and oblique modes in a three-dimensional room:

$$n_{\mathrm{ax}}(f) \cong \frac{L}{2c} \tag{4.15}$$

$$n_{\mathrm{tan}}(f) \cong \frac{\pi S}{c^2} f - \frac{L}{2c} \tag{4.16}$$

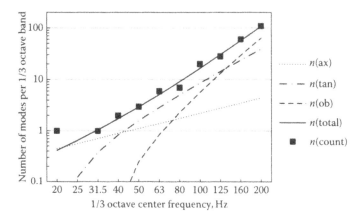

Figure 4.6 Modal density as a function of frequency in a rectangular room with dimensions 7.7 m, 5.5 m and 3.5 m. Dots indicate the actual number of modes in the third octave bands from 20 Hz to 200 Hz (compare with Table 4.1). The curves show the number of modes estimated for axial, tangential, oblique and total modes by Equations 4.13 and 4.15 through 4.17.

$$n_{ob}(f) \cong \frac{4\pi V}{c^3} f^2 - \frac{\pi S}{2c^2} f + \frac{L}{8c} \qquad (4.17)$$

Adding the modal densities of the three groups of modes yields again Equation 4.13.

Each normal mode is in itself a resonant system that can be associated with a bandwidth around the natural frequency. From Equation 4.6, it follows that the squared sound pressure of mode n is

$$p_n^2 = \frac{|A_n|^2}{\left(\omega^2 - \omega_n^2\right)^2 + 4\omega^2 \delta_n^2} \qquad (4.18)$$

The bandwidth B_r is defined as the frequency interval in which the acoustic energy is at least half the maximum energy at the natural frequency, the 3 dB bandwidth (Figure 4.7). It follows from Equation 4.18 that the maximum energy is at $\omega = \omega_n$ and the half energy is at $\omega_{1,2} = \omega_n \pm \delta_n$. The bandwidth in angular frequency is $\Delta\omega = \omega_2 - \omega_1 \approx 2\delta_n$ and, thus, the bandwidth B_r in Hz can be related to the reverberation time, T:

$$B_r = \frac{\Delta\omega}{2\pi} \cong \frac{\delta}{\pi} = \frac{2.2}{T} \qquad (4.19)$$

The relation between damping constant and reverberation time follows from the definition of T as the time for the energy to decay by 60 dB after the source is interrupted. A resonant system with damping constant δ has the decay $10 \lg (\exp(-2\delta_T)) = -60$ dB, and so $\delta_T = 3/\lg (e) = 2.2\pi$.

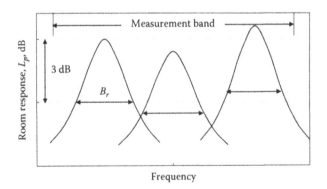

Figure 4.7 Frequency response of three modes within a frequency band of measurement. The modal overlap index is $M = 0.7$. (Adapted from Bies, D.A. and Hansen, C.H., *Engineering Noise Control, Theory and Practice*, Unwin Hyman Ltd., London, 1988, Figure 7.3.)

In a room, the natural frequencies of the normal modes are quite spread at low frequencies, but at higher frequencies, they are very close together. The number of modes within the bandwidth of the modes is called the modal overlap index, M:

$$M = B_r \cdot n(f) \approx \frac{8.8 \cdot \pi \cdot V}{T \cdot c^3} f^2 \qquad (4.20)$$

The modal overlap index in a room increases rapidly with increasing frequency, which is clearly seen in the approximation where Equations 4.14 and 4.19 are inserted. Figure 4.7 shows an example of low modal overlap. There are three modes within the measurement band, which is 4.2 times the modal bandwidth. Thus, $M = 4.2/3 \approx 0.7$.

4.1.5 The Schroeder limiting frequency

Due to the high number of modes in a room at high frequencies and in order to avoid too complicated methods, it is necessary to apply statistical considerations, e.g. to assume the same energy in all modes within a certain frequency band. As a minimum requirement, it has been suggested by Schroeder (1962) that the modal overlap should be $M = 3$, although this cannot be considered a very sharp borderline. Using Equation 4.20, this leads to a lower limiting frequency for statistical methods, the so-called 'Schroeder's limiting frequency', f_g:

$$f_g = 2000 \sqrt{\frac{T}{V}} \qquad (4.21)$$

where T is the reverberation time (s) and V is the volume (m³).

As an example, a room with volume 50 m³ and reverberation time 0.5 s has a limiting frequency of $f_g = 200\,\text{Hz}$. Thus, below this frequency, a weak coupling between the room modes can be assumed, or – in other words – the sound field below 200 Hz may be dominated by single room modes.

4.2 STATISTICAL ROOM ACOUSTICS

4.2.1 The diffuse sound field

In this section, the acoustical behaviour of a room is treated from a statistical point of view, based on energy balance considerations. It is assumed that the modal density is high enough, so that the influence of single modes in the room can be neglected. It is also assumed that the reflection density is high enough so that the phase relations between individual reflections can be neglected. This means that the reflections in the room are assumed to be uncorrelated and their contribution can be added on an energy basis.

The diffuse sound field is defined as a sound field in which

- The energy density is the same everywhere.
- All directions of sound propagation occur with the same probability.

It is obvious that the direct sound field near a sound source is not included in the diffuse sound field. Neither are the special interference phenomena that are known to give increased energy density near the room boundaries and corners. The diffuse sound field is an ideal sound field that does not exist in any room. However, in many cases, the diffuse sound field can be a good and very practical approximation to the real sound field.

4.2.2 Incident sound power on a surface

In a plane propagating sound wave, the relation between RMS sound pressure p_1 and sound intensity I_1 is

$$p_1^2 = I_1 \cdot \rho c$$

In a diffuse sound field, the RMS sound pressure p_{diff} is the result of sound waves propagating in all directions, and all having the sound intensity I_1. By integration over a sphere with the solid angle $\psi = 4\pi$, the RMS sound pressure in the diffuse sound field is

$$p_{\mathrm{diff}}^2 = \int_{\psi=4\pi} I_1 \cdot \rho c \, \mathrm{d}\psi = 4\pi \cdot I_1 \cdot \rho c \tag{4.22}$$

In the case of a plane wave with the angle of incidence θ relative to the normal of the surface, the incident sound power per unit area on the surface is

$$I_\theta = I_1 \cos\theta = \frac{p_{\mathrm{diff}}^2}{4\pi\rho c} \cos\theta \tag{4.23}$$

where p_{diff} is the RMS sound pressure in the diffuse sound field. This is just the sound intensity in the plane propagating wave multiplied by the cosine, which is the projection of a unit area as seen from the angle of incidence (Figure 4.8).

The total incident sound power per unit area is found by integration over all angles of incidence covering a half sphere in front of the surface (Figure 4.8b).

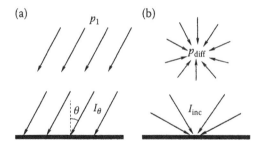

Figure 4.8 (a) Plane wave at oblique incidence on a surface and (b) diffuse incidence on a surface.

The integration covers the solid angle $\psi = 2\pi$ and Figure 4.9 shows the definition of the angles.

$$I_{inc} = \int_{\psi=2\pi} I_\theta \, d\psi = \frac{1}{4\pi} \int_0^{2\pi} \int_0^{\pi/2} \frac{p_{diff}^2}{\rho c} \cos\theta \sin\theta \, d\theta \, d\phi$$

$$= \frac{1}{4\pi} \cdot 2\pi \cdot \frac{p_{diff}^2}{\rho c} \int_0^1 \sin\theta \, d(\sin\theta) = \frac{1}{2} \cdot \frac{p_{diff}^2}{\rho c} \cdot \frac{1}{2} \qquad (4.24)$$

$$I_{inc} = \frac{p_{diff}^2}{4\rho c}$$

It is noted that this is four times less than in the case of a plane wave of normal incidence.

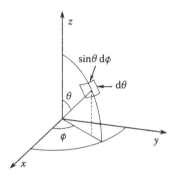

Figure 4.9 Definition of angles of incidence in a diffuse sound field.

4.2.3 Equivalent absorption area

The absorption coefficient α is defined as the ratio of the non-reflected sound energy to the incident sound energy on a surface. It can take values between 0 and 1; $\alpha = 1$ means that all incident sound energy is absorbed in the surface. An example of a surface with absorption coefficient, $\alpha = 1$ is an open window. Sabine in his original paper on reverberation from 1898 suggested the use of open-window unites for the absorption area (Paper one in collected papers, Sabine, 1922).

The product of area and absorption coefficient of a surface material is the equivalent absorption area of that surface, i.e. the area of open windows giving the same amount of sound absorption as the actual surface. The equivalent absorption area of a room is

$$A = \sum_i S_i \alpha_i = S_1 \alpha_1 + S_2 \alpha_2 + \cdots = S \alpha_m \qquad (4.25)$$

where S is the total surface area of the room and α_m is the mean absorption coefficient. The unit of A is m². In general, the equivalent absorption area may also include sound absorption due to the air and due to persons or other objects in the room.

4.2.4 Energy balance in a room

The total acoustic energy in a room is the sum of potential energy and kinetic energy, or twice the potential energy, since the time average of the two parts must be equal. The total energy E is the energy density multiplied by the room volume V:

$$E = \left(w_{\text{pot}} + w_{\text{kin}} \right) V = 2 w_{\text{pot}} V = \frac{p^2}{\rho c^2} V \qquad (4.26)$$

Here and in the following, p denotes the RMS sound pressure in the diffuse sound field (called p_{diff} in Section 4.2.2). The energy absorbed in the room is the incident sound power per unit area (Equation 4.24) multiplied by the total surface area and the mean absorption coefficient, i.e. the equivalent absorption area (Equation 4.25),

$$P_{a,\text{abs}} = I_{\text{inc}} S \alpha_m = I_{\text{inc}} A = \frac{p^2}{4\rho c} A \qquad (4.27)$$

If P_a is the sound power of a source in the room, the energy balance equation of the room is

$$P_a - P_{a,\text{abs}} = \frac{dE}{dt}$$

$$P_a - \frac{p^2}{4\rho c} A = \frac{V}{\rho c^2} \frac{d}{dt}(p^2) \tag{4.28}$$

With a constant sound source, a steady state situation is reached after some time, and the right side of the equation is zero. Hence, the absorbed power equals the power emitted from the source, and the steady state sound pressure in the room is

$$p_s^2 = \frac{4P_a}{A} \rho c \tag{4.29}$$

This equation shows that the sound power of a source can be determined by measuring the sound pressure generated by the source in a room, provided that the equivalent absorption area of the room is known. It also shows how the absorption area in a room has a direct influence on the sound pressure in the room. In some cases, it is more convenient to express Equation 4.29 in terms of the sound pressure level L_p and the sound power level L_W,

$$L_p \cong L_W + 10 \lg\left(\frac{4A_0}{A}\right)(dB) \tag{4.30}$$

where $A_0 = 1\,m^2$ is a reference area. The approximation comes from neglecting the term with the constants and reference values:

$$10 \lg \frac{\rho c P_{ref}}{A_0 p_{ref}^2} = 10 \lg \frac{1.204 \times 343 \times 10^{-12}}{1 \times (20 \times 10^{-6})^2} = 0.14\,dB \cong 0\,dB$$

4.2.5 Reverberation time: Sabine's formula

If the sound source is turned off after the sound pressure has reached the stationary value, the first term in the energy balance equation (Equation 4.28) is zero, and the RMS sound pressure is now a function of time:

$$\frac{A}{4\rho c} p^2(t) + \frac{V}{\rho c^2} \frac{d}{dt}\left(p^2(t)\right) = 0 \tag{4.31}$$

The solution to this equation can be written as

$$p^2(t) = p_s^2 e^{-\frac{cA}{4V}(t - t_0)} \tag{4.32}$$

where p_s^2 is the mean square sound pressure in the steady state and $t_0 = 1\,s$ is the time when the source is turned off. It can be seen that the mean square sound pressure, and hence, the sound energy, follow an exponential decay function. On a logarithmic scale, the decay is linear, and this is called the decay curve (Figure 4.10).

If instead the source is turned on at time $t=0$, the sound build up in the room follows a similar exponential curve, also shown in Figure 4.10.

$$p^2(t) = p_s^2 \left(1 - e^{-\frac{cA}{4V}t} \right)$$

(4.33)

The reverberation time T is defined as the time it takes for the sound energy in the room to decay to one millionth of the initial value, i.e. a 60 dB decay of the sound pressure level. Hence, for $t=T$

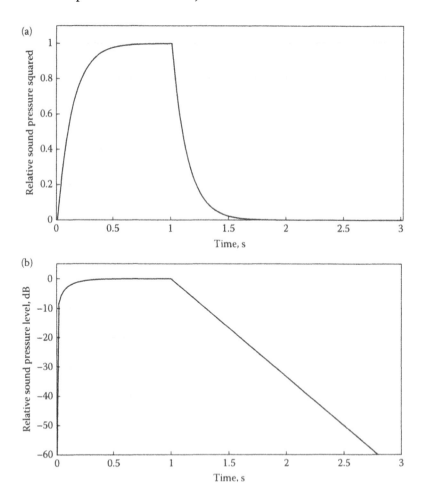

Figure 4.10 Build up and decay of sound in a room. Here, the source is turned on at $t = 0$ and turned off at $t_0 = 1$ s. The reverberation time is 1.8 s. (a) Linear scale (sound pressure squared), and (b) logarithmic scale (dB).

$$p^2(t) = p_s^2 10^{-6} = p_s^2 \, e^{\frac{cA}{4V}T}$$

Hence, the reverberation time is

$$T = 6 \cdot \ln(10) \cdot \frac{4V}{cA} = \frac{55.3V}{cA} \qquad (4.34)$$

This is *Sabine's formula* named after Wallace C. Sabine (1868–1919). He introduced the reverberation time concept in a pioneering paper published first time 1898 in Proceedings of the American Institute of Architects. (Paper one in collected papers, Sabine, 1922.). He was the first to demonstrate that T is inversely proportional to the equivalent absorption area A.

Note: Sabine's formula is often written as $T = 0.16 \ V/A$. However, this implies that V must be in m^3 and A in m^2.

There is no doubt that Sabine's formula is the most well-known formula in room acoustics, and it is widely used to calculate reverberation time. However, the preconditions for the formula to be valid are often neglected:

• The sound field should be a three-dimensional, diffuse sound field.
• The absorption should be evenly distributed on all surfaces.

None of these conditions are fulfilled in ordinary rooms like classrooms and offices that typically have a sound absorbing ceiling and highly reflecting walls.

In Figure 4.10 we see the rapid build-up of sound energy after the source is turned on. The point of –3 dB (half of the asymptotic final energy) is reached after 5 % of the reverberation time, –1 dB after 11 % and –0.1 dB after 27 %. This means that using a stationary sound source, the sound field can be considered stationary after one-third of the reverberation time, and measurements can start.

4.2.6 Stationary sound field in a room: Reverberation distance

A reverberation room is a special room with long reverberation time and a good diffusion. In such a room, the diffuse sound field is a good approximation, and the results for stationary conditions (Equation 4.29) and for sound decay (Equation 4.34) can be applied to measure the sound power of a sound source:

$$P_a = \frac{p_s^2}{4\rho c} \cdot \frac{55.3V}{cT} \qquad (4.35)$$

The reverberation time and the average sound pressure level in the reverberation room are measured, and the sound power level is calculated from

$$L_W = L_p + 10 \lg \frac{p_{ref}^2 \cdot 55.3 \cdot V}{P_{ref} \cdot 4\rho c^2 \cdot T}$$

$$= L_p + 10 \lg \frac{V}{V_0} - 10 \lg \frac{T}{t_0} - 14 \, dB \tag{4.36}$$

where $V_0 = 1 \, m^3$ and $t_0 = 1 \, s$.

In most ordinary rooms, the diffuse sound field is not a good approximation. Each of the following conditions may indicate that the sound field is not diffuse:

- An uneven distribution of sound absorption on the surfaces, e.g. only one surface is highly absorbing
- A lack of diffusing or sound scattering elements (e.g. furniture) in the room
- The ratio of longest to shortest room dimension is higher than three
- The volume is very large, say more than $5000 \, m^3$

A rather simple modification to the stationary sound field is to separate the direct sound. The sound power radiated by an omnidirectional source is the sound intensity at the distance r in a spherical sound field multiplied by the surface area of a sphere with radius r

$$P_a = I_r \cdot 4\pi r^2 \tag{4.37}$$

Thus, the sound pressure squared of direct sound in the distance r from the source is

$$p_{dir}^2 = \frac{P_a}{4\pi r^2} \rho c \tag{4.38}$$

The stationary sound is described by Equation 4.29:

$$p_s^2 = \frac{4P_a}{A} \rho c$$

The reverberation distance r_{rev} is defined as the distance where $p_{dir}^2 = p_s^2$ when an omnidirectional point source is placed in a room. It is a descriptor of the amount of absorption in a room since the reverberation distance depends only on the equivalent absorption area:

$$r_{rev} = \sqrt{\frac{A}{16\pi}} = 0.14\sqrt{A} \tag{4.39}$$

At a distance closer to the source than the reverberation distance, the direct sound field dominates, which is called the direct field. At longer distances, the reverberant sound field dominates, and in this so-called far field, the stationary, diffuse sound field may be a usable approximation.

An expression for the combined direct and diffuse sound field can derived by the simple addition of the squared sound pressures of the two sound fields. However, since direct sound is treated separately, it should be extracted from the energy balance equation, which was used to describe the diffuse sound field. To do this, the sound power of the source should be reduced by a factor of $(1 - \alpha_m)$, which is the fraction of the sound power emitted to the room after the first reflection. Hence, the squared sound pressure in the total sound field is

$$p_{\text{total}}^2 = p_{\text{dir}}^2 + p_s^2 (1 - \alpha_m) = p_s^2 \left(\frac{r_{\text{rev}}^2}{r^2} + 1 - \alpha_m \right) \tag{4.40}$$

$$p_{\text{total}}^2 = P_a \cdot \rho c \left(\frac{1}{4\pi r^2} + \frac{4}{A} (1 - \alpha_m) \right) \tag{4.41}$$

The absorption area A divided by $(1 - \alpha_m)$ is sometimes called the *room constant*.

Normal sound sources like a speaking person, a loudspeaker, or a musical instrument radiate sound with different intensities in different directions. The directivity factor Q is the ratio of the intensity in a certain direction to the average intensity,

$$Q = I \cdot \frac{4\pi r^2}{P_a} \tag{4.42}$$

Hence, the squared sound pressure of the direct sound is

$$p_{\text{dir}}^2 = \frac{Q \cdot P_a}{4\pi r^2} \rho c \tag{4.43}$$

This leads to a general formula for the sound pressure level as a function of the distance from a sound source in room.

$$L_p \cong L_W + 10 \lg \frac{4 A_0}{A} + 10 \lg \left(Q \frac{r_{\text{rev}}^2}{r^2} + 1 - \alpha_m \right) (\text{dB}) \tag{4.44}$$

where $A_0 = 1 \, \text{m}^2$. This formula is displayed in Figure 4.11 for the case of an omnidirectional sound source ($Q = 1$). In a reverberant room with little sound absorption (e.g., $\alpha_m < 0.1$), the sound pressure level in the far field will be approximately as predicted by the diffuse field theory, i.e. the last

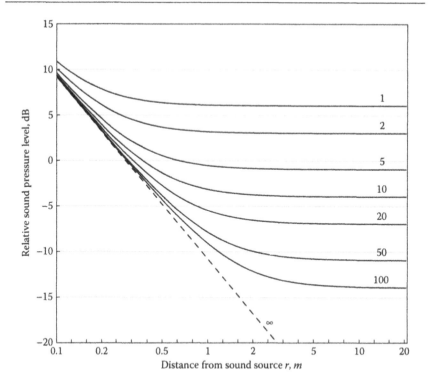

Figure 4.11 Relative sound pressure level as a function of distance in a room with an approximately diffuse sound field. The source has a directivity factor of one. The parameter on the curves is the room constant $A/(1 - \alpha_m)$ in m².

term will be close to zero. In the case of a highly directive sound source like a trumpet $(Q \gg 1)$, the direct field can be extended to distances much longer than the reverberation distance. In the latter situation, the last term in Equation 4.44 raises the sound pressure level above the diffuse field value.

In large rooms with medium or high sound absorption (e.g., $\alpha_m > 0.2$), the sound pressure level will continue to decrease as a function of the distance because the diffuse field theory is not valid in such a room. Instead, the slope of the spatial decay curve may be taken as a measure of the degree of acoustic attenuation in a room. Hence, in large industrial halls, the attenuation in dB per doubling of the distance may be a better descriptor than the reverberation time.

A reverberation room can be used to determine the sound power of a sound source by measuring the average sound pressure level in the room. If measurements are made in positions that avoid the direct sound, the last term in Equation 4.44 becomes more correctly (Vorländer 1995):

$$10\lg(1-\alpha_m) \rightarrow 10\lg\left(e^{-\frac{A}{S}}\right) \cong -4.34 \cdot \frac{A}{S}(dB)$$

where A is the equivalent absorption area, S is the total surface area and taking the air attenuation into account (see Section 4.3.2). For correct results at low frequencies, it is also necessary to apply the Waterhouse correction (Equation 3.31)), and, thus, the sound power level is determined from

$$L_W = L_p + 10\lg\frac{A}{4A_0} + 4.34.\frac{A}{S} + 10\lg\left(1+\frac{cS}{8Vf}\right)(dB) \qquad (4.45)$$

This is how the sound power of sound sources can be measured according to ISO 3741. In addition, the possible variation of temperature and static pressure may also be taken into account for very high precision.

4.3 GEOMETRICAL ROOM ACOUSTICS

4.3.1 Sound rays and a general reverberation formula

In geometrical acoustics, rays are used to describe sound propagation. The concept of rays implies that the wavelength and the phase of the sound are neglected, and only the direction of sound energy propagation is treated.

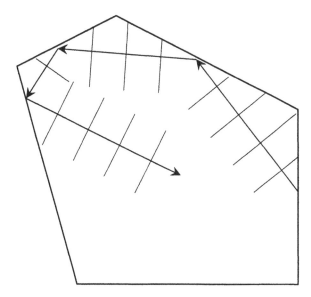

Figure 4.12 A plane wave travelling as a ray from wall to wall in a room.

Sound decay shall now be studied by following a plane wave travelling as a ray from wall to wall (Figure 4.12). The energy of the wave is gradually decreased due to absorption at the surfaces, all of which are assumed to have the mean absorption coefficient α_m.

The ray representing a plane wave may start in any direction and it is assumed that the decay of energy in the ray is representative of the decay of energy in the room. The room may have any shape.

By each reflection, the energy is reduced by a factor $(1 - \alpha_m)$. The initial sound pressure is p_0 and after n reflections, the squared sound pressure is

$$p^2(t) = p_0^2 \cdot (1 - \alpha_m)^n = p_0^2 \cdot e^{n \cdot \ln(1 - \alpha_m)} \tag{4.46}$$

The distance the ray travels from one reflection to the next is l_i and the total distance travelled by the ray up to the time t is

$$\sum_i l_i = c \cdot t = n \cdot l_m \tag{4.47}$$

where l_m is the mean free path. Hence, the squared sound pressure is

$$p^2(t) = p_0^2 \cdot e^{\frac{c}{l_m} \cdot \ln(1 - \alpha_m) \cdot t} \tag{4.48}$$

When the squared sound pressure has dropped to 10^{-6} of the initial value, the time t is by definition the reverberation time T:

$$10^{-6} = e^{\frac{c}{l_m} \cdot \ln(1 - \alpha_m) \cdot T} \Rightarrow -6 \cdot \ln(10) = \frac{c}{l_m} \cdot \ln(1 - \alpha_m) \cdot T$$

This leads to an interesting pair of general reverberation formulae:

$$T = \frac{13.8 \cdot l_m}{-c \cdot \ln(1 - \alpha_m)} \approx \frac{13.8 \cdot l_m}{c \cdot \alpha_m} \tag{4.49}$$

The last approximation is valid if $\alpha_m < 0.3$, i.e. only in rather reverberant rooms. The approximation comes from

$$-\ln(1 - \alpha_m) = \ln\left(\frac{1}{1 - \alpha_m}\right) = \alpha_m + \frac{\alpha_m^2}{2} + \frac{\alpha_m^3}{3} + \cdots$$

With the assumption that all directions of sound propagation appear with the same probability, it can be shown from Kosten (1960) that the mean free path in a three-dimensional room is

$$l_m = \frac{4V}{S}\left(\text{three-dimensional}\right) \qquad (4.50)$$

where V is the volume and S is the total surface area.

Similarly, the mean free path in a two-dimensional room can be derived. This could be the narrow air space in a double wall, or structure-borne sound in a plate. The height or thickness must be small compared to the wavelength. In this case, the mean free path is

$$l_m = \frac{\pi S_x}{U}\left(\text{two-dimensional}\right) \qquad (4.51)$$

where S_x is the area and U is the perimeter. The one-dimensional case is just the sound travelling back and forth between two parallel surfaces with the distance $l = l_m$.

Insertion of Equation 4.50 in the last part of Equation 4.35 gives Sabine's formula (Equation 4.34), whereas insertion in the first part of Equation 4.49 leads to the so-called *Eyring's formula* for reverberation time in a room:

$$T = \frac{55.3 \cdot V}{-c \cdot S \cdot \ln\left(1 - \alpha_m\right)} \qquad (4.52)$$

In a reverberant room $(\alpha_m < 0.3)$, it gives the same result as Sabine's formula, but in highly absorbing rooms, Eyring's formula is theoretically more correct. In practice, the absorption coefficients are not the same for all surfaces and the mean absorption coefficient is calculated as in Equation 4.20:

$$\alpha_m = \frac{1}{S} \cdot \sum_i S_i \alpha_i \qquad (4.53)$$

In the extreme case of an anechoic room $(\alpha_m = 1)$, Eyring's formula gives correctly a reverberation time of zero, whereas Sabine's formula is obviously wrong, giving the value $T = 55.3\ V/(c\ S)$. However, in normal rooms with a mixture of different absorption coefficients, it is recommended to use Sabine's formula.

4.3.2 Sound absorption in air

A sound wave travelling through air is attenuated by a factor m, which depends on the temperature and the relative humidity (RH) of the air. The unit of the air attenuation factor is m^{-1}. If this attenuation is included in Equation 4.48, the squared sound pressure in the decay is

$$p^2(t) = p_0^2 \cdot e^{\frac{c}{l_m}\cdot\ln(1-\alpha_m)\cdot t} \cdot e^{-mct} = p_0^2 \cdot e^{\frac{ct}{l_m}(\ln(1-\alpha_m)-m\cdot l_m)} \tag{4.54}$$

The general reverberation formula then becomes

$$T = \frac{13.8 \cdot l_m}{c\left(-\ln(1-\alpha_m) + m\cdot l_m\right)} \approx \frac{13.8 \cdot l_m}{c\left(\alpha_m + m\cdot l_m\right)} \tag{4.55}$$

In the three-dimensional case with Equation 4.50, we then have

$$T = \frac{55.3 \cdot V}{c(-S\cdot\ln(1-\alpha_m) + 4mV)} \approx \frac{55.3 \cdot V}{c(S\cdot\alpha_m + 4mV)} \tag{4.56}$$

These two expressions are Eyring's and Sabine's formula, respectively, with the air absorption included. By comparison with Equation 4.34, the equivalent absorption area including air attenuation is

$$A = \sum_i S_i\alpha_i + 4mV \tag{4.57}$$

Some typical values of m are found in Table 4.3.

4.3.3 Reflection density in a room

The image source principle can easily be applied to higher order reflections in a rectangular room. An infinite number of image rooms make a grid, and each cell in the grid is an image room containing an image source. The principle is shown for the two-dimensional case in Figure 4.13.

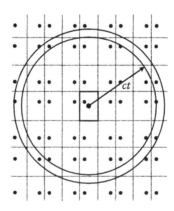

Figure 4.13 Rectangular room with a sound source and image sources, shown in two dimensions. Image sources located inside the circle with radius ct will contribute reflections up to time t.

If an impulse sound is emitted, the number of reflections that will arrive within the time t can be calculated as the volume of a sphere with radius ct divided by the room volume V:

$$N(t) = \frac{\frac{4}{3}\pi(ct)^3}{V} \tag{4.58}$$

The reflection density is then the number of reflections within a small time interval dt, and by differentiation:

$$\frac{dN}{dt} = 4\pi\frac{c^3}{V}t^2 \tag{4.59}$$

The reflection density increases with the time squared, so the higher order reflections are normally so dense in arrival time that it is impossible to distinguish separate reflections. If Equation 4.59 is compared to Equation 4.14, it is striking to observe the analogy between reflection density in the time domain and modal density in the frequency domain.

4.4 CALCULATION OF REVERBERATION TIME

4.4.1 Using Sabine's formula

Sabine's formula (Equation 4.34) is the most well-known and simple method for calculation of reverberation time in a room:

$$T = \frac{55.3V}{cA} \cong \frac{0.16V}{A} \tag{4.60}$$

with volume V in m³ and A in m². The equivalent absorption area is calculated as in Equation 4.57, but in addition to absorption from surfaces and air, the absorption from persons or other items in the room should be included, if relevant:

$$A = \sum_i S_i\alpha_i + \sum_j n_j A_j + 4mV \tag{4.61}$$

Here n_j is the number of items, each contributing with an absorption area A_j. Examples of absorption coefficients of common materials are given in Table 4.2. The air attenuation for different values of RH at 20 °C can be taken from Table 4.3. It is noted that high RH means low air attenuation.

Table 4.2 Typical values of the absorption coefficient α for some common materials

Material	Frequency (Hz)					
	125	250	500	1000	2000	4000
Brick, bare concrete	0.01	0.02	0.02	0.02	0.03	0.04
Parquet floor on studs	0.16	0.14	0.11	0.08	0.08	0.07
Needle-punch carpet	0.03	0.04	0.06	0.10	0.20	0.35
Window glass	0.35	0.25	0.18	0.12	0.07	0.04
Curtain draped to half its area, 100 mm air space	0.10	0.25	0.55	0.65	0.70	0.70

Table 4.3 Examples of air attenuation coefficient m (m^{-1}) at a temperature of 20 °C

Relative humidity (%)	Frequency (kHz)			
	1	2	4	8
40	0.0011	0.0026	0.0072	0.0237
50	0.0010	0.0024	0.0061	0.0192
60	0.0009	0.0023	0.0056	0.0162
70	0.0009	0.0021	0.0053	0.0143
80	0.0008	0.0020	0.0051	0.0133

Source: Harris, C.M., J. Acoust. Soc. Am., 40, 148, 1966.

4.4.2 Reverberation time in non-diffuse rooms

In a room with the sound absorption unequally distributed on the surfaces, the assumption of a diffuse sound field is not fulfilled and, thus, Sabine's formula will not be reliable. The measured reverberation time may be either shorter or longer than predicted by Sabine's formula.

A shorter reverberation time will appear in a room in which the first reflections are directed towards the most absorbing surface. In an auditorium, the most absorbing surface is typically the floor with the audience.

In a rectangular room without sound scattering surfaces or elements, there is a possibility of prolonged decay in certain directions. In order to give an idea of the problem, it is possible to calculate the different reverberation times associated with one-dimensional decays in each of the three main directions using the general reverberation formula (Equation 4.49).

$$T \approx \frac{13.8 \cdot l_m}{c \cdot \alpha_m} \approx 0.04 \cdot \frac{l_m}{\alpha_m} \quad (l_m \text{ in m}) \tag{4.62}$$

As an example, the room in Figure 4.14 is considered. The ceiling has a high absorption coefficient ($\alpha = 0.8$), but all other surfaces are acoustically hard ($\alpha = 0.1$).

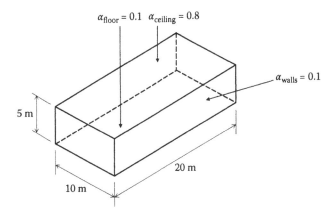

Figure 4.14 A rectangular room with indicated absorption coefficients.

Volume $V = 5 \times 10 \times 20 = 1000\,\mathrm{m^3}$
Surface area $S = 700\,\mathrm{m^2}$
Equivalent absorption area $A = 200 \times 0.8 + 500 \times 0.1 = 210\,\mathrm{m^2}$
Mean absorption coefficient $\alpha_m = A/S = 210/700 = 0.30$
Mean absorption coefficient (height) $\alpha_m = (0.8 + 0.1)/2 = 0.45$
Mean free path (three-dim.) $l_m = 4V/S = 4 \times 1000/700 = 5.7\,\mathrm{m}$
Mean free path (two-dim.) $l_m = \pi S_x/U = \pi \times 200/60 = 10.5\,\mathrm{m}$

The results are shown in Table 4.4. A two-dimensional reverberation in the horizontal plane between the walls has also been calculated (4.2 s). The one-dimensional decays are the extreme cases with the longest reverberation time being 20 times the shortest one, 8.0 s and 0.4 s, respectively.

The real decay that is measured in the room will be a mixture of these different decays, and the reverberation time will be considerably longer than predicted from Sabine's formula. Eyring's formula is even worse. The measured decay curve will be bent, and thus, the measuring result depends on which part of the decay curve is considered for the evaluation of reverberation time.

In a room with long reverberation time due to non-diffuse conditions and at least one sound-absorbing surface, introducing some sound scattering elements

Table 4.4 Calculation of the one-dimensional reverberation times of the rectangular room in Figure 4.14

Direction	l_m (m)	α_m	T (s)
Three-dimensional (Sabine)	5.7	0.30	0.8
Three-dimensional (Eyring)	5.7	0.30	0.6
Two-dimensional (horizontal)	10.5	0.10	4.2
One-dimensional (length)	20	0.10	8.0
One-dimensional (width)	10	0.10	4.0
One-dimensional (height)	5	0.45	0.4

in the room can have a significant effect. It could be furniture or machines on the floor or some diffusers on the walls. This will make the sound field more diffuse, and the reverberation time will be reduced, i.e. it will come closer to the Sabine value. In other words: The sound absorption available in the room becomes more efficient when scattering elements are introduced to the room.

Note: In the one-dimensional case, it is strictly not correct to use the arithmetic average of the absorption coefficients for calculating the reverberation time of the axial mode. By inspection of Equation 4.46, it can be seen that the mean absorption coefficient should be calculated from

$$(1 - \alpha_m) = \sqrt{(1 - \alpha_1)(1 - \alpha_2)} \qquad (4.63)$$

Hence, if one of the surfaces is is totally absorbing, $\alpha_m = 1$ and, therefore, the reverberation time is zero.

4.5 MEASUREMENT OF REVERBERATION TIME

The reverberation time in a room can be measured with a noise signal or with an impulse. Actually, the modern measurement technique is to measure the impulse response using a sine-sweep or an maximum length sequence signal, see also (see Chapter 12). Both methods (interrupted noise and integrated squared impulse response) are described in the measurement standard ISO 3382-2. The traditional method uses pink noise or white noise emitted by a loudspeaker and a microphone to measure the sound pressure level as a function of time after the source is switched off. This gives a decay curve. A typical example is shown in Figure 4.15.

From the microphone, the signal is led to a frequency filter, which is either an octave filter or a one-third octave filter. If the sound in the room is adequately diffuse and a sufficiently large number of modes are excited, the decay curve is

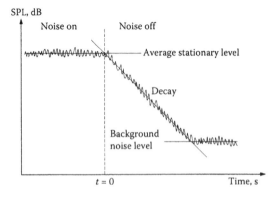

Figure 4.15 Typical decay curve measured with noise interrupted at time $t=0$. SPL, sound pressure level.

close to a straight line between the excitation level and the background level. The dynamic range is seldom more than around 50 dB and the whole range of the measured decay curve is not used. The lower part of the decay curve is influenced by the background noise and the upper part may be influenced by the direct sound, which gives a steeper start of the curve. Hence, the part of the decay curve used for evaluation begins 5 dB below the average stationary level and ends normally 25 dB or 35 dB below the same level. The evaluation range is, thus, either 20 dB or 30 dB and the slope is determined by fitting a straight line or automatically by calculating the slope of a linear regression line. From the slope of the decay curve in dB per second, the reverberation time is calculated, being the time for a 60 dB drop following the straight line. According to ISO 3382-2, the preferred evaluation range is 20 dB and the result is denoted as T_{20}. Similarly, T_{30} is used if the 30 dB evaluation range has been applied. If the sound field in the room is sufficiently diffuse, the decay curves are reasonably straight and T_{20} and T_{30} are approximately the same.

Sometimes, the decay curves are not nice and straight and it is difficult to measure a certain reverberation time; T_{20} and T_{30} yield different results. One reason can be that it is a measurement at low frequencies in a small room and maybe only a few modes are excited within the frequency band of the measurement. In this case, interference between the modes can cause very irregular decay curves.

The reverberation time is measured in a number of source and receiver positions, often two source positions and three microphone positions, i.e. six combinations are used. If the interrupted noise method is used, several decays should be measured and averaged in each position because the noise signal is a stochastic signal and, thus, the measured decay curves are always a little different. However, if the integrated squared impulse response method is used, a single measurement in each position is sufficient; the decay curve obtained with this method corresponds to averaging an infinite number of decays with the interrupted noise method (Schroeder, 1965). Figure 4.16 shows an example of a measured squared impulse response (the curve with fluctuations) and the corresponding integrated decay curve.

Annex B of ISO 3382-2 contains two parameters that are useful for describing the quality of reverberation time measurements. One is the degree of non linearity of the part of the decay curve within the evaluation range. Since the evaluation of the decay curve is made by a least-squares fit of a straight line, the correlation coefficient r is known. The non linearity parameter ξ is defined as the permillage deviation from perfect linearity:

$$\xi = 1000\left(1 - r^2\right) \ (\permil) \tag{4.64}$$

Values higher than 10 ‰ indicate a decay curve being far from a straight line, and the value of the reverberation time derived from the curve may be suspicious.

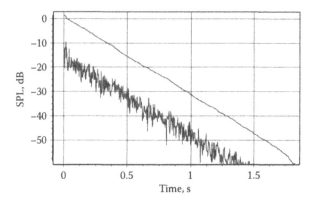

Figure 4.16 Typical decay curve measured using the integrated squared impulse response method. The lower curve is the squared impulse response and the upper curve is the decay curve.

The other quality parameter is the degree of curvature C defined as

$$C = 100\left(\frac{T_{30}}{T_{20}} - 1\right) \quad (\%) \tag{4.65}$$

Values higher than 10 % indicate a bent decay curve; the decay may be a combination of room modes with very different reverberation times.

REFERENCES

D.A. Bies and C.H. Hansen (1988). *Engineering Noise Control, Theory and Practice,* Unwin Hyman Ltd., London.

C.M. Harris (1966). Absorption of sound in air versus humidity and temperature. *Journal of the Acoustical Society of America* 40, 148–159.

ISO 3382-2 (2008). *Acoustics – Measurement of room acoustic parameters – Part 2: Reverberation time in ordinary rooms,* International Organization for Standardization, Geneva, Switzerland.

ISO 3741 (2010). *Acoustics – Determination of sound power levels and sound energy levels of noise sources using sound pressure – Precision methods for reverberation test rooms,* International Organization for Standardization, Geneva, Switzerland.

C.W. Kosten (1960). The mean free path in room acoustics. *Acustica* 10, 245–250.

H. Kuttruff (1973). *Room Acoustics,* Applied Science Publishers, London.

A.R. Molares (2010). *A Monte Carlo approach to the analysis of uncertainty in acoustics.* PhD Dissertation, Escola Técnica Superior de Enxeñeiros de Telecomunicación Universidade de Vigo, Spain.

A.D. Pierce (1989). *Acoustics. An Introduction to Its Physical Principles and Applications.* 2nd Edition, The Acoustical Society of America, New York.

M.R. Schroeder (1962). Frequency correlation functions of frequency responses in rooms. *Journal of the Acoustical Society of America* 34, 819–1823.

M.R. Schroeder (1965). New method of measuring reverberation time, *Journal of the Acoustical Society of America* 37, 409–412.

J.H. Rindel (2015). Modal energy analysis of nearly rectangular rooms at low frequencies. *Acta Acustica United with Acustica* 101, 1211–1221.

J.H. Rindel (2016). A note on modal reverberation times in rectangular rooms. *Acta Acustica United with Acustica*, 102, 600–603.

W.C. Sabine (1922). *Collected Papers on Acoustics*. Harvard University Press. Published 1964 by Dover Publications Inc., New York.

M. Vorländer (1995). Revised relation between the sound power and the average sound pressure level in rooms and consequences for acoustic measurements. *Acustica* 81, 332–342.

Chapter 5

Introduction to sound insulation

This chapter describes the basic principles of sound insulation with simplified assumptions. The methods for measuring airborne and impact sound insulation are briefly described. Also presented are the methods of objective evaluation of the measurement results. As a supplement to the fundamental theoretical models, the principles are illustrated through examples of measurement results on typical constructions.

5.1 AIRBORNE SOUND INSULATION

5.1.1 The sound transmission loss

A sound wave incident on a wall or any other surface separating two adjacent rooms is partly reflected back to the source room, partly dissipated as heat within the material of the wall, propagates partly to other connecting structures and is partly transmitted into the receiving room. Note that it is important to distinguish between sound insulation and sound absorption. Sound insulation has to do with the reduction of sound energy when sound is transmitted through a barrier or a wall, whereas sound absorption has to do with the sound energy that is lost (to heat) in some material. A material that is a good sound absorber is not necessarily a good material for sound insulation and vice versa. In fact, a good sound absorber is typically light and porous, whereas a good sound insulator is heavy and airtight.

If the power incident on the wall is P_1 and the power transmitted into the receiving room is P_2, the *sound transmission coefficient* τ is defined as the ratio of transmitted to incident sound power:

$$\tau = \frac{P_2}{P_1} \tag{5.1}$$

However, the sound transmission coefficients are typically very small numbers, and it is more convenient to use the *sound reduction index R* with the unit decibel (dB). It is defined as:

$$R = 10\lg\frac{P_1}{P_2} = 10\lg\frac{1}{\tau} = -10\lg\tau \text{ (dB)} \tag{5.2}$$

Another name for the same term is *sound transmission loss*, which will be used throughout this book.

5.1.2 Sound insulation of a partition between two rooms

Consider the sound insulation between two rooms. With the assumption of diffuse sound fields in both rooms, it is possible to derive a simple relation between the transmission loss and the sound pressure levels in the two rooms. The rooms are called the source room and the receiving room, respectively. In the source room is a sound source that generates the average sound pressure p_1. The sound power (see Equation 4.24) incident on the wall is (Figure 5.1),

$$P_1 = I_{\text{inc}}S = \frac{p_1^2 S}{4\rho c} \tag{5.3}$$

The area of the wall is S. In the receiving room, the sound pressure p_2 is generated from the sound power P_2 radiated into the room (see Equation 4.29),

$$p_2^2 = \frac{4P_2}{A_2}\rho c \tag{5.4}$$

A_2 denotes the absorption area in the receiving room. Insertion in Equation 5.2 gives

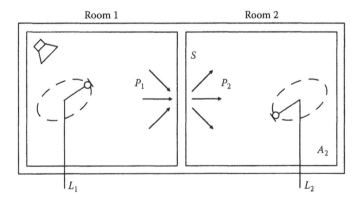

Figure 5.1 Airborne sound transmission from source room (1) to receiving room (2).

$$R = 10 \lg \frac{p_1^2 S}{4 \rho c P_2} = 10 \lg \frac{p_1^2 S}{p_2^2 A_2} = L_1 - L_2 + 10 \lg \frac{S}{A_2} \text{ (dB)} \qquad (5.5)$$

L_1 and L_2 are the spatially averaged sound pressure levels in the source and receiving rooms, respectively. This important result is the basis for transmission loss measurements.

5.1.3 Measurement of sound insulation

Sound insulation is measured in one-third octave bands covering the frequency range from 100 Hz to 3150 Hz. In recent years, the international standards for measurement of sound insulation have been revised, and it is recommended to extend the frequency range down to 50 Hz and up to 5000 Hz. One reason for this is that the low frequencies 50 Hz to 100 Hz are very important for the subjective evaluation of the sound insulation properties of lightweight constructions. Lightweight constructions have been more commonly used in new building technology, whereas heavy constructions have traditionally been used for sound insulation.

The sound pressure levels are measured as the average from a number of microphone positions or as the average from microphones slowly moving on a circular path (if mechanically operated) or on other paths (if manually operated). The results are usually averaged over two different source positions. More details can be found in Chapter 12.

In addition to the average sound pressure levels in the source and receiving rooms, it is also necessary to measure the reverberation time in the receiving room to calculate the absorption area. Sabine's equation is used for this (see Equation 4.34):

$$A_2 = \frac{55.3 V_2}{c T_2} \qquad (5.6)$$

where V_2 and T_2 are the volume and reverberation time of the receiving room, respectively, and c is the speed of sound. Only under special laboratory conditions can the transmission loss of a wall be measured without influence from other transmission paths. In a normal building, sound will not only be transmitted through the separating construction but the flanking constructions will also influence the result, as discussed later in Section 5.5.

For measurements of sound insulation in buildings, the *apparent sound reduction index* is

$$R' = L_1 - L_2 + 10 \lg \frac{S}{A_2} \text{ (dB)} \qquad (5.7)$$

The apostrophe after the symbol indicates that flanking transmission can be assumed to influence the result. Thus, R' is typically a few dB less than

R for the same partition wall measured in a laboratory. For the characterization of sound insulation in buildings, the R' has a problem – the area S of the partition must be known, and this is not always the case. For example, the source and receiving rooms may be located with an offset (staggered rooms) so that the area S does not exist or is very small? In the latter case, Equation 5.7 would yield a very low result, not corresponding to the perceived sound insulation.

To overcome this problem in field measurements, the area S can be set to a constant value of $10\,\mathrm{m}^2$, and the result using Equation 5.7 is called the *normalized level difference* D_n. An alternative measure that overcomes the problem with the partition area is the *standardized level difference* D_{nT}:

$$D_{nT} = L_1 - L_2 + 10\lg\frac{T_2}{T_0} \quad (\mathrm{dB}), \quad T_0 = 0.5\,\mathrm{s} \tag{5.8}$$

This is the level difference between the source and receiving rooms if the reverberation time of the receiving room is 0.5 s at all frequencies. This reverberation time is typical for a furnished living room in a dwelling. According to ISO 16283-1, the standardized level difference D_{nT} provides a straightforward link to the subjective impression of airborne sound insulation.

the relation between D_{nT} and R' is approximately: From Equations 5.6 through 5.8,

$$D_{nT} = R' + 10\lg\frac{V_2}{S\cdot d_0} \quad (\mathrm{dB}), \quad d_0 = 3\,\mathrm{m} \tag{5.9}$$

where d_0 is a reference distance, V_2 is the volume of the receiving room and S is the area of the partition (if any). In the case of a rectangular receiving room where one of the walls is the partition wall, the relation depends on the depth of the receiving room perpendicular to the partition wall, $d_2 = V_2/S$.

If $d_2 < 3\,\mathrm{m}$; $D_{nT} < R'$
If $d_2 \approx 3\,\mathrm{m}$; $D_{nT} \approx R'$
If $d_2 > 3\,\mathrm{m}$; $D_{nT} > R'$

As a consequence, D_{nT} depends on the measurement direction if the size of the source room and the receiving room is different. For testing to show compliance with a minimum airborne sound insulation requirement, the smaller room should be used as the receiving room to get the lowest result.

5.1.4 Multi-element partitions and apertures

A partition is often divided into elements with different sound insulation properties, e.g. a wall with a door. Each element is described by the area S_i and the transmission coefficient τ_i. If the sound intensity incident on the surfaces of the source room is denoted by I_{inc}, the total incident sound power on the partition is

$$P_1 = \sum_{i=1}^{n} S_i I_{\text{inc}} = S I_{\text{inc}}$$

The total area of the partition is called S. The total sound power transmitted through the partition is

$$P_2 = \sum_{i=1}^{n} \tau_i S_i I_{\text{inc}}$$

Thus, the transmission coefficient of the partition is

$$\tau_{\text{res}} = \frac{P_2}{P_1} = \frac{1}{S}\sum_{i=1}^{n} \tau_i S_i \qquad (5.10)$$

The same result can also be written in terms of the transmission losses R_i of each element

$$R_{\text{res}} = -10\lg\tau_{\text{res}} = -10\lg\left(\frac{1}{S}\sum_{i=1}^{n} S_i 10^{-0,1R_i}\right) \qquad (5.11)$$

In the simple case of only two elements, the graph in Figure 5.2 may be used.

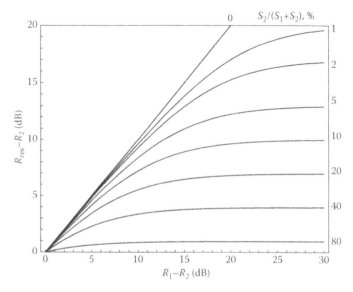

Figure 5.2 Estimating the transmission loss of a multi-element partition.

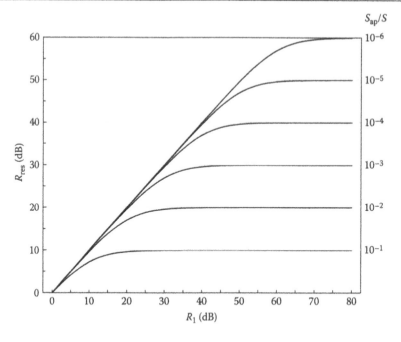

Figure 5.3 Estimating the transmission loss of a construction with an aperture.

An aperture in a wall is a special example of an element with different transmission properties. As an approximation, it can be assumed that the transmission coefficient of the aperture is 1. If the area of the aperture S_{ap} is also very small compared to the total area, the resulting transmission loss of the wall with aperture will be:

$$R_{res} = -10\lg\left(\frac{1}{S}\left(S_1 10^{-0,1R_1} + S_{ap}\right)\right) \cong -10\lg\left(10^{-0,1R_1} + \frac{S_{ap}}{S}\right) \qquad (5.12)$$

Figure 5.3 can illustrate the result. The relative area of the aperture defines an upper limit of the sound insulation that can be achieved.

5.2 SINGLE LEAF CONSTRUCTIONS

5.2.1 Sound transmission through a solid material

Consider a solid material shaped as a large plate with thickness h. The material is characterized by the density ρ_m and the speed of longitudinal waves c_L (Equation 1.20). The surface of the material defines two transition planes, where the sound waves change from one medium to another. It is assumed that the medium on either side is air with density ρ and the speed of sound c (also longitudinal waves). The symbols and notation are explained in Figure 5.4.

The sound pressure is equal on either side of the two transition planes:

$$p_i + p_r = p_1 + p_4$$

$$p_t = p_2 + p_3 \tag{5.13}$$

Also, the particle velocity is equal on either side of the two transition planes:

$$u_i - u_r = u_1 - u_4$$

$$u_t = u_2 - u_3 \tag{5.14}$$

The characteristic impedance in the surrounding medium (air) is denoted by Z_0 and in solid material it denoted by Z_m. Thus, the ratio of sound pressure to particle velocity in each of the plane propagating waves is:

$$\frac{p_i}{u_i} = \frac{p_r}{u_r} = \frac{p_t}{u_t} = Z_0 = \rho c$$

$$\frac{p_1}{u_1} = \frac{p_2}{u_2} = \frac{p_3}{u_3} = \frac{p_4}{u_4} = Z_m = \rho_m c_L \tag{5.15}$$

Using Equation 5.15 in Equation 5.14 leads to:

$$p_i - p_r = \frac{Z_0}{Z_m}(p_1 - p_4)$$

$$p_t = \frac{Z_0}{Z_m}(p_2 - p_3) \tag{5.16}$$

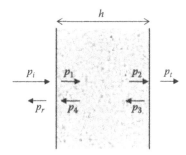

Figure 5.4 Thick wall with incident, reflected and transmitted sound waves. p_1, p_2, p_3 and p_4 denote the sound pressures of the longitudinal waves in the solid material, incident and reflected at the boundaries. p_i, p_r and p_t denote the incident, reflected and transmitted sound pressures in the air.

Assuming propagation from one side of the material to the other without losses means that there is only a phase difference between the pressures at the two intersections:

$$p_2 = p_1 e^{-jk_m h}$$

$$p_4 = p_3 e^{-jk_m h} \tag{5.17}$$

$k_m = \omega/c_L$ is the angular wave number for longitudinal sound propagation in the solid material.

From Equations 5.13, 5.16 and 5.17, the ratio between the sound pressures p_i and p_t and, thus, the transmission loss at normal incidence can be expressed by:

$$R_0 = 10\lg\left|\frac{p_i}{p_t}\right|^2 = 10\lg\left(\cos^2(k_m h) + \frac{1}{4}\left(\frac{Z_0}{Z_m} + \frac{Z_m}{Z_0}\right)^2 \sin^2(k_m h)\right) \tag{5.18}$$

The resulting transmission loss is shown in Figure 5.5, where some dips can be observed at high frequencies. They occur at frequencies where the thickness is equal to half a wavelength in the solid material, or a multiple of half wavelengths. However, the dips are very narrow and they are mainly of theoretical interest.

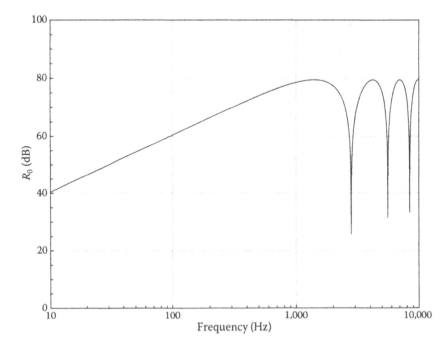

Figure 5.5 Transmission loss at normal incidence of sound on a 600 mm thick concrete wall.

Two special cases can be studied. First, the case of a thin wall: $Z_m \gg Z_0$ and $k_m h \ll 1$

$$R_0 \cong 10 \lg \left[1 + \left(\frac{Z_m}{2Z_0} \right)^2 \sin^2(k_m h) \right] \cong 10 \lg \left(1 + \left(\frac{\omega \rho_m h}{2 \rho c} \right)^2 \right) \qquad (5.19)$$

The other special case is a very thick wall: $Z_m \gg Z_0$ and $k_m h \gg 1$

$$R_0 \cong 10 \lg \left(\frac{Z_m}{2Z_0} \right)^2 \cong 20 \lg \left(\frac{\rho_m c_L}{2 \rho c} \right) \qquad (5.20)$$

The cross-over frequency from Equation 5.19 to Equation 5.20 is the frequency f_h at which $k_m h = 1$:

$$f_h = \frac{c_L}{2\pi h} \qquad (5.21)$$

This is the frequency at which the thickness is approximately one-sixth of the longitudinal wavelength λ_L in the material:

$$h = \frac{c_L}{2\pi f} = \frac{\lambda_L}{2\pi}$$

The result for the thin wall (Equation 5.19) is the so-called mass law, which will be derived in a different way in the next section. The result for a very thick wall (Equation 5.20) means that there is an upper limit on the sound insulation that can be achieved by a single-leaf construction, and this limit depends on the density of the material. For wood, it is 68 dB, for concrete 80 dB and for steel 94 dB. These numbers should be reduced by 5 dB in the case of random incidence, instead of normal incidence – see Section 5.2.3.

5.2.2 The mass law

Consider a thin wall with the mass per unit area m (see Figure 5.6). The application of Newton's second law (force=mass × acceleration) gives:

$$\Delta p = p_i + p_r - p_t = m \frac{\mathrm{d}v_n}{\mathrm{d}t} = j\omega m v_n \qquad (5.22)$$

where v_n is the velocity of the wall vibrations (in the direction normal to the wall). *Wall impedance Z_w* is introduced:

$$Z_w = \frac{\Delta p}{v_n} = j\omega m \qquad (5.23)$$

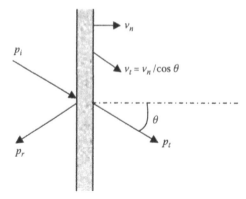

Figure 5.6 Thin wall with sound pressures and particle velocities.

The wall impedance will be more complicated if the bending stiffness of the wall is also taken into account (see Section 5.2.4).

The particle velocities in the sound waves are called *u* with the same indices as the corresponding sound pressures. Due to the continuity requirement, the normal component of the velocity on both sides of the wall is:

$$v_n = u_t \cos \theta = (u_i - u_r) \cos \theta \qquad (5.24)$$

which leads to:

$$p_t = p_i - p_r = Z_0 \frac{v_n}{\cos \theta} \qquad (5.25)$$

The sound transmission loss R_θ at a certain angle of incidence θ is:

$$R_\theta = 10 \lg \left| \frac{p_i}{p_t} \right|^2 = 10 \lg \left| 1 + \frac{Z_w \cos \theta}{2Z_0} \right|^2 \text{ (dB)} \qquad (5.26)$$

In the special case of normal sound incidence ($\theta=0$), the insertion of Equation 5.23 gives the important *mass law* of sound insulation:

$$R_0 = 10 \lg \left| 1 + j \frac{\omega m}{2\rho c} \right|^2 \cong 20 \lg \left(\frac{\pi f m}{\rho c} \right) \text{(dB)} \qquad (5.27)$$

Since $m = \rho_m h$, this result is the same as that derived earlier in a different way (Equation 5.19). The mass law for sound insulation was first suggested in a dissertation by Berger (1911).

5.2.3 Sound insulation at random incidence

The transmission coefficient at the angle of incidence θ is, from Equation 5.26,

$$\tau(\theta) = \frac{1}{1+\left(\dfrac{\omega m}{2\rho c}\right)^2 \cos^2 \theta} \qquad (5.28)$$

Random incidence means that the sound field on the source side of the partition is approximately a diffuse sound field. In a diffuse sound field, the incident sound power P_1 on a surface is found by integration over the solid angle $\psi = 2\pi$, assuming the same sound intensity I_1 in all directions. Since, in each direction, the transmitted sound power is equal to the incident sound power multiplied by the transmission coefficient, the ratio between transmitted and incident power is:

$$\tau = \frac{P_2}{P_1} = \frac{\displaystyle\int_{\psi=2\pi} \tau(\theta) I_1 S \, d\psi}{\displaystyle\int_{\psi=2\pi} I_1 S \, d\psi} = \frac{\displaystyle\int_0^{\pi/2} \tau(\theta) \cos\theta \sin\theta \, d\theta}{\displaystyle\int_0^{\pi/2} \cos\theta \sin\theta \, d\theta}$$

$$= 2 \int_0^{\pi/2} \tau(\theta) \cos\theta \sin\theta \, d\theta = \int_0^1 \tau(\theta) d(\cos^2\theta)$$

$$= \int_0^1 \frac{d(\cos^2\theta)}{1+\left(\omega m/2\rho c\right)^2 \cos^2\theta} = \left(\frac{2\rho c}{\omega m}\right)^2 \ln\left(1+\left(\omega m/2\rho c\right)^2\right)$$

$$R = -10\lg\tau = R_0 - 10\lg\left(0{,}23 R_0\right) \text{ (dB)} \qquad (5.29)$$

where R_0 is the mass law for normal incidence, Equation 5.27. This is the theoretical result for random incidence, and for typical values (R_0 between 30 dB and 60 dB), it means that R is 8 dB to 11 dB lower than R_0. However, this is not true in real life, and it can be shown that the aforementioned result is related to partitions of infinite size. Taking the finite size into account, the result is approximately:

$$R \cong R_0 - 5 \text{ dB} \qquad (5.30)$$

The transmission loss at random incidence is approximately the normal incidence transmission loss less 5 dB. This is in good agreement with measuring results on real walls. The sound transmission through finite-size walls will be described in Chapter 8.

5.2.4 The critical frequency

The bending stiffness of a beam with height h is the product of Young's modulus E and the moment of inertia, $h^3/12$. In a plate, E must be replaced by $E/(1-\mu^2)$, and the bending stiffness per unit length of a plate with thickness h is:

$$B = \frac{Eh^3}{12(1-\mu^2)} \tag{5.31}$$

where E is Young's modulus of the material and μ is Poisson's ratio ($\mu \cong 0.3$ for most rigid materials).

The speed of propagation of bending waves in a plate with bending stiffness per unit width B and mass per unit area m is (see Equation 1.25):

$$c_B = \sqrt{\omega}\sqrt[4]{\frac{B}{m}} = c\sqrt{\frac{f}{f_c}} \tag{5.32}$$

f_c is introduced as the *critical frequency* and is defined as the frequency at which the speed of the bending waves equals the speed of sound in air, $c_B = c$.

The critical frequency is:

$$f_c = \frac{c^2}{2\pi}\sqrt{\frac{m}{B}} \tag{5.33}$$

A sound wave with the angle of incidence θ propagates across the wall with the *phase speed* $c/\sin\theta$, i.e. the phase speed is in general higher than c (Figure 5.7). If the bending wave speed happens to be equal to the phase speed of the incident sound wave, this is called *coincidence*:

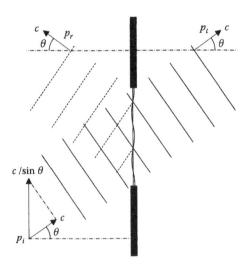

Figure 5.7 Thin wall with bending wave and indication of speed of propagation along the wall.

$$c_B = c / \sin\theta$$

The coincidence leads to a significant dip in the sound transmission loss. The coincidence dip will be at the *coincidence frequency*, which is higher than or equal to the critical frequency:

$$f_{co} = f_c / \sin^2\theta \tag{5.34}$$

The coincidence effect was first described by Lothar Cremer (1905–1990) in a pioneering paper (Cremer 1942). Taking the bending stiffness into account, the wall impedance (Equation 5.23) is replaced by:

$$Z_w = j\omega m \left(1 - \left(\frac{f}{f_c}\right)^2 \sin^4\theta\right) \tag{5.35}$$

Insertion in the general form, Equation 5.26, leads to the sound transmission loss at a certain angle of incidence:

$$R_\theta = R_0 + 20 \lg|\cos\theta| + 20 \lg\left|1 - (f/f_c)^2 \sin^4\theta\right| \text{ (dB)} \tag{5.36}$$

The frequency course of the transmission loss at different angles of incidence is displayed in Figure 5.8. At normal incidence ($\theta=0$), the transmission loss increases monotonically with frequency, but at all other angles of incidence, there is a coincidence dip at the coincidence frequency (Equation 5.34). At grazing incidence, i.e. when θ approaches 90°, the coincidence dip is at the critical frequency, but at 45°, the coincidence dip is one octave higher at $f=2 f_c$. Another interesting observation from Equation 5.36 is that the transmission loss is extremely low near grazing incidence. If we go back to the original formula, Equation 5.26, we see that $R_\theta=0$ dB at 90°. This is a theoretical result that obviously disagrees with reality. In Chapter 8, we shall derive a more realistic solution by taking the finite area into account.

The three-dimensional contour plot in Figure 5.9 shows the transmission loss of a single panel as a function of frequency along one axis and as a function of the angle of incidence along the other axis. At normal incidence, there is a monotone increase with frequency and no coincidence dip. At other angles of incidence, there is a dip at the coincidence frequency given by Equation 5.34.

5.2.5 A simple model of sound insulation of single constructions

The simple model of sound insulation is based on the mass law as given in Equation 5.27, but with the influence of the dip around the critical

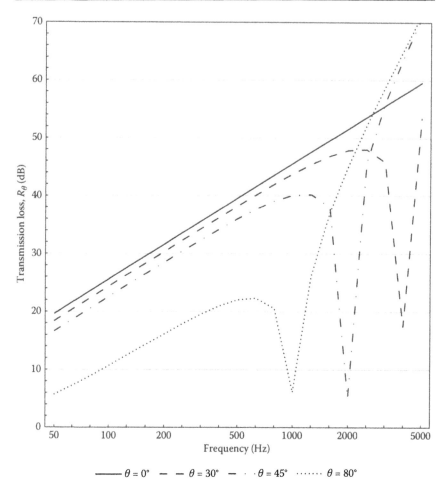

Figure 5.8 Transmission loss of a single panel as a function of frequency for different angles of incidence. The panel has $m=25$ kg/m^2 and $f_c=1000$ Hz.

frequency. Only bending waves are assumed. This model does not include the influence of a finite area, the normal modes at low frequencies, nor the shear waves at high frequencies. These topics will be dealt with in Chapter 8.

The following results are valid for sound insulation between rooms with approximately diffuse sound fields. In the frequency range below the critical frequency, $f < f_c$, the sound insulation is mass-controlled and the approximate transmission loss is:

$$R \cong R_0 + 20\lg\left|1-\left(f/f_c\right)^2\right| - 5 \text{ dB} \tag{5.37}$$

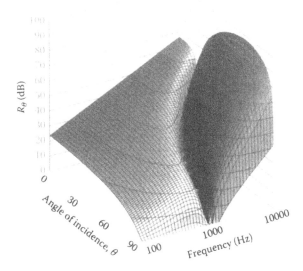

Figure 5.9 Contour plot of the transmission loss of a single panel as a function of angle of incidence and frequency. The step between contours is 10 dB. The panel has m=25 kg/m² and f_c=1000 Hz.

In the frequency range above the critical frequency, $f \geq f_c$, the sound insulation is stiffness-controlled and the resonant transmission due to the structural modes dominates the transmission. The approximate transmission loss in this region is:

$$R \cong R_0 + 10 \lg \frac{2\eta f}{\pi f_c} \text{ (dB)} \qquad (5.38)$$

where η is the loss factor (see Equation 2.28). This result will be derived and explained later in Chapter 7.

A sketch of the transmission loss as a function of frequency is shown in Figure 5.10. Well below the critical frequency, the course increases by 6 dB per octave, but it changes towards the dip at the critical frequency as described by Equation 5.37. The maximum transmission loss below f_c will be at $f = f_c / \sqrt{3} \cong 0.58 \cdot f_c$. Insertion of this frequency and Equation 1.28 in Equation 5.37 leads to:

$$R_{\max,f<f_c} = 10 \lg \left(\frac{c^2}{\rho^2} \frac{\rho_m^3}{E} \left(1 - \mu^2\right) \right) - 8.5 \text{ dB} \qquad (5.39)$$

This is a theoretical maximum, and for some building materials (Table 1.4), this can be calculated: around 33 dB for gypsum board; 35 dB for glass,

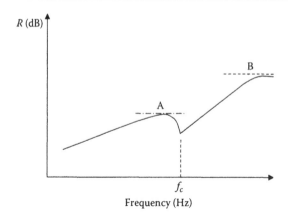

Figure 5.10 Simplified sound insulation of a single-leaf construction, f_c is the critical frequency and the horizontal dotted line at A indicates the maximum below f_c (Equation 5.39). The dotted line at B indicates the upper limit due to longitudinal waves (Equation 5.20).

brick and porous concrete; 37 dB for concrete; 44 dB for steel and 59 dB for lead. In reality, the maximum value may be a few dB lower because resonant transmission contributes to the total sound transmission, especially when the physical dimensions are relatively small, like windows, or when the plate is supported by laths, like gypsum boards. Hence, a maximum around 30 dB to 32 dB is typical for single layers of gypsum board and glass.

5.2.6 Examples of single constructions

Some examples of measured transmission loss of single constructions will illustrate the physical principles explained so far.

Figure 5.11 shows the effect of increasing the thickness h of a glass pane. This means a shift of the frequency course towards lower frequencies such that the critical frequency goes down, $f_c \sim 1/h$ (see Equation 1.28). Below f_c, the mass law predicts an increase of R with 6 dB per doubling of h. The weak point of a single glass pane is the dip around the critical frequency, and for a thin glass (4 mm), this dip is at a sufficiently high frequency, so it is not a serious problem. But increasing the thickness means that the frequency region with weak sound insulation moves down into more important frequencies – in general, solid glass should not be more than 6 mm to 8 mm thick.

If a heavier glass is needed, it should be a laminated glass, which is made from two or more layers of glass that is glued together with a special soft material. An example of the transmission loss of a laminated glass is shown in Figure 5.12. The dip around the critical frequency is almost removed due to the losses inherent in the lamination.

The previous examples indicate that it may be difficult to take full advantage of the mass law. However, if we could increase the mass without

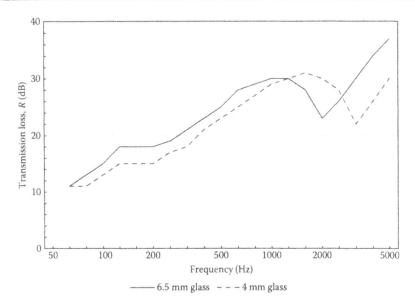

Figure 5.11 Influence of thickness on sound insulation of a single construction. (Adapted from Lewis, P.T., *Building Acoustics*, Oriel Press, London, 1971, Figure 7.18.)

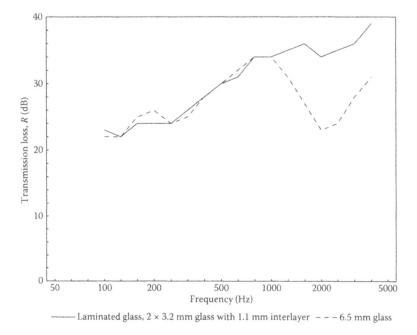

Figure 5.12 Transmission loss of a laminated and a solid glass pane with same mass per unit area. (Adapted from Marsh, J.A., *Building Acoustics*, Oriel Press, London, 1971, Figure 5.3.)

increasing the bending stiffness, this might be interesting. The example in Figure 5.13 is a door leaf with cylindrical cavities, which can be filled with sand. This doubles the mass without changing the bending stiffness, and the improvement in sound insulation is much better than the theoretically expected 6 dB. In addition to the weight, the sand probably also increases the internal loss factor.

The opposite solution is also possible – reducing the bending stiffness without significantly changing the mass. In Figure 5.14, we see the effect of carving slits halfway through a plywood panel. The reduced bending stiffness means lower speed of propagation of bending waves, a higher critical frequency and less radiation of sound from resonant modes.

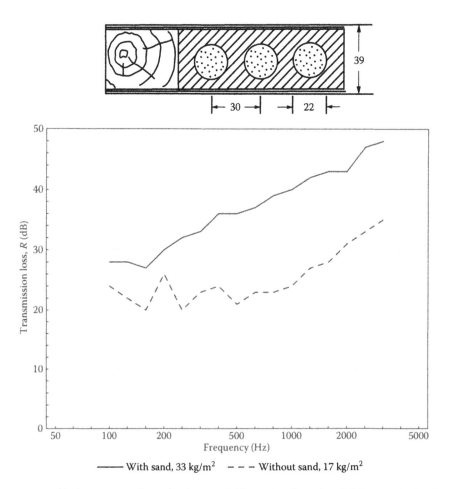

——— With sand, 33 kg/m² – – – Without sand, 17 kg/m²

Figure 5.13 Transmission loss of a door leaf, 39 mm hardboard with 22 mm cylindrical cavities per 30 mm. (Adapted from Gösele, K., Schalldämmung von Türen, Berichte aus der Bauforschung, Heft 63, 1969, Figure 10.)

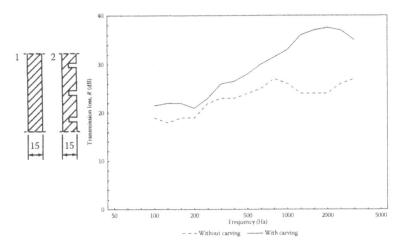

Figure 5.14 Transmission loss of a 15 mm plywood plate with or without slits. (Adapted from Fasold, W. and Sonntag, E., *Bauakustik. Bauphysikalisches Entwurfslehre,* Band 4, VEB Verlag für Bauwesen, Berlin, 1978, Figure 252.)

A simple and widely used method to increase the mass without too much trouble with the critical frequency is simply to layer two or more sheets on top of each other. With two layers instead of one, the bending stiffness is doubled and so is the mass, i.e. the ratio is constant and the critical frequency is the same as for the single sheet. Figure 5.15 shows the transmission loss of one or two layers of 13 mm gypsum board, with the critical

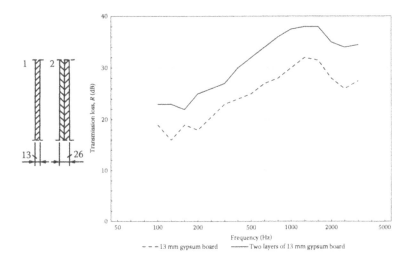

Figure 5.15 Transmission loss of one or two layers of 13 mm gypsum board. (Adapted from Fasold, W. and Sonntag, E., *Bauakustik. Bauphysikalisches Entwurfslehre,* Band 4, VEB Verlag für Bauwesen, Berlin, 1978, Figure 253.)

frequency unchanged at 2500 Hz. Three or even four layers are sometimes used when very high sound insulation is required. It is important that the layers are not glued together, but free to slide against each other. In praxis, the plates may be screwed together and they still behave as layered plates.

Good sound insulation requires that the partition wall is airtight. In the case of porous concrete and for some types of brick walls, this means that the surface must be treated with plaster, preferably on both sides. The effect of plaster on the sound insulation of a porous concrete wall is shown in Figure 5.16.

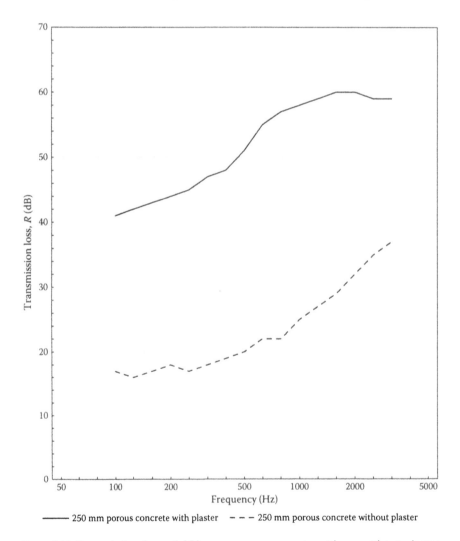

———— 250 mm porous concrete with plaster   – – – 250 mm porous concrete without plaster

Figure 5.16 Transmission loss of 250 mm porous concrete with or without plaster. (Adapted from Kristensen, J., Lydisolation: Teori, måling, vurdering og bestemmelser. SBI-notat 24 (In Danish), Danish Building Research Institute, Aalborg University Copenhagen, 1973, Figure 14.)

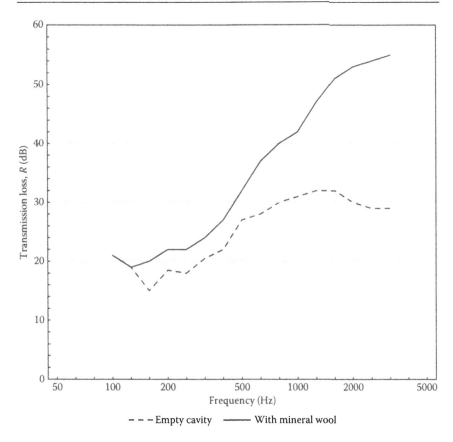

Figure 5.17 Transmission loss of a 10 mm gypsum board with or without sound absorbing treatment on both sides. (Adapted from Fasold, W. and Sonntag, E., *Bauakustik. Bauphysikalisches Entwurfslehre*, Band 4, VEB Verlag für Bauwesen, Berlin, 1978.)

Sometimes, the partition wall should also provide sound absorption in the rooms. The treatment of the surface with a sound absorbing material may influence the sound transmission. An example of the sound insulation of a 10 mm gypsum board with or without sound absorbing treatment on both sides is seen in Figure 5.17. The sound absorbing treatment is 30 mm mineral wool covered with a perforated plate.

5.3 DOUBLE LEAF CONSTRUCTIONS

5.3.1 Impedance model for sound transmission

Consider a double construction with two plates separated by distance d (Figure 5.18). The wall impedance of the two plates is denoted by Z_1 and

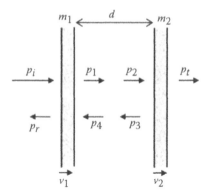

Figure 5.18 A double construction showing sound pressures and wall velocities. p_1, p_2, p_3 and p_4 denote the sound pressures in the cavity, incident and reflected at the boundaries. p_i, p_r and p_t denote the incident, reflected and transmitted sound pressures. v_1 and v_2 are the velocities of the two walls when excited by sound.

Z_2, respectively. As for the single construction in Equation 5.12, the movement of each wall is:

$$p_i + p_r - (p_1 - p_4) = Z_1 v_1$$

$$p_2 + p_3 - p_t = Z_2 v_2 \tag{5.40}$$

The velocity of each wall equals the particle velocity on either side:

$$v_1 = u_i - u_r = \frac{1}{Z_0}(p_i - p_r)$$

$$v_1 = u_1 - u_4 = \frac{1}{Z_0}(p_1 - p_4)$$

$$v_2 = u_2 - u_3 = \frac{1}{Z_0}(p_2 - p_3)$$

$$v_2 = u_t = \frac{1}{Z_0}p_t \tag{5.41}$$

Assuming propagation from one side of the cavity to the other without losses means that there is only a phase difference between the pressures at the two intersections:

$$p_2 = p_1 e^{-jkd}$$

$$p_4 = p_3 e^{-jkd} \tag{5.42}$$

From Equations 5.40 through 5.42, the ratio between the sound pressures p_i and p_t can be derived and, thus, the transmission loss can be expressed by:

$$R_0 = 10 \lg \left| \frac{p_i}{p_t} \right|^2$$

$$= 10 \lg \left| \left(1 + \frac{Z_1 + Z_2}{2Z_0} \right) \cos(kd) + j \left(1 + \frac{Z_1 + Z_2}{2Z_0} + \frac{Z_1 Z_2}{2Z_0^2} \right) \sin(kd) \right|^2 \quad (5.43)$$

If only the mass of each wall is taken into account, the wall impedances are:

$$Z_1 = j \omega m_1$$

$$Z_2 = j \omega m_2 \quad (5.44)$$

Neglecting the smaller parts and inserting $Z_0 = \rho c$ together with Equation 5.44 yields:

$$R_0 \cong 10 \lg \left[\left(\frac{\omega (m_1 + m_2)}{2 \rho c} \sin(kd) \right)^2 + \left(\frac{\omega (m_1 + m_2)}{2 \rho c} \cos(kd) - \frac{\omega^2 m_1 m_2}{2(\rho c)^2} \sin(kd) \right)^2 \right] \quad (5.45)$$

This result will be discussed and simplified in the following section.

5.3.2 The mass-air-mass resonance frequency

The transmission loss is minimum when the last term in Equation 5.45 is zero, for example:

$$\tan(kd) = \frac{m_1 + m_2}{m_1 m_2} \frac{\rho c}{\omega} \quad (5.46)$$

For a cavity that is narrow compared to the wavelength $(kd \ll 1)$, we get:

$$\tan(kd) \cong kd = \frac{\omega d}{c}$$

The solution to Equation 5.46 is the mass-air-mass resonance frequency $f_0 = \omega_0 / 2\pi$:

$$f_0 \cong \frac{c}{2\pi} \sqrt{\frac{\rho}{d} \left(\frac{1}{m_1} + \frac{1}{m_2} \right)} \quad (5.47)$$

If the depth d of the cavity is comparable to the wavelength, there are many solutions to Equation 5.46, and they are approximately $kd = n\pi$, where n is a natural number. The dips in the sound insulation occur at frequencies at which the cavity depth equals one or more half wavelengths: $d = n\ \lambda/2$.

However, more important than these dips is the shift from low- to high-frequency behaviour of the air cavity. The *cross-over frequency* is the frequency f_d at which $kd = 1$:

$$f_d = \frac{c}{2\pi d} \tag{5.48}$$

This is quite similar to the result found for the sound transmission through a solid material (Equation 5.21). Only, in this case, the transmission is through air. The spring-like behaviour of the air cavity changes from that of a simple spring below the cross-over frequency to that of a transmission channel at higher frequencies.

5.3.3 Sound insulation of double constructions

The result (Equation 5.45) can be simplified in different ways depending on the frequency range. In the frequency range below the resonance frequency, $f < f_0$,

$$R_0 \approx 20 \lg\left(\frac{\omega(m_1 + m_2)}{2\rho c}\right) = R_{(1+2)} \tag{5.49}$$

This means that the construction behaves as a single construction with the mass per unit area $(m_1 + m_2)$. In the frequency range above the resonance frequency and below the cross-over frequency, $f_0 < f < f_d$,

$$R_0 \approx 20 \lg\left(\frac{\omega^3 m_1 m_2 d}{2\rho^2 c^3}\right) \approx R_1 + R_2 + 20 \lg(2kd) \tag{5.50}$$

In this range, a much better sound insulation can be obtained, and the transmission loss depends on the product of the three parameters, m_1, m_2 and d. At frequencies above f_d where the cavity is wide compared to the wavelength, $\sin(kd)$ is replaced by its maximum value 1, and for $f \geq f_d$:

$$R_0 \approx 20 \lg\left(\frac{\omega^2 m_1 m_2}{2(\rho c)^2}\right) \approx R_1 + R_2 + 6\ \text{dB} \tag{5.51}$$

In this high-frequency range, d is no longer an important parameter.

A sketch of the transmission loss as a function of frequency is shown in Figure 5.19. We recall that the transmission loss of a single leaf will exhibit a dip at the critical frequency, and if the two leafs have different critical frequencies, there will be two corresponding dips in the transmission loss of

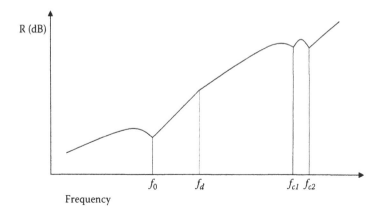

Figure 5.19 Sound insulation of a double leaf construction, f_0 is the resonance frequency and f_d is the cross-over frequency of the cavity. The critical frequencies of the two plates are f_{c1} and f_{c2}, respectively. Note that $f_{c1} = f_{c2}$ if it is a symmetric construction.

the double construction. To include this in the calculations, Equations 5.37 and 5.38 should be used to represent R_1 and R_2 in Equations 5.50 and 5.51.

A more accurate approximation of Equation 5.45 valid at frequencies below f_d is:

$$R_0 = R_{(1+2)} + 10\lg\left[\sin^2(kd) + \left(\cos(kd) - \left(\frac{\omega}{\omega_0}\right)^2 \cdot \mathrm{sinc}(kd)\right)^2\right]$$

$$\cong R_{(1+2)} + 10\lg\left(1 - \left(\frac{f}{f_0}\right)^2\right)^2 \tag{5.52}$$

5.3.4 Examples of double constructions

Some examples of measured transmission loss of double constructions will illustrate the physical principles described here.

Figure 5.20 shows the effect of increasing the distance d between the glasses in a double window. With 12 mm cavity, the resonance frequency is 250 Hz, which is very unfortunate in relation to traffic noise. Increasing the distance to 85 mm yields a much better sound insulation at all frequencies, as the resonance frequency is lower and the resonance dip not so pronounced. The critical frequency of a 4 mm glass is 3 kHz, and this is noted in the measured frequency course.

In a double window, it can be a good idea to use glass of different thicknesses because the dips at the critical frequencies, then, are less

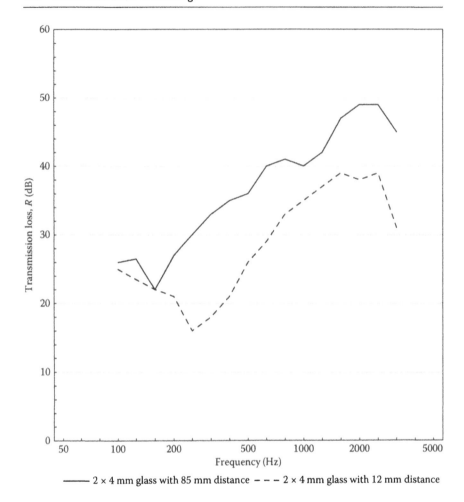

Figure 5.20 Sound insulation of a double window with two 4 mm glass panes, and the influence of the cavity depth. (Adapted from Ingemansson, S., *Ljudisolerande fönsterkonstruktioner* (In Swedish), Stockholm, Sweden, 1968, Figure 8.)

pronounced compared to the dip in a symmetric construction. The example in Figure 5.21 is a double window with a 100 mm cavity and glass thicknesses of 4 mm and 6.5 mm. The advantage of the asymmetric solution is obvious.

In a drywall, it is important that the cavity is acoustically damped, e.g. with mineral wool. Without this, the sound in the cavity can resonate and significantly reduce the sound insulation. The example in Figure 5.22 shows the result for a wall with two layers of 13 mm gypsum board on steel profiles per 600 mm. The cavity depth is 70 mm and this is either empty or

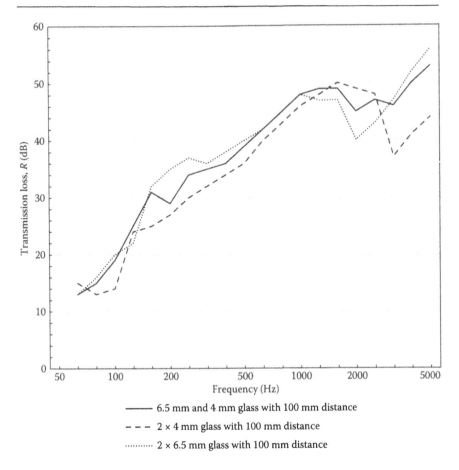

Figure 5.21 Influence of symmetry on the sound insulation of a double construction. (Adapted from Lewis, P.T., *Building Acoustics*, Oriel Press, London, 1971, Figure 7.19.)

filled with 50 mm mineral wool. The effect of the mineral wool is significant in the entire frequency range.

Asymmetric double constructions have a special application in relation to heavy- or medium-weight single walls. Such walls have a relatively low critical frequency and, thus, the sound radiation from resonant modes in the wall is very important for sound transmission. If the structural vibrations are transferred to a thin plate with high critical frequency, the sound radiation is reduced and, thus, the sound insulation is significantly improved, as shown in the example in Figure 5.23. In fact, it is possible to improve the sound insulation even more by applying a similar light construction to the other side of the wall.

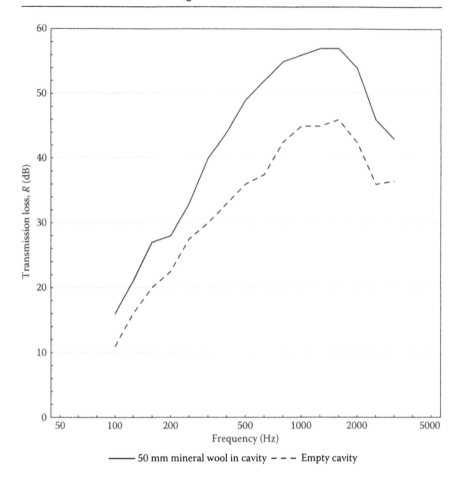

Figure 5.22 The influence of absorption material in the cavity on the sound insulation of a double construction.

5.4 SLITS AND LEAKS

In Figure 5.3 and Equation 5.12, it was shown that an aperture in a wall leads to an upper limit of the sound insulation. As a first approximation, this limit is independent of the frequency, which in practice means that the high frequency sound insulation is limited while the lower frequencies can remain more or less unaffected by a small slit or a leakage. In the following, we shall see a few examples of how slits and leaks influence the sound insulation.

The effect of the width of a slit is seen in Figure 5.24. While transmission loss of the wall without slits increase with frequency, a 1 mm wide slit means an upper limit around 40 dB. In addition, we see very efficient transmission around 2000 Hz and 4000 Hz. This can be explained by the depth of the slit that equals one or two half wavelength at these frequencies.

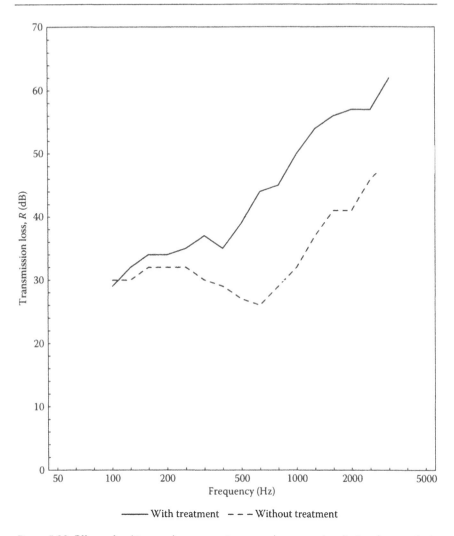

Figure 5.23 Effect of a thin panel construction to reduce sound radiation from a single wall, shown here as a 68 mm lightweight concrete wall. The thin panel is 3 mm of hardboard on 25 mm wooden laths with 25 mm mineral wool in the cavity. (Adapted from Brandt, O., *Akustisk planering*. (In Swedish), V. Pettersons Bokindustri AB, Stockholm, Sweden, 1958, Figure 3.24b.)

In Figure 5.25, the 3 mm slit from the previous figure is filled with felt, and we see that such a filling can be efficient; the transmission loss with felt in the slit is nearly the same as that without any slit in Figure 5.24.

Doors can be particularly difficult in terms of sound insulation, because it sometimes becomes necessary to avoid any sealing lists at the floor, e.g. in hospitals. In such cases, it is possible to apply special moving sealing lists that close when the door is closed. However, it is also possible to leave the

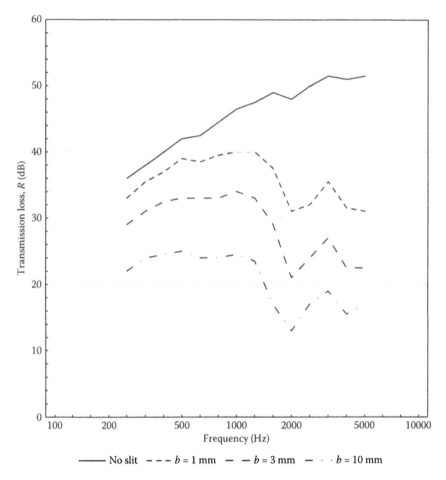

Figure 5.24 Transmission loss of a 1 m² wall with a 1 m long slit. The depth is 75 mm and the width of the slit varies from 0 mm to 10 mm. (Adapted from Gösele, K., Schalldämmung von Türen, Berichte aus der Bauforschung, Heft 63, 1969, Figure 17.)

slit under the door open, if there is an efficient sound absorbing lining in the door, like a resonator. In the measured result shown in Figure 5.26, the underside of the door has an opening into a 100 mm deep cavity filled with mineral wool.

Sound insulation of doors is also particularly difficult in practice due to the risk of leaks and because the sealing between the frame and the surrounding wall is important. Good workmanship is important here.

Openable windows are also an issue, where leaks can easily lead to poor sound insulation. It is important that a soft sealing list is applied on all sides of the sash, and if it is possible to have two sets of continuous sealing that is even better that a single sealing (Figure 5.27).

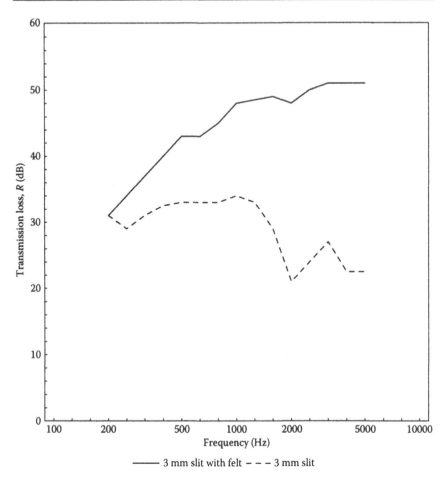

— 3 mm slit with felt – – – 3 mm slit

Figure 5.25 Effect of filling a slit with felt. Transmission loss of a 1 m² wall with a 1 m long and 75 mm deep slit. The width of the slit is 3 mm. (Adapted from Gösele, K., Schalldämmung von Türen, Berichte aus der Bauforschung, Heft 63, 1969, Figure 24.)

5.5 FLANKING TRANSMISSION

The transmission of sound from a source room to a receiver room can be via flanking constructions like the floor, the ceiling or the façade. When all relevant transmission paths are considered, the sound insulation is described by the *apparent sound transmission loss* (*apparent sound reduction index*):

$$R' = 10\lg\frac{P_1}{P_2 + P_3} = L_1 - L_2 + 10\lg\frac{S}{A_2} \text{ (dB)} \qquad (5.53)$$

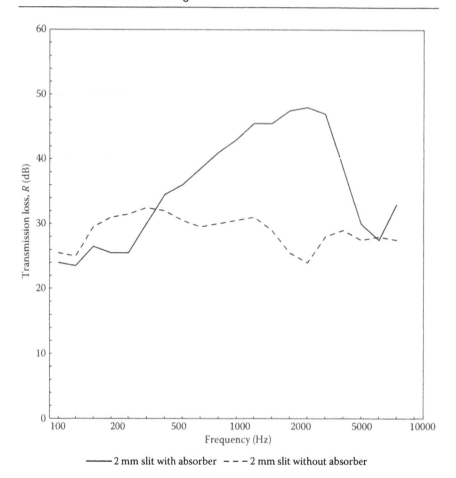

——— 2 mm slit with absorber – – – 2 mm slit without absorber

Figure 5.26 Effect of a Helmholtz resonator to reduce sound transmission through a slit. (Adapted from Gösele, K., Schalldämmung von Türen, Berichte aus der Bauforschung, Heft 63, 1969, Figure 25.)

where P_1 is the incident sound power on the partition, P_2 is the sound power transmitted through the partition wall to the receiver room and P_3 is the sound power radiated to the receiver room from the flanking surfaces and other flanking paths:

$$P_3 = \sum_i P_{F,i} \qquad (5.54)$$

Each single flanking transmission path i can be characterized by the *flanking transmission loss, $R_{F,i}$*:

$$R_{F,i} = 10 \lg \frac{P_1}{P_{F,i}} \text{ (dB)} \qquad (5.55)$$

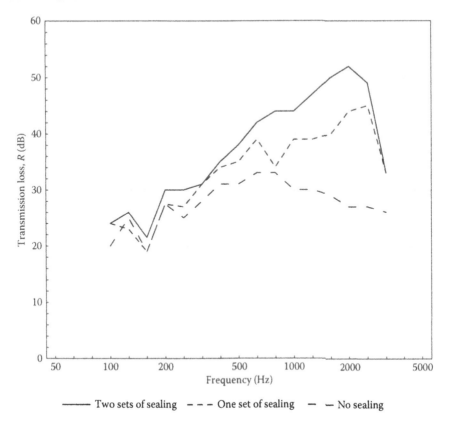

Figure 5.27 Effect of sealing in an openable double window with two panes of 4 mm glass and 85 mm cavity depth. (Adapted from Ingemansson, S., Ljudisolerande fönsterkonstruktioner (In Swedish), Stockholm, Sweden, 1968, Figure 6.)

It is convenient to keep the incident sound power P_1 on the partition wall as a reference for all the flanking transmission losses. In this way, it is very simple to add all the contributions. The apparent transmission loss is calculated from:

$$R' = -10\lg\left(10^{-0,1R} + \sum_i 10^{-0,1R_{F,i}} \right) \text{(dB)} \tag{5.56}$$

In the typical case of horizontal transmission through a wall, there will be 12 flanking paths, namely, three possible paths for each of the four surrounding flanking constructions (Figure 5.28).

Flanking transmission can be controlled in different ways. Three basic principles include the following:

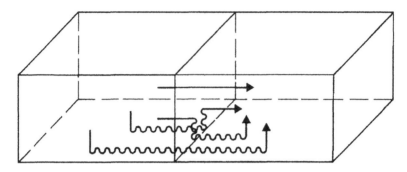

Figure 5.28 Direct transmission and three flanking transmission paths via the floor. (Reproduced from Kristensen, J. and Rindel, J.H., *Bygningsakustik – Teorti og praksis*. SBI-Anvisning 166 (In Danish), Danish Building Research Institute, Hørsholm, Denmark, 1989. With permission.)

- Use of heavyweight constructions. The heavy mass of the flanking construction means that the excitation of structural vibrations is relatively weak. If the transmission loss of the flanking construction itself is at least as good as that of the partition wall, the flanking transmission should not be a problem.
- Interruption of the flanking construction, either by a slit or by elastic interlayers. This calls for careful design of the details in the junction. Some examples of good and bad solutions are shown in Figures 5.29 and 5.30.

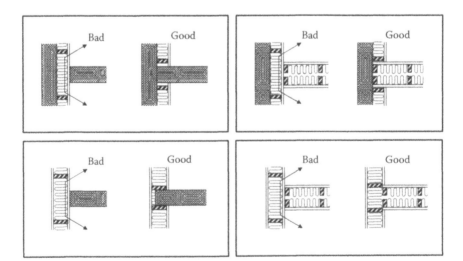

Figure 5.29 Examples of junctions between external and internal walls in horizontal view. Solutions are shown in pairs, those with a flanking transmission issue to the left.

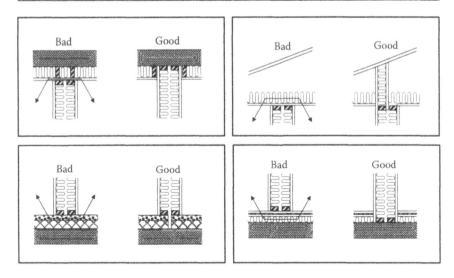

Figure 5.30 Examples of junctions between roof and internal wall (upper part) and between floor and internal wall (lower part) in vertical view. Solutions are shown in pairs, those with a flanking transmission issue to the left.

- Covering the flanking surface with a board having a high critical frequency, i.e. preferably above 2000 Hz, the higher the better. This can be a secondary surface with one or two layers of plasterboard. The reduction of flanking transmission is efficient on both surfaces of the flanking structure, in the source room as well as in the receiving room.

5.6 ENCLOSURES

A noise source is supposed to radiate sound power P_a. Consider that the noise source is totally covered by an enclosure with surface area S, absorption coefficient α on the inside and the enclosure is made from a plate with transmission loss R or transmission coefficient τ. The average sound pressure in the enclosure p_{encl} can be estimated, if a diffuse sound field is assumed:

$$p_{encl}^2 = \frac{4P_a}{\alpha S} \rho c \tag{5.57}$$

The sound power incident on the inner surface of the enclosure is (still with the assumption of a diffuse sound field):

$$P_{inc} = \frac{p_{encl}^2 S}{4 \rho c} \tag{5.58}$$

The sound power transmitted through the enclosure is then:

$$P_{out} = \tau P_{inc} = \frac{\tau}{\alpha} P_a \qquad (5.59)$$

The *insertion loss* of the enclosure is the difference in radiated sound power level without and with the enclosure:

$$\Delta L = 10 \lg \frac{P_a}{P_{out}} = 10 \lg \frac{\alpha}{\tau} = R + 10 \lg \alpha \ (\text{dB}) \qquad (5.60)$$

This result cannot be considered to be very accurate. The assumption of a diffuse sound field inside the enclosure is especially doubtful. However, the result is not bad as a rough estimate for the design of an enclosure. It is clearly seen from Equation 5.60 that both transmission loss and absorption coefficient are important for an efficient reduction of noise by an enclosure.

5.7 IMPACT SOUND INSULATION

5.7.1 Impact sound pressure level

The noise generated from footsteps on floors is characterized by the impact noise level. It is measured using a standardized tapping machine according to ISO 16283-2 in buildings or ISO 10140-3 in the laboratory. The main data for the tapping machine are the following:

- The noise is generated by steel hammers with a fall height of 40 mm.
- Each steel hammer has a mass of 500 g.
- The number of taps per second is 10.

The tapping machine is placed on the floor in the source room in a number of positions. The calibrated sound pressure level L_2 is measured in the room below – or any other room in the building (see Figure 5.31). The reverberation time in the receiving room must also be measured in order to calculate the absorption area A_2. The impact sound pressure level is the sound pressure level in dB re 20 µPa that would be measured if the absorption area is $A_0 = 10 \, \text{m}^2$:

$$L_n = L_2 + 10 \lg \frac{A_2}{A_0} \ (\text{dB}) \quad A_0 = 10 \ \text{m}^2 \qquad (5.61)$$

The frequency range is the same as for airborne sound insulation, i.e. the 16 one-third octave bands from 100 Hz to 3150 Hz. However, it is recommended that the frequency range be extended down to 50 Hz, especially in the case of lightweight floors and heavyweight constructions with a floating floor.

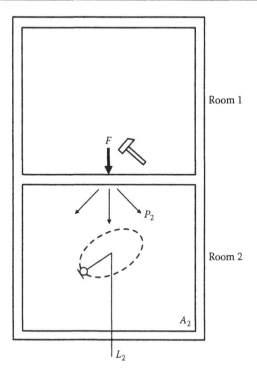

Figure 5.31 Principle of measuring the impact sound pressure level from a floor to a receiving room (Room 2).

The *normalized impact sound pressure level* L_n is used for measurements in the laboratory to characterize specimens of floor constructions. However, for field measurements in buildings, the *standardized impact sound pressure level* L'_{nT} is sometimes preferred.

$$L'_{n,T} = L_2 - 10\lg\frac{T_2}{T_0} \text{ (dB)} \quad T_0 = 0.5 \text{ s} \tag{5.62}$$

Just like D_{nT} for airborne sound insulation, L'_{nT} is corrected for the reverberation time in the receiving room if different from 0.5 s. According to ISO 16283-2, L'_{nT} provides a straightforward link to the subjective impression of impact sound insulation.

The relationship between the standardized and the normalized impact sound pressure level is approximately:

$$L'_{nT} = L'_n - 10\lg\frac{V_2}{V_0} \text{ (dB)}, \quad V_0 = 30 \text{ m}^3 \tag{5.63}$$

where V_2 is the volume of the receiving room. For typical volumes in a dwelling around 50 m³ to 80 m³, the difference will be −2 dB to −4 dB.

5.7.2 Historical note

The tapping machine originated from the German standard DIN 4110 (1938), but the impact sound pressure level was measured in Phon as a broad-band level applying a frequency weighting filter B. The practice to measure impact sound in one-third octave bands was developed in Denmark (Ingerslev et al. 1947) – they used the frequency range from 100 Hz to 2000 Hz. An alternative method was developed in America, using a tapping machine with different specifications and measuring in seven octave bands from 63 Hz to 4 kHz (Lindahl and Sabine 1940).

5.7.3 Sum of airborne and impact sound insulation

Figure 5.32 shows the measured impact sound pressure level of a 230 mm clinker concrete floor, either bare or with the traditional Danish flooring, a 22 mm wooden floor on joists on resilient pads. The flooring makes a big difference on the impact sound, and the improvement of the impact sound pressure level is denoted by ΔL. It increases from about 200 Hz with

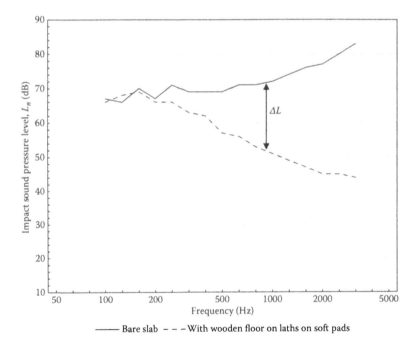

———— Bare slab – – – With wooden floor on laths on soft pads

Figure 5.32 Impact sound pressure level of a 200 mm clinker concrete slab with 30 mm concrete finish, measured without and with a 22 mm wooden floor on laths (38 mm×57 mm) on 10 mm fiberboard pads. (Data from Lydteknisk Institut, *Building Acoustic Laboratory Measurements*. Bygningsakustiske laboratoriemålinger, in Danish), Lyngby, Denmark, LL 906/80.

9 dB per octave; a slight amplification is seen at 125 Hz, which is due to a mass-spring resonance of the floor. Figure 5.33 shows the airborne sound insulation of the same construction, and it is seen that the improvement of the transmission loss due to the flooring ΔR is almost negligible.

For massive homogeneous floors, there is a relationship between the airborne and impact sound insulation. In Chapter 10, it will be shown that for homogeneous floors with a hard surface, the relationship is as simple as:

$$R + L_n = 38 + 30 \lg f \tag{5.64}$$

With a floating floor or a soft flooring, the relationship changes above the resonance frequency f_0 of the flooring:

$$R + L_n = 38 + 30 \lg f_0 - \Delta L + \Delta R \qquad (f > f_0) \tag{5.65}$$

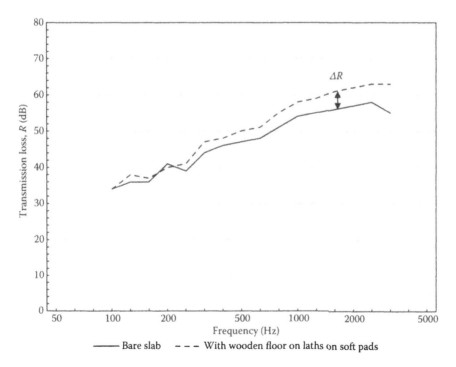

——— Bare slab – – – With wooden floor on laths on soft pads

Figure 5.33 Transmission loss of a 200 mm clinker concrete slab with 30 mm concrete finish, measured without and with a 22 mm wooden floor on laths (38 mm × 57 mm) on 10 mm fiberboard pads. (Data from Lydteknisk Institut, *Building Acoustic Laboratory Measurements*. Bygningsakustiske laboratoriemålinger, in Danish), Lyngby, Denmark, LL 884/80.

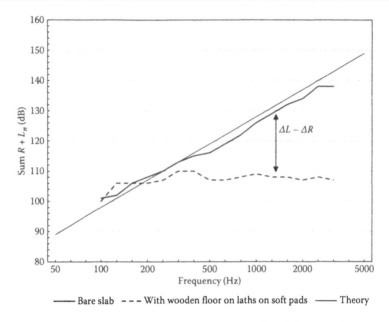

Figure 5.34 Example of sum curves obtained by adding R and L_n, using the measured results from Figures 5.32 and 5.33.

Using the example of the measured sound insulation in Figures 5.32 and 5.33, the sum curve with and without the wooden floor is shown in Figure 5.34. The agreement with the theoretical curves is quite satisfactory. The wooden floor on joists on resilient pads is efficient in attenuating the impact sound at frequencies above 200 Hz.

5.8 SINGLE-NUMBER RATING OF SOUND INSULATION

5.8.1 Weighted sound reduction index and the sound transmission class

The single-number rating of sound insulation is practical for several purposes:

- To characterize in a single number the measured result of a building construction
- For quick comparison of the sound insulation obtained with different constructions
- To specify requirements for sound insulation

The *weighted sound reduction index* R_w is based on a standardized reference curve that is defined in one-third octaves in the frequency range from

100 Hz to 3150 Hz. The method is defined in the international standard ISO 717-1. The reference curve has three straight lines with a slope of 9 dB per octave from 100 Hz to 400 Hz, 3 dB per octave from 400 Hz to 1250 Hz and 0 dB per octave from 1250 Hz to 3150 Hz.

The measured transmission loss is compared to the reference curve, and the difference is calculated in each of the 16 one-third octave bands. The sum of *unfavourable deviations* (deficiencies) is calculated. An unfavourable deviation is the deviation between the reference curve and the measured curve when the measured sound insulation is *lower* than the value of the reference curve. Deviations where the measured value is better than the reference curve do not count.

The reference curve is shifted up or down in steps of 1 dB, and the correct position of the reference curve is found when the sum of unfavourable deviations is as large as possible, but do not exceed 32 dB. This limit of 32 dB is chosen because it means an average deviation of 2 dB in the 16 frequency bands covered by the curve. No credit is given for (favourable) deviations above the reference curve. The value of the reference curve at 500 Hz is taken as the single-number value of the measuring result. The application of the method on an example is shown in Figure 5.35.

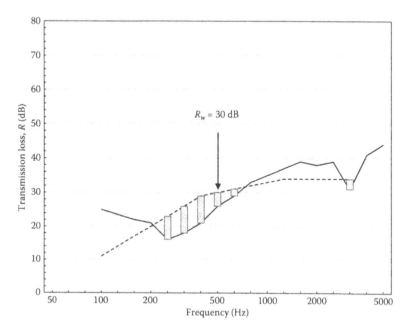

Figure 5.35 Determination of the weighted sound reduction index. Full line is the measured transmission loss, in this example, a double window measured in the laboratory. The dotted curve is the reference curve, which has been shifted to the position 30 dB at 500 Hz. The sum of the unfavourable deviations is 7.0+8.0+8.0+4.0+2.0+3.0=32.0 dB. The result is R_w=30 dB.

The same procedure is applied to field measurement results of airborne sound insulation, e.g. as the *weighted apparent sound reduction index* R'_w and the *weighted standardized level difference* $D_{nT,w}$.

The American standard ASTM E413-16 defines a very similar single-number rating called *sound transmission class* (STC). The method and the reference curve are the same as in ISO 717-1, but there are the following differences:

- The frequency range is 125 Hz to 4000 Hz, i.e. there are still 16 one-third octave bands but they are shifted one band up; the 100 Hz value of the reference curve is not used and the reference curve is extended horizontally up to 4000 Hz.
- The measured transmission loss results are rounded to whole dB values before comparison with the reference curve.
- An additional criterion is that the maximum unfavourable deviation is 8 dB; in the case of a dip in the frequency curve of sound insulation, this may be decisive for the single-number value.
- The result is written STC *xx* (without dB), where *xx* is the single-number value in dB.

The R_w and STC values are usually very close, within 1 or 2 points, but larger differences may occur in the case of a resonance dip.

5.8.2 Weighted impact sound pressure level and the impact insulation class

The *weighted impact sound pressure level* $L_{n,w}$ is based on a standardized reference curve that is defined in one-third octaves in the frequency range 100 Hz to 3150 Hz. The method is defined in the international standard ISO 717-2. The reference curve has three straight lines with a slope of 0 dB per octave from 100 Hz to 315 Hz, −3 dB per octave from 315 Hz to 1000 Hz and −9 dB per octave from 1000 Hz to 3150 Hz.

The measured impact sound pressure level is compared to the reference curve, and the sum of *unfavourable deviations* (deficiencies) is calculated. An unfavourable deviation is the deviation between the reference curve and the measured curve if the measured impact sound pressure level is *higher* than the value of the reference curve.

The reference curve is shifted up or down in steps of 1 dB, and the correct position of the reference curve is found when the sum of unfavourable deviations is as large as possible, but do not exceed 32 dB. The value of the reference curve at 500 Hz is taken as the single-number value of the measuring result. The application of the method on an example is shown in Figure 5.36.

The same procedure is applied to laboratory and field measurement results of impact sound insulation, e.g. as the *weighted standardized impact sound pressure level* $L'_{nT,w}$.

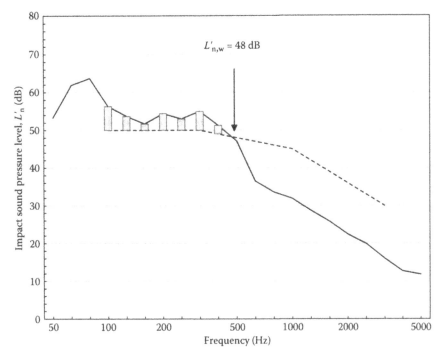

Figure 5.36 Determination of the weighted impact sound pressure level. The full line is the measured impact sound pressure level, in this example, a field measurement of a concrete slab with a floating floor. The dotted curve is the reference curve, which has been shifted to the position 48 dB at 500 Hz. The sum of the unfavourable deviations is $6.3 + 3.7 + 1.7 + 4.4 + 3.0 + 5.0 + 2.3 = 26.4$ dB. The result is $L'_{n,w} = 48$ dB.

The American standard ASTM E989-06 defines a very similar single-number rating called *impact insulation class* (IIC). The method, the frequency range and the reference curve are the same as in ISO 717-2, but there is a significant difference in the presentation of the result: The IIC value is equal to 110 dB $- L_{n,w}$. This implies that a high impact sound insulation gives a high IIC value, but a low value of $L_{n,w}$. The IIC is characterized by the following:

- The frequency range is 100 Hz to 3150 Hz, i.e. the same 16 one-third octave bands as used for $L_{n,w}$.
- The measured impact sound pressure levels are rounded to whole dB values before comparison with the reference curve.
- An additional criterion is that the maximum unfavourable deviation is 8 dB; in the case of a peak in the frequency curve of impact sound pressure level, this may be decisive for the single-number value.
- The result is written IIC xx (without dB), where xx is the single-number value in dB.

The importance of frequencies below 100 Hz, especially in connection with lightweight floor constructions and floating floors on heavyweight constructions, has made it relevant to extend the frequency range for measurements of impact sound insulation down to 50 Hz, or even further down to 20 Hz. More about this is explained in Chapter 12.

5.8.3 Alternative reference curve for evaluating the impact sound insulation

There has been some controversy about the rating curves, especially the one for impact sound. An alternative rating curve was suggested by Bodlund (1985) based on a thorough analysis and comparison with results from a socio-acoustic investigation. The so-called Bodlund curve is a straight line with a slope of 1 dB per one-third octave from 50 Hz to 1000 Hz. Although this covers only 14 frequency bands, the evaluation method is the same as for the ISO curve, i.e. the sum of unfavourable deviations should be as large as possible without exceeding 32 dB. An example of the application is shown in Figure 5.37, using the same measured data as in Figure 5.36.

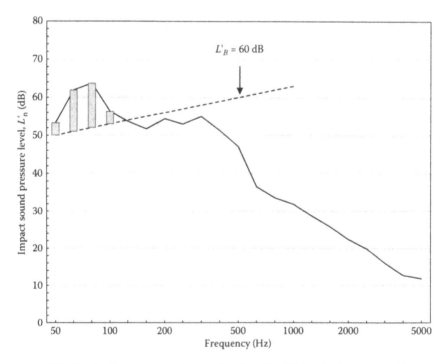

Figure 5.37 The Bodlund curve and its application. The full line is the measured impact sound pressure level, in this example, a field measurement of a concrete slab with a floating floor. The dotted curve is the reference curve, which has been shifted to the position 60 dB at 500 Hz. The sum of the unfavourable deviations is $3.3 + 10.9 + 11.7 + 3.3 = 29.2$ dB. The result is $L'_B = 60$ dB.

This alternative evaluation method for impact sound is often applied or referred to in research. The reason is that the results of this evaluation method have proved to be highly correlated with subjective response to footfall noise from neighbours in multi-unit houses. If $P_{satisfied}$ is the percentage of persons rating the impact sound insulation quite or nearly quite satisfactory, the corresponding Bodlund-weighted impact sound pressure level, according to Hveem et al. (1996, 34–35), is:

$$L'_B = 74,7 - 0,247 \cdot P_{satisfied} \text{ (dB)} \tag{5.66}$$

It has been found that the result, using the Bodlund curve, is highly correlated with the ISO method when extended with the spectrum adaptation term going down to 50 Hz (the spectrum adaptation term is explained in Section 5.8.6). As an example, we find that more than 80 % should be satisfied with the impact sound if $L'_B < 55$ dB, or approximately $L'_{n,w} + C_{I,50-2500} < 49$ dB, because there is an average difference of 6.4 dB between the two ratings (Hveem et al. 1996, 35).

5.8.4 Previous methods for rating sound insulation

The idea of using a shifted reference curve for evaluating airborne and impact sound insulation comes from Germany (DIN 4109, 1959) and was adapted in ISO/R 717 in 1968. It was called the *airborne sound insulation index* I_a and was similar to R_w with the exception, that maximum unfavourable deviation in any one-third octave band was limited to 8 dB. Hence, $I_a \cong R_w$. In the 1982 edition of ISO 717, the symbol was changed to R_w and the 8 dB-rule was removed.

Similarly, for the rating of impact sound insulation, the *impact sound index* I_i was defined in ISO/R 717. The rating curve was the same as shown in Figure 5.31, but the impact sound pressure levels should either be measured in 1/1 octave bands or measured in one-third octave bands and then corrected +5 dB before comparison with the reference curve. Hence, $I_i \cong L_{n,w} + 5$ dB. In the 1982 edition of ISO 717, the 5 dB correction and the 8 dB-rule was removed and the symbol was changed to $L_{n,w}$.

5.8.5 Historical note

It is interesting to note that ISO/R 717, which is specifically meant for measurements in dwellings, only mentions the normalized measures for sound insulation R' and L_n, not the standardized measures that adjust the result to 0.5 s reverberation time. Since some countries were using the standardized measures, it is not surprising that there was some controversy about the proposal among the ISO member counties; 20 countries approved the recommendation, but six countries opposed the approval (Belgium, Denmark, France, Italy, Norway and USA).

Before the first ISO standard on evaluation of sound insulation, many countries used a fixed reference curve to define minimum requirements in the building code. In Britain, the *aggregate adverse deviation* (AAD) was used as single-number rating of airborne sound insulation. This measure is the sum of all deviations below a fixed two-segment reference curve and ignores bands where measured performance is above the reference curve. An approximate conversion is $R'_w \cong STC \cong -0.113\,AAD + 56.7\,dB$ (Bradley 1982). Hence, AAD = 15 corresponds to $R'_w \cong 55\,dB$ and AAD = 60 corresponds to $R'_w \cong 50\,dB$.

5.8.6 Spectrum adaptation terms

Historically, two different methods were developed for single-number rating of airborne sound insulation. One was the comparison with a weighting curve as described earlier, and the other method was the French '*Indice d'affaiblissement acoustique R vis-à-vis d'un bruit rose*', R_{rose}. This is calculated as the difference between two A-weighted sound pressure levels: (1) a sound with pink spectrum from 100 Hz to 5000 Hz, and (2) the sound with a pink spectrum reduced in each one-third octave band by the sound reduction index. For the evaluation of windows and facades, a generalized traffic noise spectrum was applied instead of the pink noise spectrum, and the term was '*Indice d'affaiblissement acoustique R vis-à-vis d'un bruit de traffic routier*', R_{route}.

The revision of ISO 717 published in 1996 was the result of a compromise, merging the German and the French methods into one. The reference curve and R_w was kept, but new spectrum adaptation terms C and C_{tr} were defined in such a way that

$$R_{rose} \cong R_w + C \text{ and } R_{route} \cong R_w + C_{tr}$$

Single-number values that include a spectrum adaptation term characterize the sound insulation as the A-weighted level difference for noise with either a pink noise spectrum or a road traffic noise spectrum. The latter has more energy at the low frequencies and is primarily intended for the evaluation of windows and façades. The frequency range used in this method is more flexible, and in addition to the frequencies used for R_w, (100 Hz to 3150 Hz), the range can be extended to include higher frequencies (4000 Hz and 5000 Hz) and lower frequencies (50 Hz, 63 Hz and 80 Hz).

The spectrum adaptation terms are calculated by the following equation:

$$C_j = -10\lg \sum_{i=1}^{n} 10^{(L_{ij} - X_i)/10} - X_w \text{ (dB)} \tag{5.67}$$

where

C_j is the spectrum adaptation term, e.g. C, C_{tr}, $C_{50-5000}$, $C_{tr,50-3150}$.

j is index for the sound spectrum, either 1 or 2.

i is index for the one-third octave band in the relevant range, up to n.

L_{ij} is the level of the spectrum j in the frequency band i (Table 5.1).

X_i is the sound insulation at frequency band i, e.g. R or D_{nT}, given to 0.1 dB.

X_w is the single-number quantity calculated with the shifted reference curve, e.g. R_w.

For simplicity, $D_{nT,w,50}$ is sometimes used instead of the more correct $D_{nT,w}+C_{50-3150}$ and, similarly, $R_{w,50}$ is used as abbreviation for $R_w+C_{50-3150}$.

Table 5.1 Spectra for calculation of the spectrum adaptation terms according to ISO 717-1

Frequency (Hz)	Spectrum 1 From 50 Hz or 100 Hz up to 3150 Hz $L_{j,i}$ (dB)	Spectrum 1 From 50 Hz or 100 Hz up to 5000 Hz $L_{j,i}$ (dB)	Spectrum 2 Any frequency range $L_{j,i}$ (dB)
50	−40	−41	−25
63	−36	−37	−23
80	−33	−34	−21
100	−29	−30	−20
125	−26	−27	−20
160	−23	−24	−18
200	−21	−22	−16
250	−19	−20	−15
315	−17	−18	−14
400	−15	−16	−13
500	−13	−14	−12
630	−12	−13	−11
800	−11	−12	−9
1000	−10	−11	−8
1250	−9	−10	−9
1600	−9	−10	−10
2000	−9	−10	−11
2500	−9	−10	−13
3150	−9	−10	−15
4000		−10	−16
5000		−10	−18

Note: Spectrum 1 is A-weighted pink noise and spectrum 2 is A-weighted traffic noise.

Note that there is no need to calculate the weighted value with the shifted reference curve if only the extended frequency range is relevant. For example, $R_{w,50}$ is calculated directly from:

$$R_{w,50} = -10\lg \sum_{i=1}^{n} 10^{(L_{ij}-R_i)/10} \text{ (dB)} \tag{5.68}$$

The principle of applying an assumed noise spectrum and characterizing the sound insulation by the A-weighted level difference is particularly useful for windows and façade elements, and many different traffic noise spectra are available for this purpose. The collection of spectra in Nordtest ACOU 061 contains seven spectra representing various kinds of road traffic noise, railway noise and aircraft noise. Nordtest Spectrum 1 is for mixed urban road traffic at 50 km/h and about 10 % heavy vehicles, and this is the traffic noise spectrum adapted in ISO 717-1.

The ASTM E1332 describes a similar method, using a spectrum representing a mixture of road, rail and air traffic. The one-third octave bands from 80 Hz to 4000 Hz inclusive are used, and the calculated A-weighted level difference is called the outdoor-indoor transmission class (OITC). This spectrum has a substantial amount of energy at low frequencies. However, Davy (2004) brought important new information about the measurements behind the OITC spectrum, and he concluded that there is some doubt about the validity of the spectrum, especially concerning the high levels at low frequencies.

Figure 5.38 shows the traffic noise spectra from ISO 717-1 and ASTM E1332 together with a spectrum for city street traffic noise derived from newer measurements and proposed by Buratti et al. (2014). It is clearly seen that the newer spectrum for city street traffic has less energy at low frequencies than the older Nordtest/ISO spectrum; whether this is due to a general change in cars as noise sources, or to differences in the measurement conditions, is not known. However, there are certainly differences from one country to another in the typical composition of road traffic (e.g. the number of very light vehicles, scooters, frequent use of horns etc.). Further discussion of the applicability of the Nordtest/ISO traffic noise spectrum is found in the work of Mesihovic et al. (2016).

Spectrum adaptation terms are also used for impact sound insulation. This allows the frequency range to be more flexible, and in addition to the frequencies used for $L_{n,w}$ (100 Hz to 3150 Hz), the range can be extended to include lower frequencies (50 Hz, 63 Hz and 80 Hz), or even lower if desired.

The spectrum adaptation term for impact sound is calculated by the following equation:

$$C_I = 10\lg \sum_{i=1}^{n} 10^{L_i/10} - 15 - X_w \text{ (dB)} \tag{5.69}$$

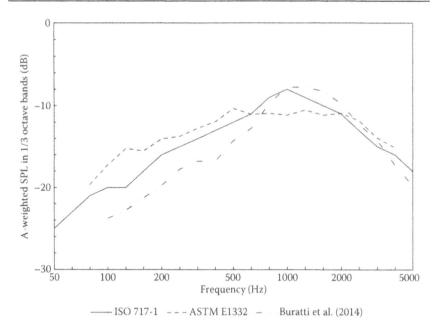

Figure 5.38 Traffic noise spectra in third octave bands relative to the A-weighted level. The ISO 717-1 spectrum used for the spectrum adaptation term C_{tr}, the ASTM E1332 spectrum used for OITC, and the city street traffic spectrum from Buratti et al. (2014).

where

C_I is the spectrum adaptation term; $C_{I, 50-2500}$ in the case of extended frequency range.

i is index for the one-third octave band in the relevant range, up to n.

L_i is the level of the measured impact sound pressure level in the frequency band i, given to 0.1 dB.

X_w is the single-number quantity calculated with the shifted reference curve, e.g. $L_{n,w}$ or $L'_{nT,w}$.

The number 15 in Equation 5.69 has been so determined that the value of C_I is approximately 0 dB for solid floors with effective floor coverings.

For simplicity, $L_{n,w,50}$ is sometimes used instead of the more correct $L_{n,w}+C_{I, 50-2500}$ and, similarly, for $L'_{nT,w,50}$. Note that there is no need to calculate the weighted value with the shifted reference curve if only the extended frequency range is relevant. For example, $L_{n,w,50}$ is calculated directly from:

$$L_{n,w,50} = 10\lg \sum_{i=1}^{n} 10^{L_i/10} - 15 \, \text{(dB)} \tag{5.70}$$

5.9 REQUIREMENTS FOR SOUND INSULATION

Requirements for sound insulation are usually given in the national building codes and may differ from one country to another. Some countries have detailed and complicated requirements for different types of buildings and rooms, while other countries have few and simple requirements. Some examples of requirements between dwellings are shown in Table 5.2. The highest level of sound insulation requirements is found in the Nordic countries and in Austria, while Portugal has significantly lower requirements. The latter is characteristic for countries in southern Europe. It appears that sound insulation requirements depend on the climate of the country, and a warm climate means that people spend most time with open windows or outdoors, while the opposite is true in a cold climate.

In 2003, the UK changed the compliance limits for England and Wales to use the traffic noise spectrum adaptation term C_{tr} for the evaluation of party walls and floors; Australia did the same a few years later. The reasoning behind this rather surprising adoption was that low-frequency sound from music, Hi-Fi equipment, etc. was recognized as an increasing problem between neighbours in multi-unit buildings, but there was at the same time a strong resistance against extending the measurements to include frequencies below 100 Hz. The traffic noise spectrum would certainly put more weight on the lower frequencies (e.g. 100 Hz to 500 Hz), but instead new problems have appeared with complaints about insufficient sound insulation at the mid and high frequencies.

Table 5.2 Examples of sound insulation requirements between dwellings in multi-unit buildings (as of 2016)

Country	Airborne	Impact
Some Nordic countries (DK, ES, FI, IS, NO)	$R'_w \geq 55 \, dB$	$L'_{n,w} \leq 53 \, dB$
Sweden	$D_{nT,w,50} \geq 52 \, dB$	$L'_{nT,w,50} \leq 56 \, dB$
Austria	$D_{nT,w} \geq 55 \, dB$	$L'_{nT,w} \leq 48 \, dB$
Germany (walls/floors)	$R'_w \geq 53/54 \, dB$	$L'_{n,w} \leq 53 \, dB$
Italy	$R'_w \geq 50 \, dB$	$L'_{n,w} \leq 63 \, dB$
France	$D_{nT,w} + C \geq 53 \, dB$	$L'_{nT,w} \leq 58 \, dB$
Portugal	$D_{nT,w} \geq 50 \, dB$	$L'_{nT,w} \leq 60 \, dB$
Spain	$D_{nT,w} + C \geq 50 \, dB$	$L'_{nT,w} \leq 65 \, dB$
Scotland	$D_{nT,w} \geq 56 \, dB$	$L'_{nT,w} \leq 56 \, dB$
England and Wales	$D_{nT,w} + C_{tr} \geq 45 \, dB$	$L'_{nT,w} \leq 62 \, dB$
Australia	$D_{nT,w} + C_{tr} \geq 45 \, dB$	$L'_{nT,w} + C_I \leq 62 \, dB$
New Zealand	STC 55 (FSTC 50)	IIC 55 (FIIC 50)
USA (2006 International Building Code)	STC 50 (FSTC 45)	IIC 50 (FIIC 45)
Canada	STC 50 (ASTC 47)	N/A

In the USA, most of the states have adapted the International Building Code (IBC), which requires STC 50 and IIC 50. However, this is the requirement for laboratory tests of the building components, and in the case of a field test, 5 dB lower values are allowed (field sound transmission class (FSTC) and field impact insulation class (FIIC)). It should be noted that the American measurement standard ASTM 336 requires that '*the flanking paths be shielded*' during measurement of transmission loss in the field. Therefore, FSTC 45 is a rather weak requirement for airborne sound insulation. Canada has no requirement for impact sound and the requirement for airborne sound insulation is similar to that in the USA, except that field measurements should be made in terms of the apparent sound transmission class (ASTC), which *includes the flanking paths* in the building.

The progress of lightweight building systems has led to increased awareness of the importance of low frequencies for evaluation of impact sound, and as a consequence, a revised building code SS 25267 was introduced in Sweden in 2015. This requires, as a minimum, the sound insulation to be evaluated from 50 Hz, and impact sound to be measured down to 20 Hz in case of sound classes above the minimum requirement.

REFERENCES

ASTM E90-09 (2009). *Standard test method for laboratory measurement of airborne sound transmission loss of building partitions and elements*, American Society for Testing and Materials, West Conshohocken, PA.

ASTM E336-16 (2016). *Standard test method for measurement of airborne sound attenuation between rooms in buildings*, American Society for Testing and Materials, West Conshohocken, PA.

ASTM E413-16 (2016). *Classification for rating sound insulation*, American Society for Testing and Materials, West Conshohocken, PA.

ASTM E492-09 (2016). *Standard test method for laboratory measurement of impact sound transmission through floor-ceiling assemblies using the tapping machine.* American Society for Testing and Materials, West Conshohocken, PA.

ASTM E989-06 (2012). *Standard classification for determination of impact insulation Class (IIC)*, American Society for Testing and Materials, West Conshohocken, PA.

ASTM E1007-16 (2016). *Standard test method for field measurement of tapping machine impact sound transmission through floor-ceiling assemblies and associated support structures*, American Society for Testing and Materials, West Conshohocken, PA.

ASTM E1332-16 (2016). *Standard classification for rating outdoor-indoor sound attenuation*, American Society for Testing and Materials, West Conshohocken, PA.

R. Berger (1911). *Über die schalldurchlässigkeit.* (On the Sound Transmission, in German). Dissertation, Munich, Germany.

K. Bodlund (1985). Alternative reference curves for evaluation of the impact sound insulation between dwellings. *Journal of Sound and Vibration* 102, 381–402.

J.S. Bradley (1982). *Subjective Rating of the Sound Insulation of Party Walls.* Building Research Note 196. National Research Council Canada, Division of Building Research.

O. Brandt (1958). *Akustisk planering.* (Acoustical planning, in Swedish). V. Pettersons Bokindustri AB, Stockholm, Sweden.

C. Buratti, E. Belloni, E. Moretti (2014). Façade noise abatement prediction: New spectrum adaptation terms measured in field in different road and railway traffic conditions. *Applied Acoustics 76,* 238–248.

L. Cremer (1942): Theorie der Schalldämmung dünner Wände bei schrägem Einfall. (Theory of sound insulation of thin walls at oblique incidence, in German). *Akustische Zeitschrift 7,* 81–104.

J. Davy (2004). Insulating buildings against transportation noise. *Proceedings of ACOUSTICS 2004,* 3–5 November 2004, Gold Coast, Australia, 447–453.

DIN 4109 (1959). *Schallschutz im Hochbau,* Entwurf. (In German). Berlin 1959. (Final edition 1962).

DIN 4110 (1938). *Technische Bestimmung für die Zulassung neuer Bauweisen.* (In German), 2nd edition. Berlin.

W. Fasold, E. Sonntag (1978). *Bauakustik. Bauphysikalisches Entwurfslehre,* Band 4, 3rd edition. VEB Verlag für Bauwesen. Berlin.

K. Gösele (1969). *Schalldämmung von Türen* (Sound insulation of doors, in German). Berichte aus der Bauforschung, Heft 63.

S. Hveem, A. Homb, K. Hagberg, and J.H. Rindel (1996). *Low-frequency footfall noise in multi-storey timber frame buildings.* NKB Work and Committee Report 1996:12 E. Nordic Committee on Building Regulations, Acoustics Group, Helsinki.

S. Ingemansson (1968). *Ljudisolerande fönsterkonstruktioner.* (Sound insulating window constructions, in Swedish). Report from the Building Research, 3/68, Stockholm, Sweden.

F. Ingerslev, A. Kjerbye Nielsen, and A. Falck Larsen (1947). The measuring of impact sound transmission through floors. *Journal of the Acoustical Society of America 19,* 981–987.

ISO 10140-2 (2010). *Acoustics – Laboratory measurement of sound insulation of building elements – Part 2: Measurement of airborne sound insulation,* International Organization for Standardization, Geneva, Switzerland.

ISO 10140-3 (2010). *Acoustics – Laboratory measurement of sound insulation of building elements – Part 3: Measurement of impact sound insulation,* International Organization for Standardization, Geneva, Switzerland.

ISO 16283-1 (2014). *Acoustics – Field measurements of sound insulation in buildings and of building elements – Part 1: Airborne sound insulation,* International Organization for Standardization, Geneva, Switzerland.

ISO 16283-2 (2015). *Acoustics – Field measurements of sound insulation in buildings and of building elements – Part 2: Impact sound insulation,* International Organization for Standardization, Geneva, Switzerland.

ISO 16283-3 (2016). *Acoustics – Field measurements of sound insulation in buildings and of building elements – Part 3: Façade sound insulation,* International Organization for Standardization, Geneva, Switzerland.

ISO 717-1 (1996). *Acoustics – Rating of sound insulation in buildings and of building elements – Part 1: Airborne sound insulation,* International Organization for Standardization, Geneva, Switzerland.

ISO 717-2 (1996). *Acoustics – Rating of sound insulation in buildings and of building elements – Part 2: Impact sound insulation*, International Organization for Standardization, Geneva, Switzerland.

ISO/R717 (1968). *Rating of sound insulation for dwellings*, International Organization for Standardization, Geneva, Switzerland.

J. Kristensen (1973). *Lydisolation: Teori, måling, vurdering og bestemmelser*, (Sound insulation: Theory, measurement, evaluation and requirements, in Danish). SBI-notat 24 (In Danish). Danish Building Research Institute, Aalborg University Copenhagen.

J. Kristensen, J.H. Rindel (1989). *Bygningsakustik – Teori og praksis* (Building acoustics – Theory and practice, in Danish) SBI-Anvisning 166. Danish Building Research Institute, Hørsholm, Denmark.

P.T. Lewis (1971). Real windows, Chapter 7, in *Building Acoustics* (Eds. T. Smith, P.E. O'Sullival, B. Oakes, R.B. Conn). British Acoustical Society, Special Volume No. 2, Oriel Press, London.

R. Lindahl and H.J. Sabine (1940). Measurement of impact sound transmission through floors. *Journal of the Acoustical Society of America* 11, 401–405.

J.A. Marsh (1971). The airborne sound insulation of glass, Chapter 5, in *Building Acoustics* (Eds. T. Smith, P.E. O'Sullival, B. Oakes, R.B. Conn). British Acoustical Society, Special Volume No. 2, Oriel Press, London.

NT ACOU 061 (1987). *Windows: Traffic Noise Reduction Indices*, Nordtest Method. Helsinki, Finland.

SS 25267 (2015). *Acoustics – Sound Classification of Spaces in Buildings*, Swedish Standards Institute, Stockholm.

Lydteknisk Institut. *Building Acoustic Laboratory Measurements* (Bygningsakustiske laboratoriemålinger, in Danish). Lyngby, Denmark.

M. Mesihovic, J.H. Rindel, and I. Milford (2016). The need for updated traffic noise spectra, used for calculation of sound insulation of windows and facades. *Proceedings of InterNoise 2016*, Hamburg, Germany, 3890–3897.

Chapter 6

Sound radiation from plates

The sound radiation from plates is described in this chapter. Both forced and resonant vibrations are considered, and the application of Rayleigh's method of radiation calculation is demonstrated in the case of radiation from forced bending waves. The influence of the finite area of a plate is also shown. Forced vibrations are the part of the vibrations that are directly due to the surrounding sound field exciting the plate. In contrast, resonant vibrations are the free vibrations caused by reflections of the forced vibrations from the boundaries.

6.1 NORMAL MODES IN A RECTANGULAR PLATE

To study radiation of the reverberant part of the vibration field in a plate, we need some basic information about this field. The modal density for the plate is especially needed, as this is an important parameter for the statistical energy analysis (SEA). The reverberant vibration field will be analysed with the aid of SEA in Chapter 7.

The normal modes are the free vibration patterns that can exist in a finite system without excitation or after an excitation has stopped. They are also responsible for building up the field when an excitation starts. Each normal mode is associated with a natural frequency (resonance frequency) and – for a given excitation – a modal amplitude. The normal modes are orthogonal to each other, meaning among other things that the energy is determined from the sum of the squared amplitudes of the normal modes.

For a plate, the normal modes shall fulfil the bending wave equation (Equation 1.24) – if we can assume to be in the low-frequency region, i.e. well below the cross-over frequency fs (Equation 1.29) – and simultaneously fulfil the boundary conditions. In the present case, we are only interested in the general trends (valid for all boundary conditions), so it is practical to use the simplest of all situations – the simply supported rectangular plate. The boundary condition of a simple support is that the displacement and the moment reaction at the boundary is zero – the plate is, however, free to rotate at the boundary. The moment reaction is related to the second derivative of the displacement (or vibration velocity) field. The modes we are looking for are, therefore, to have zero value at the boundary, and the second-order derivative should also be zero at this location. A function fulfilling these conditions is sin $(k_x x)$ (convince yourself). It is also easy to show that this function fulfils the bending wave equation (once again, convince yourself). These two facts prove that the normal modes we are looking for are sine functions. The wave number k is, however, still not fixed. If there is a simple support both at $x=0$ and $x=l_x$, then k_x must equal

$$k_{x,m} = \frac{m\pi}{l_x} \tag{6.1}$$

where m is an arbitrary integer. (OBS: In other connections m is the mass per unit area.) However, we are talking about a plate. Thus, the y-direction must also be included. This direction will have the same type of field, sin $(k_y y)$ and

$$k_{y,n} = \frac{n\pi}{l_y} \tag{6.2}$$

where l_y is the plate length in the y-direction and n is an arbitrary integer.
    A normal mode can then be written as

$$\phi_{m,n}(x,y) = \sin\left(\frac{m\pi}{l_x}x\right)\sin\left(\frac{n\pi}{l_y}y\right) \tag{6.3}$$

Figure 6.1 shows an example of a normal mode in a plate. Some other examples are shown in Figure 6.2. Normal modes can be used to find solutions for the (forced) bending wave equation in the form

$$w(x,y) = \sum_{m,n=1}^{\infty} A_n \phi_{m,n}(x,y) \tag{6.4}$$

However, that is not what we shall do here. Instead, the starting point will be the *Pythagoras* relationship between the free bending wave numbers and the mode numbers m and n,

$$k_b^2 = k_{m,x}^2 + k_{n,y}^2 = \left(\frac{m\pi}{l_x}\right)^2 + \left(\frac{n\pi}{l_y}\right)^2 \tag{6.5}$$

If the bending wavelength is used instead of the wave numbers, the relationship that shall be fulfilled is

$$\left(\frac{2}{\lambda_B}\right)^2 = \left(\frac{m}{l_x}\right)^2 + \left(\frac{n}{l_y}\right)^2 \tag{6.6}$$

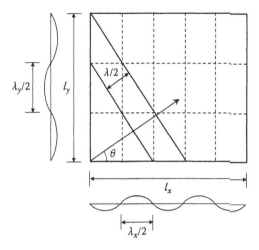

Figure 6.1 Example of a normal mode in a simply supported rectangular plate. In this case, the mode numbers are $(m, n)=(5, 3)$. The corresponding half wavelength and direction of a propagating bending wave are shown.

where, as before, m and n are natural numbers (Figure 6.1). The numbers represent the number of half wavelengths along the x- and y-axis, respectively. Some examples are shown in Figure 6.2. With simply supported edges, the lowest natural frequency is the one with $(m, n)=(1, 1)$.

In general, it is a little complicated to calculate the natural frequencies of the normal modes, because of the transition from bending to shear

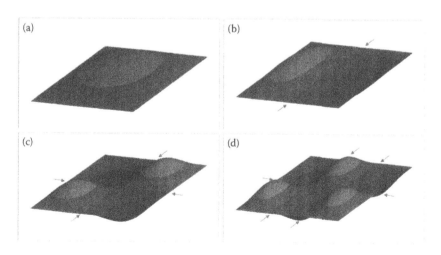

Figure 6.2 Examples of normal modes in a simply supported rectangular plate, illustrated as maximum deflection. The modal numbers are (a) (1,1), (b) (2,1), (c) (2,2), (d) (3,2). The arrows indicate the nodal lines.

waves (see Equation 1.27). At low frequencies, however, it may normally be acceptable to assume bending waves, i.e. the frequency is well below the cross-over frequency for shear waves (Equation 1.29). Then, the wavelength in the panel is $\lambda_B = c_B / f = c / \sqrt{f \cdot f_c}$ and, thus, the natural frequency of the mode with numbers (m, n) is:

$$f_{mn} = \frac{c^2}{4 f_c} \left[\left(\frac{m}{l_x} \right)^2 + \left(\frac{n}{l_y} \right)^2 \right] \tag{6.7}$$

Here c is the speed of sound in air and f_c is the critical frequency (Equation 1.28). Considering the distribution of the natural frequencies, it can be found that, on average, the number of natural frequencies below a certain frequency, i.e. for wavelengths longer than λ_B, is:

$$N = \pi \cdot l_x \cdot l_y \frac{1}{\lambda_B^2} - \left(l_x + l_y \right) \frac{1}{\lambda_B} + \frac{1}{4} \tag{6.8}$$

Assuming bending waves and introducing the area $S = l_x \cdot l_y$ and the perimeter of the panel $U = 2(l_x + l_y)$, the modal density yields:

$$\frac{\Delta N}{\Delta f} = \frac{\pi S}{c^2} f_c - \frac{U}{4c} \sqrt{\frac{f_c}{f}} \cong \frac{\pi S}{c^2} f_c \tag{6.9}$$

Similarly, by assuming shear waves, $\lambda_B = c_s / f$ and the modal density yields:

$$\frac{\Delta N}{\Delta f} = \frac{2 \pi S}{c_s^2} f - \frac{U}{2 c_s} \cong \frac{2 \pi S}{c_s^2} f \tag{6.10}$$

We notice that the modal density in a plate is constant in the bending wave region, but increases with frequency in the shear wave region.

The modal density of a homogeneous, isotropic plate depends not only on the length and width but also on the thickness. If the plate is thin (thickness less than approximately one-sixth of the wavelength of bending waves), bending waves can be assumed; but if the plate is thick compared to the wavelength, shear waves will occur instead. If the bandwidth is one-third octave, $\Delta f = 0.23 \, f$ and, thus, the approximate number of modes within the band is:

$$\Delta N_{1/3} \cong \begin{cases} \dfrac{0.23 \pi}{c^2} S f_c f & (f < f_s / 2) \quad \text{(bending)} \\[4mm] \dfrac{0.46 \pi}{c_s^2} S f^2 & (f > f_s / 2) \quad \text{(shear)} \end{cases} \tag{6.11}$$

Here f_c is the critical frequency, $f_s = f_c (c_s/c)^2$ is the cross-over frequency for transition to shear waves and c_s is the speed of shear waves. The two formulas in Equation 6.11 are shown graphically in Figure 6.3. The curves cross at the frequency $f_s/2$, so this is the cross-over frequency from bending to shear when the modal density is considered. In other words, the cross-over for modal density is one octave lower than the cross-over for speed of propagation.

As an example, the first modes have been calculated for a 10 m² brick wall, 250 mm thick (Table 6.1). The actual number of modes in each one-third octave band is displayed in Figure 6.4 and compared with the statistical estimate (Equation 6.11). The shear wave cross-over frequency is $f_s = 4$ kHz, so for modal density, the bending wave assumption holds up to 2 kHz. It is noted that up to 250 Hz, many one-third octave bands are without any modes and the first modes from f_{11} to f_{22} are widely separated. Actually, this interval is two octaves $(f_{22} = 4 \, f_{11})$ and there may be only two or three modes within that range. Thus, the statistical estimates in Equation 6.11 should only be used for frequencies above $f_{22} = 4 \, f_{11}$, that is 366 Hz in this example.

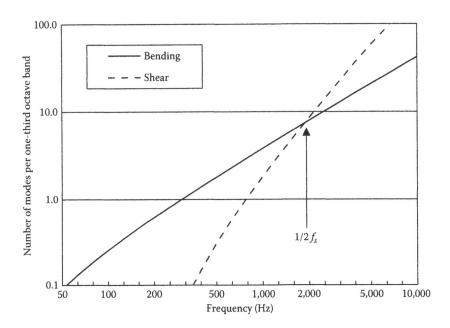

Figure 6.3 Number of normal modes per one-third octave bands in a 250 mm thick brick wall, calculated for bending waves and shear waves with Equation 6.11. $f_c = 72$ Hz, $f_s \approx 4000$ Hz, $f_{11} = 91.4$ Hz. Below 315 Hz, there is statistically less than one mode per band. Note that the ordinate is logarithmic.

Table 6.1 First 14 modes calculated for a 250 mm
thick brick wall, 4 m × 2.5 m

m	n	f_{mn} (Hz)
1	1	91.4
2	1	168
1	2	289
3	1	297
2	2	366
4	1	477
3	2	494
1	3	617
4	2	674
2	3	694
5	1	708
3	3	823
5	2	905
6	1	990

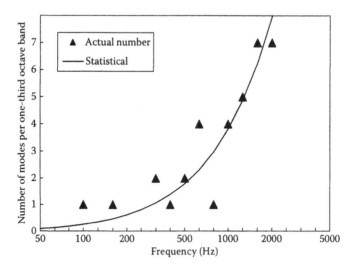

Figure 6.4 Number of normal modes per one-third octave bands in a 250 mm thick brick
wall, 4 m × 2.5 m. The first normal mode is f_{11} = 91.4 Hz. The statistical modal
density calculated from Equation 6.11 is also shown. Bending waves only.

6.2 SOUND RADIATION FROM A PROPAGATING TRANSVERSE WAVE

Consider a large plate located in the *xy*-plane with sound radiation into
the side where $z > 0$. A bending wave is assumed to travel in the plate as a

plane wave in the x-direction. Hence, the velocity of the vibration can be written as:

$$v = |v| e^{-jk_B x} \tag{6.12}$$

The time factor $e^{j\omega t}$ is omitted here and in the following. The angular wave number of the bending wave is $k_B = \omega/c_B$. In the air above the plate, there is no variation in the y-direction, so the sound pressure can be written in the form:

$$p = |p| e^{-j(k_x x + k_z z)} \tag{6.13}$$

where k_x and k_z are the wave numbers that represent the sound propagation in the x- and z-directions, respectively. When this general solution is inserted in the Helmholtz equation:

$$\nabla^2 p + k^2 p = 0 \tag{6.14}$$

it is found that the wave numbers are related as:

$$k^2 = k_x^2 + k_z^2 \tag{6.15}$$

The z-component of the particle velocity in the air field is:

$$u_z = -\frac{1}{j\omega\rho}\frac{\partial p}{\partial z} = \frac{|p| k_z}{\omega\rho} e^{-j(k_x x + k_z z)} \tag{6.16}$$

At the surface of the plate ($z=0$), this must equal the normal velocity of the plate (Equation 6.12) and thus:

$$|v| = \frac{|p| k_z}{\omega\rho} \text{ and } k_x = k_B \tag{6.17}$$

From Equation 6.15, it follows that

$$k_z = \sqrt{k^2 - k_B^2} \tag{6.18}$$

Insertion in Equation 6.13 yields the following result for the sound pressure:

$$p = \frac{\rho c |v|}{\sqrt{1 - (k_B / k)^2}} e^{-jk_B x} e^{-j\sqrt{k^2 - k_B^2} z} \tag{6.19}$$

This result is discussed here for three cases.

The first case is $k_B < k$ (meaning that $c_B > c$ and $f > f_c$). The solution represents a sound field propagation in both the x- and z-directions. The angle of propagation (Figure 6.5c) is determined by the following:

$$\sin\theta = \frac{c}{c_B} = \frac{k_B}{k} = \sqrt{\frac{f_c}{f}} \qquad (6.20)$$

In this case, the sound pressure can be written in terms of the angle of radiation, θ:

$$p = \frac{\rho c |v|}{\cos\theta} e^{-j(x\sin\theta + z\cos\theta)} \qquad (6.21)$$

The second case is $k_B > k$ (meaning that $c_B < c$ and $f < f_c$). The sound pressure (Equation 6.19) yields:

$$p = \frac{j\rho c |v|}{\sqrt{(k_B / k)^2 - 1}} e^{-jk_Bx} e^{-i\sqrt{k_B^2 - k^2}\,z} \qquad (6.22)$$

This represents a sound field that only propagates in the x-direction, but not in the z-direction; the amplitude of the sound pressure decreases exponentially with the distance from the plate. There is an acoustic near field, but no sound radiation (Figure 6.5a).

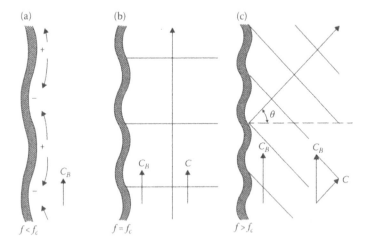

Figure 6.5 A propagating bending wave in a plate cannot radiate sound if the speed is below c (case a). If the speed equals c, the sound is radiated along the surface of the plate (case b). A plane wave is radiated in the direction θ if the speed exceeds c (case c). (Reproduced from Kristensen, J., Rindel J.H., Bygningsakustik – Teori og praksis (Building acoustics – Theory and practice, in Danish). SBI Anvisning 166. Danish Building Research Institute, Hørsholm, Denmark, 1989. With permission.)

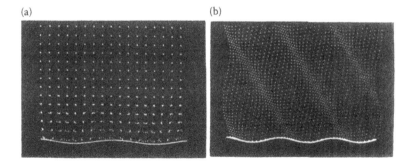

(a)   (b)

Figure 6.6 Simulated particle movements in the air due to a bending wave travelling from left to right. (a) $c_B < c$. (b) $c_B > c$. (Reproduced from Heckl, M. and Heckl, M., Darstellung von abstrahlvorgängen im film. *Fortschritte der Akustik – DAGA 1976*, VDI-Verlag, Düsseldorf, 165–168, 1976. With permission.)

The third case is the transition from non-radiation to radiation, which is at $k_B = k$, i.e. when the bending wave number equals the wave number in air, or the speed of the propagating bending wave equals the speed of sound (Figure 6.5b). This is called *coincidence*. Also note that when coincidence occurs, the sound radiation is extremely high since $p \to \infty$ as the denominator of Equation 6.22 becomes zero.

A simulation of the sound field above a propagating bending wave without and with sound radiation is shown in Figure 6.6.

6.3 THE RADIATION EFFICIENCY

The result in the previous section is for structures with infinite extent. However, real structures are finite and this has an important effect on the sound radiation. The rest of this chapter is about finite structures. In order to study the radiation of these structures, radiation efficiency is introduced.

The concept of radiation efficiency was first introduced by Gösele (1953). Radiation efficiency is a measure of the radiated sound power from a source relative to the sound power radiated from an equivalent piston source that generates a plane wave. The sound power from a very large piston with a harmonic vibration $v = |v|\, e^{j\omega t}$ and the area S is:

$$P_{\text{piston}} = \tfrac{1}{2}\rho c S |v|^2 = \rho c S \left\langle \tilde{v}^2 \right\rangle \tag{6.23}$$

Here $\tilde{v}$ denotes the RMS (root mean squared) value of the velocity, and the brackets mean the average taken over the surface of the vibrating source.

If the actual radiated sound power is P from a surface with area S and velocity v, the radiation efficiency σ is defined as:

$$\sigma = \frac{P}{\rho c S \left\langle \tilde{v}^2 \right\rangle} \tag{6.24}$$

The radiation efficiency has no dimension and it may take values between 0 and around 4. Values higher than unity are possible because bending waves in a plate may radiate more efficiently than the piston, as we shall see later.

The radiation efficiency will be used to describe the radiated sound power from various kinds of vibrations. When sound is transmitted through a plate, the radiated sound power can be assigned to two different kinds of vibrations in the same plate: *forced* and *resonant* vibrations. As these two vibration fields have different physical origins and different wave numbers associated with them, the radiation efficiency and velocity magnitude will also be very different. The resonant vibration is associated with free bending wave numbers and the forced vibration is associated with the wave number in the sound field in the air. The total transmission through the wall is found by adding together the transmitted sound powers:

$$P_{tot} = P_{for} + P_{res} = \rho c S \left\langle \tilde{v}_{for}^2 \right\rangle \sigma_{for} + \rho c S \left\langle \tilde{v}_{res}^2 \right\rangle \sigma_{res} \tag{6.25}$$

6.4 SOUND RADIATION CALCULATED BY RAYLEIGH'S METHOD

In this section, a general method for calculating the sound radiation from a vibrating surface is introduced – the so-called Rayleigh's method to be used in the following sections. (Rayleigh, 1896, §278, Equation 3). The method is described in Lord Rayleigh's "The Theory of Sound" first published 1877, and the steps are the follows:

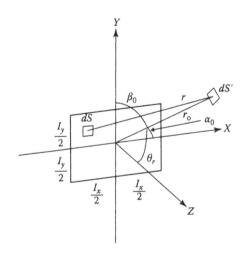

Figure 6.7 A sound-radiating surface surrounded by a large, rigid baffle.

- Knowing the velocity pattern of the vibrating panel, the velocity potential (and thus the sound pressure) is found in an arbitrary point by integration over the surface of the panel.
- The squared sound pressure is integrated over the surface of an infinite half-sphere.
- The result is the radiated sound power.

The surface is assumed to be rectangular with the dimensions lx and ly and the position in a coordinate system is shown in Figure 6.7. A small area dS can be considered as a point source with the volume velocity $v\,dS$. Since the radiation into a half sphere is considered, the sound pressure in the distance r is

$$p = j\omega p \frac{v\,dS}{2\pi r} e^{-jkr} \tag{6.26}$$

Here and in the following, the time factor $e^{-j\omega t}$ is omitted. The total sound pressure at the point of observation can be found by integration over the surface of the plate:

$$p = \frac{j\omega p}{2\pi} \int_S v \frac{e^{-jkr}}{r} \, dS \tag{6.27}$$

The distance from the centre of the plate to the point of observation is r_0 and it is assumed that this is large compared to the dimensions of the plate. In Equation 6.27, r appears in two places – in the divisor, the approximation $r \approx r_0$ is sufficient, whereas in the exponent, a more accurate approximation is needed:

$$r \approx r_0 - (x \cos \alpha_0 + y \cos \beta_0) \tag{6.28}$$

Here x and y are the coordinates of the source point dS, and the angles α_0 and β_0 are as defined in Figure 6.7. With these approximations, the equation becomes:

$$p = \frac{j\omega p}{2\pi r_0} e^{-jkr_0} \int_S v e^{jk(x \cos \alpha_0 + y \cos \beta_0)} \, dS \tag{6.29}$$

With this equation, the sound pressure can be calculated in any direction (α_0, β_0) if the velocity v of the plate is known as a function of the position (x, y).

The radiated sound intensity at the point of observation is $I = \frac{1}{2}|p|^2/\rho c$, and the total sound power radiated into the half sphere $S' = 2\pi r_0^2$ is:

$$P = \int_{S'} I\,dS' = \frac{r_0^2}{2\rho c} \int_{2\pi} |p|^2 \, d\psi \tag{6.30}$$

6.5 SOUND RADIATION FROM FORCED VIBRATIONS IN A RECTANGULAR PLATE

We are now ready to study sound radiation and the radiation efficiency of a finite plate. The theoretical derivation and experimental verification of the results were done by Rindel (1975). To start with, consider the case of forced vibrations (the resonant vibration case will be discussed in Section 6.6).

It is assumed that the vibrations of the plate are generated by a plane sound wave incident from the side $z < 0$. The angles of incidence relative to the axis of the coordinate system are α, β and θ, respectively (Figure 6.8).

The sound pressure of the incident sound can be described as:

$$p_i = |p_i| e^{-jk(x\cos\alpha + y\cos\beta + z\cos\theta)} \tag{6.31}$$

Accordingly, the forced vibrations of the plate can be written as:

$$v = |v| e^{-jk(x\cos\alpha + y\cos\beta)} \tag{6.32}$$

At this point of our discussion, it is not important whether there is a phase shift between sound pressure and velocity. For the sake of simplicity, it is assumed that the velocity amplitude is constant within the area of the plate and zero on the rigid baffle surrounding the plate. Then, the sound pressure of the radiated sound field is found by inserting Equation 6.32 in Equation 6.29:

$$p = \frac{j\omega\rho}{2\pi r_0} |v| e^{-jkr_0} \int_{-\frac{1}{2}l_y}^{\frac{1}{2}l_y} \int_{-\frac{1}{2}l_x}^{\frac{1}{2}l_x} e^{jkx(\cos\alpha_0 - \cos\alpha)} e^{jky(\cos\beta_0 - \cos\beta)} \, dx \, dy$$

$$= \frac{j\omega\rho}{2\pi r_0} |v| e^{-jkr_0} \cdot \frac{e^{j\frac{1}{2}kl_x(\cos\alpha_0 - \cos\alpha)} - e^{-j\frac{1}{2}kl_x(\cos\alpha_0 - \cos\alpha)}}{jk(\cos\alpha_0 - \cos\alpha)}$$

$$\cdot \frac{e^{j\frac{1}{2}kl_y(\cos\beta_0 - \cos\beta)} - e^{-j\frac{1}{2}kl_y(\cos\beta_0 - \cos\beta)}}{jk(\cos\beta_0 - \cos\beta)}$$

Even if this result looks complicated, we note that the variables are separated. With the use of Euler's formula (Equation 1.40), the result can be rewritten in the simple form:

$$p = \frac{j\omega\rho}{2\pi r_0} |v| S e^{-jkr_0} \cdot \frac{\sin X}{X} \cdot \frac{\sin Y}{Y}$$

$$= \frac{j\omega\rho}{2\pi r_0} |v| S e^{-jkr_0} \cdot \text{sinc}(X) \cdot \text{sinc}(Y) \tag{6.33}$$

where S is the area of the plate and

$$X = \tfrac{1}{2}kl_x(\cos\alpha_0 - \cos\alpha)$$

$$Y = \tfrac{1}{2}kl_y(\cos\beta_0 - \cos\beta) \tag{6.34}$$

The function sinc $(X) = \sin X/X$ is symmetric (same value for positive and negative X), it has a maximum of 1 for $X=0$ and $\to 0$ for $X \to \infty$ (Figure 1.5). This means that the sound radiation is maximum in the direction $\alpha_0 = \alpha$ and $\beta_0 = \beta$, i.e. in the same direction as the incident sound, as could be expected. The result also shows that smaller amounts of sound are radiated in other directions; especially for small values of kl_x and kl_y, i.e. if the plate dimensions are small compared to the wavelength. If the ratio of dimension to wavelength is very small, the radiation approaches an omnidirectional pattern in a half sphere.

The *characteristic dimension a* is introduced as a general measure of the size of a plate:

$$a = \frac{2S}{U} = \frac{l_x l_y}{l_x + l_y} \tag{6.35}$$

where S is the area and U is the perimeter. $2a$ is in fact the harmonic average of l_x and l_y (see Equation 1.33), and a must be between 0.5 and 1 times the shorter dimension of the plate. In the special case of a square plate, $a = 1/2\sqrt{S}$. If the sides are very different in length, the characteristic dimension does not exceed the shorter dimension.

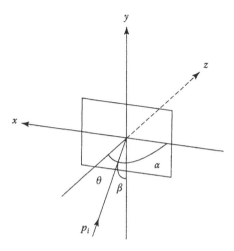

Figure 6.8 Definition of the angles of incidence of the plane sound wave generating the forced vibrations in the plate.

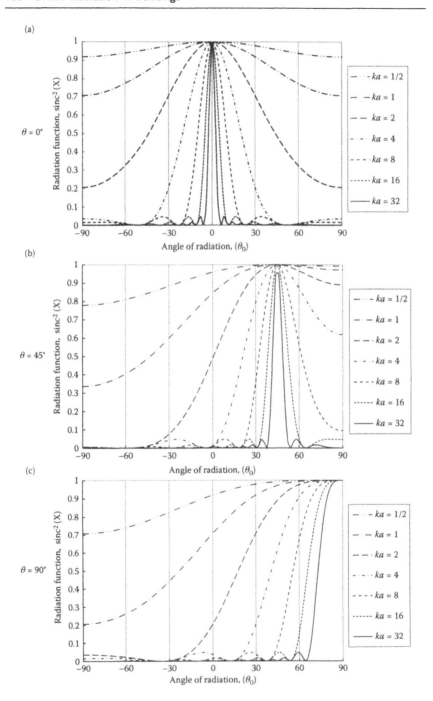

Figure 6.9 Radiation function sinc²(X) showing the sound radiation pattern from forced bending waves. Three examples of the angle of incidence, (a) 0°, (b) 45°, (c) 90°, are shown.

The radiation pattern is shown in Figure 6.9 for the two-dimensional case, i.e. $\beta_0=\beta=\pi/2$. It is seen that the radiation has little directivity for low ka, but gets more directive for high values of ka. In the case of grazing incidence, $\theta=90°$, the radiation covers a relatively wide angle even for high values of ka.

The total radiated sound power into the half sphere $z > 0$ is found by inserting Equation 6.33 in Equation 6.30:

$$P = \frac{1}{2\rho c} \int_{2\pi} \frac{\omega^2 \rho^2}{4\pi^2} |v|^2 S^2 \left(\text{sinc}X \cdot \text{sinc}Y\right)^2 d\psi$$

$$= \frac{\rho c}{8\pi^2} \cdot k^2 S^2 |v|^2 \int_{2\pi} \left(\text{sinc}X \cdot \text{sinc}Y\right)^2 d\psi \tag{6.36}$$

This integral cannot be solved analytically; so, instead, numerical approximations can be made. The radiation efficiency is denoted by σ_θ when the radiation is due to a forced bending wave generated by a sound field with the angle of incidence θ:

$$\sigma_\theta = \frac{P}{\frac{1}{2}\rho c S |v|^2} = \frac{k^2 S}{4\pi^2} \int_{2\pi} \left(\text{sinc}X \cdot \text{sinc}Y\right)^2 d\psi \tag{6.37}$$

The result of a numerical integration is shown in Table 6.2 and Figure 6.10 as a function of the angle of incidence θ in the case of a square plate.

Table 6.2 Calculated radiation efficiency for forced vibrations in a square plate

ka	0°	15°	30°	45°	60°	75°	90°	Random
0.5	−8.3	−8.3	−8.3	−8.4	−8.5	−8.6	−8.6	−8.50
0.75	−5.0	−5.1	−5.2	−5.4	−5.6	−5.7	−5.7	−5.53
1	−3.0	−3.0	−3.3	−3.6	−3.9	−4.1	−4.2	−3.79
1.5	−0.6	−0.8	−1.1	−1.7	−2.2	−2.6	−2.7	−2.01
2	0.3	0.2	−0.2	−0.8	−1.3	−1.8	−2.0	−1.15
3	−0.1	0.2	0.5	0.4	−0.2	−0.7	−0.9	−0.14
4	−0.3	0.0	0.7	1.0	0.6	0.1	−0.2	0.41
6	0.0	0.2	0.6	1.6	1.7	1.1	0.7	1.16
8	0.1	0.2	0.5	1.6	2.4	1.8	1.4	1.61
12	0.0	0.2	0.6	1.4	3.1	2.9	2.3	2.18
16	0.0	0.2	0.6	1.5	3.2	3.7	3.0	2.57
24	0.0	0.2	0.6	1.5	2.9	4.7	3.9	3.04
32	0.0	0.2	0.6	1.5	3.0	5.4	4.6	3.36
48	0.0	0.2	0.6	1.5	3.0	6.0	5.5	3.77
64	0.0	0.2	0.6	1.5	3.0	6.1	6.2	4.04

Source: Sato, H., *J. Acoust. Soc. Jpn.*, 29, 509–516, 1973.

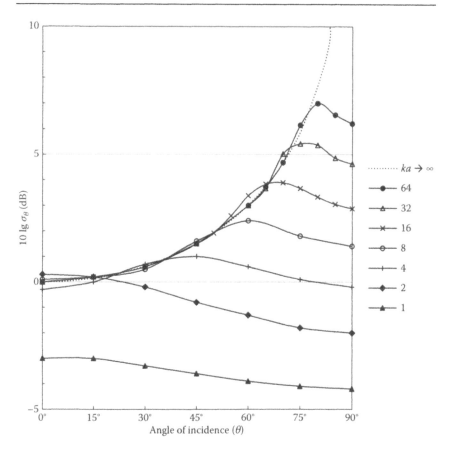

Figure 6.10 Radiation efficiency for forced vibrations in a square plate as a function of the angle of incidence for different values of *ka*. (Adapted from Sato, H., *J. Acoust. Soc. Jpn.*, 29, 509–516, 1973.)

It is particularly interesting to note the behaviour when the angle of incidence approaches grazing incidence; the radiated power is limited, instead of growing to infinity as it would theoretically for an infinite plate.

Table 6.2 also gives the result of an additional integration over all angles of incidence in a diffuse sound field, and this is the radiation efficiency for random incidence.

The limiting case of an infinitely large plate relative to the wavelength is:

$$\sigma_\theta \to \frac{1}{\cos\theta} \quad \text{for} \quad ka \to \infty \tag{6.38}$$

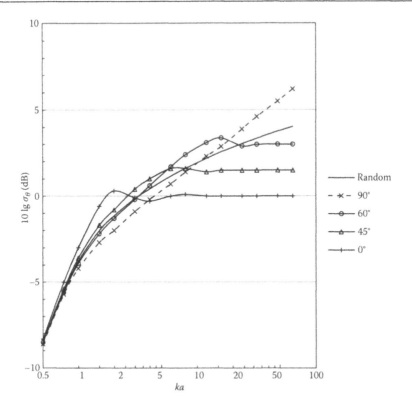

Figure 6.11 Radiation efficiency for a square plate with forced bending waves for different angles of incidence and for random incidence.

6.5.1 Approximate results

The following approximation to the result in Equation 6.37 was derived by
combining three asymptotic results for high frequencies, for low frequencies and for grazing incidence $\theta = 90°$ (Rindel 1993).

$$\sigma_{\text{for},\theta} \cong \left[\left(\cos^2 \theta - 0.6 \frac{\pi}{ka} \right)^2 + \pi \left(0.6 \frac{\pi}{ka} \right)^2 + \left(\frac{\pi}{2(ka)^2} \right)^4 \right]^{-1/4} \qquad (6.39)$$

This approximation is shown in Figure 6.11 as a function of ka for some
values of θ.

The radiation efficiency for random incidence is found by another
numerical integration over all angles of incidence. The result is also shown

in Figure 6.11. A numerical approximation for a square plate and valid for $ka > 0.5$ is:

$$\sigma_{for,d} \cong \frac{1}{2}(0.20 + \ln(2ka)) \tag{6.40}$$

(excitation by a diffuse sound field, so the index d). Usually, Equation 6.40 is sufficiently accurate for an estimate. However, a more exact formula (still for $ka > 0.5$) was derived by Sewell (1970):

$$\sigma_{for,d} = \frac{1}{2}\left(0.160 + \ln(2ka) + \left(16\pi(ka)^2\right)^{-1} + C(\Lambda)\right) \tag{6.41}$$

Here C is a function of the edge ratio $\Lambda = l_x/l_y$, where $l_x \geq l_y$. In the original paper by Sewell (1970), C was given in a table for values of Λ between 1 and 10. The function can be approximated by:

$$C(\Lambda) = -0.0022 \cdot \Lambda^2 + 0.0821 \cdot \Lambda - 0.08 \tag{6.42}$$

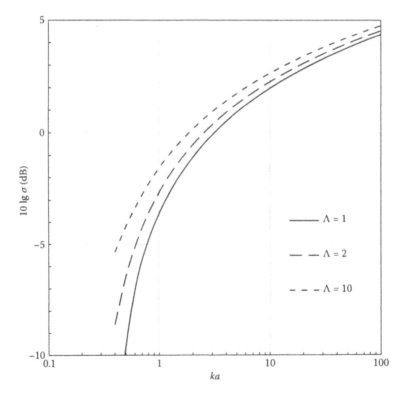

Figure 6.12 Radiation efficiency for a plate with forced bending waves excited by a diffuse sound field as a function of ka for three different edge ratios.

The radiation efficiency for diffuse sound field excitation is displayed in Figure 6.12.

In the low-frequency region, $ka < 0.5$, the radiation is weak and not directive. Thus, we can apply the low-frequency asymptotic result for a square piston in a large rigid baffle (Stenzel 1952):

$$\sigma_{\text{for}} \cong \frac{2}{\pi}(ka)^2 \qquad\qquad (6.43)$$

This is valid for $ka < 0.5$ and for any angle of incidence including random incidence.

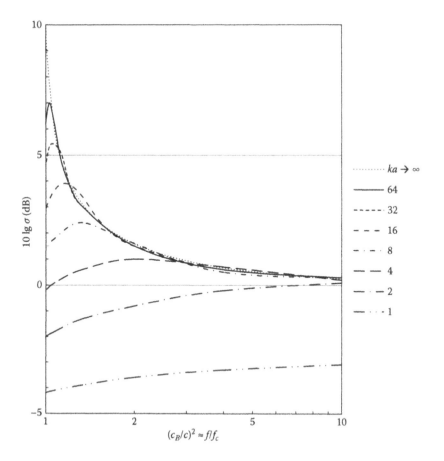

Figure 6.13 Radiation efficiency for a square plate as a function of the phase speed of bending waves relative to the speed of sound in air.

6.6 SOUND RADIATION FROM SUPERSONIC BENDING WAVES

The solution to the sound radiation problem as derived earlier is not restricted to the case of forced vibrations. There is a direct relation between the phase speed of the propagating bending wave and the angle of incidence (see Equation 6.20). This means that the radiation efficiency as displayed in Figure 6.10 can be considered as a function of the relative speed of bending waves, instead of as a function of the angle of incidence. As long as the bending wave is supersonic ($c_B > c$), it does not matter whether it is a forced or a free bending wave, the physical principle of sound radiation is the same. So, the radiation efficiency can be displayed in a new way using $(c_B/c)^2 \approx f/f_c$ as the independent parameter (Figure 6.13).

The usual assumption that sound radiation is maximum at the critical frequency is true with modifications. We can observe from Figure 6.13 that the radiation is maximum at a frequency a little higher than the critical frequency for a finite plate with high ka. However, for $ka < 4$, the maximum is shifted up about one octave or more and the peak has almost disappeared. The radiation is weak for very low ka. We also observe that for $ka > 2$, the radiation efficiency has the asymptotic value of unity (0 dB) for $f \gg f_c$.

Thin plates like a gypsum board wall or a glass pane in a window have high ka values at the critical frequency ($f_c \approx 2.5$ kHz to 4 kHz) and thus, a clear peak of radiation at the critical frequency. But thick plates like a masonry wall or a concrete slab have low ka values at the critical frequency ($f_c \approx 100$ Hz) and thus, no peak of radiation. In Section 6.7, we shall elaborate more on these findings.

6.7 SOUND RADIATION FROM RESONANT VIBRATIONS IN A RECTANGULAR PLATE

We shall now consider the sound radiation from resonant vibrations of a plate. Sound radiation is mainly dependent on two conditions: the ratio between propagation speeds in the plate and in the air (the Mach number), and the size of the plate compared to the wavelength in air. The first condition determines whether the radiation is at all efficient or not, while the second condition determines an upper limit for the radiation efficiency.

The excitation of resonant vibrations can be either mechanical impact, structural vibrations (e.g. flanking transmission) or airborne excitation from an incident sound wave. As we have seen in the beginning of this chapter, the normal modes have wavelengths $\lambda_B = c_B/f$ that are determined solely by the dimensions of the plate and the modal numbers (m, n) (see Equation 6.6). Thus, for a given frequency, the wavelength of the plate vibrations can be either smaller than or larger than the wavelength in air. In other words, it is important to determine whether the frequency is below

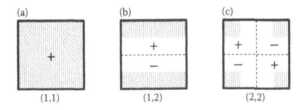

Figure 6.14 Radiation from low-numbered modes, (a) (I,I), (b) (I,2) and (c) (2,2) in a plate. The shaded areas are those radiating sound. Areas with + and − are in opposite phase.

or above the critical frequency of the plate. If $f < f_c$, we have $c_B < c$, $\lambda_B < \lambda$, and vice versa for $f > f_c$.

At the first normal mode (1,1) of a simply supported plate (Figure 6.14a), all parts of the plate move in phase and the sound radiation is similar to that from the diaphragm of a loudspeaker. At the second and third normal mode (1,2) or (2,1), the movements of the plate are divided by one nodal line (Figure 6.14b). Below the critical frequency, the pressure variations are cancelled across the modal line, but sound is radiated from a zone near the edges. If we consider the next normal mode (2,2), the sound radiation is further limited to four zones near the corners (Figure 6.14c). The radiation efficiency of these first modes is displayed in Figure 6.15 as functions of the relative speed c_B/c. All modes have radiation efficiency $\sigma \cong 1$ above

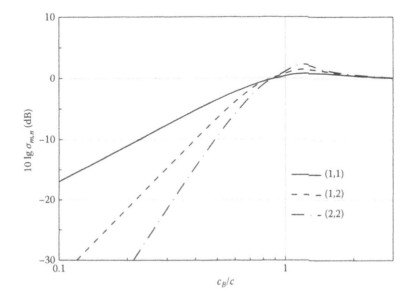

Figure 6.15 Radiation efficiency for low-numbered modes, (I,I) (I,2) and (2,2). (Principle adapted from Wallace, C.E., J. Acoust. Soc. Am., 51, 946–952, 1972, Figure 2.)

the critical frequency $(c_B/c > 1)$; but at lower frequencies, the radiation is less efficient. The sound radiation from normal modes in plates has been analysed by Wallace (1972), but here it is sufficient to look at approximate results for a square panel.

$$\sigma_{res}(1,1) = \frac{2^7}{\pi^3} \cdot \frac{Sf^2}{c^2} \cong \frac{4Sf^2}{c^2} \quad \text{mode}(1,1) \tag{6.44a}$$

$$\sigma_{res}(1,2) = \frac{2^5}{3\pi} \cdot \frac{S^2f^4}{c^4} \cong \frac{3.4S^2f^4}{c^4} \quad \text{mode}(1,2) \tag{6.44b}$$

$$\sigma_{res}(2,2) \cong \frac{8\pi}{15} \cdot \frac{S^3f^6}{c^6} \cong \frac{1.7S^3f^6}{c^6} \quad \text{mode}(2,2) \tag{6.44c}$$

Modes with higher mode numbers radiate from the edges or from the corners when the frequency is below the critical frequency, but from the entire area above the critical frequency (Figures 6.16 and 6.17). There is cancellation of sound pressures across the nodal lines of the plate as long as the wavelength in the plate is shorter than the wavelength in the air. Therefore, some mode patterns give radiation along two edges and other mode patterns only allow the four corners to radiate sound. Above the critical frequency, the entire surface contributes to the radiation and $\sigma \to 1$ at high frequencies.

The following result for the radiation efficiency was derived by Maidanik (1962) for the case of resonant vibrations in a plate generated by random incidence sound at frequencies below the critical frequency.

$$\sigma_{res} = \frac{c^2}{Sf_c^2} \cdot g_1(M) + \frac{Uc}{Sf_c} \cdot g_2(M) \quad (f_{11} < f < f_c) \tag{6.45}$$

where g_1 and g_2 are help functions:

$$g_1(M) = \begin{cases} \dfrac{8(1-2M^2)}{\pi^4 M\sqrt{1-M^2}} & (M < \sqrt{2}/2) \\ 0 & (M \geq \sqrt{2}/2) \end{cases} \tag{6.46}$$

$$g_2(M) = \frac{1}{4\pi^2} \left(\frac{(1-M^2)\ln\left((1+M)/(1-M)\right)+2M}{(1-M^2)^{3/2}} \right) \tag{6.47}$$

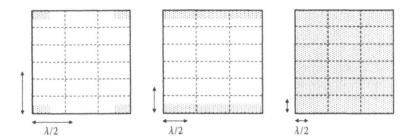

Figure 6.16 Radiation from a high-numbered mode (3,6) scaled to different frequencies. The shaded areas are those radiating sound.

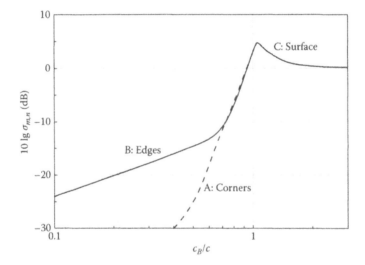

Figure 6.17 Radiation efficiency for high-numbered modes. A: Corner radiation from, e.g. mode (11, 11) or (12, 12). B: Edge radiation from, e.g. mode (1, 11) or (1, 12). C: Surface radiation from all modes when $c_B/c > 1$. (Principle adapted from Wallace, C.E., J. Acoust. Soc. Am., 51, 946–952, 1972, Figure 3.)

Here the Mach number M is used as a frequency parameter.

$$M = \frac{c_{B,\text{eff}}}{c} \approx \sqrt{\frac{f}{f_c}} \tag{6.48}$$

The parameter M is the Mach number, as it gives the ratio between the speed of the travelling bending wave and the speed of sound in air. The approximation using the critical frequency is only valid for thin plates, whereas a more general result is obtained by using the ratio between the propagation speed in the plate and in the air, $c_{B,\text{eff}}/c$.

The derivation of Equation 6.45 is based on the following assumptions:

- The modal density is high.
- All modes have the same energy.
- Contribution from 'even-odd' and 'even-even' modes is negligible compared to the contribution from 'odd-odd' modes.

The first assumptions means in reality that the result is not valid below $f_{22}=4 f_{11}$ (see Section 6.1).

Above the critical frequency, we have radiation from supersonic vibrations and for $ka \gg 1$, the radiation efficiency is:

$$\sigma_{res} = \left(1-M^{-2}\right)^{-1/2} \cong \left(1-f_c/f\right)^{-1/2} \quad (f > f_c) \tag{6.49}$$

At frequencies near the critical frequency f_c, the radiation efficiency is determined by a maximum related to the dimensions of the plate compared to the wavelength, as discussed in Section 6.6. A simple empirical approximation for the maximum is:

$$\sigma_{res} \cong \sqrt{\frac{\pi f U}{16c}} \quad (f \approx f_c) \tag{6.50}$$

Combining the aforementioned equations into one:

$$\sigma_{res} = (\sigma_1^{-4} + \sigma_2^{-4} + \sigma_3^{-4})^{-1/4} \tag{6.51}$$

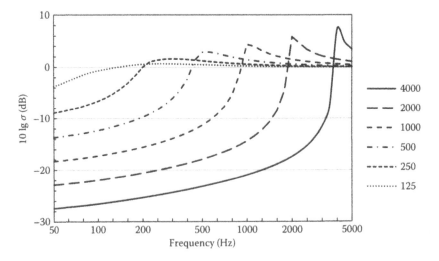

Figure 6.18 Radiation efficiency for resonant vibrations in plates with dimension 4.0 m × 2.5 m and critical frequencies from 125 Hz to 4000 Hz.

where σ_1, σ_2 and σ_3 are from Equations 6.45, 6.49 and 6.50, respectively. In this way, the transition between each equation is smoothed.

Some calculation examples are shown in Figure 6.18. The previous observations concerning a peak near the critical frequency for thin plates and a very flat radiation course for thick plates are clearly seen. For frequencies above the critical frequency, the results shown in Figure 6.18 are in fact the same as those in Figure 6.13.

6.7.1 Approximate results

Assuming bending waves, i.e. plates that are relatively thin and sufficiently large so that the first mode of vibration $f_{11} \ll f_c$, we get the following approximations for the frequency range below the critical frequency (also shown in Figure 6.19).

$$\sigma_{\text{res}} \cong \frac{Uc}{\pi^2 Sf_c} \sqrt{\frac{f}{f_c}} \qquad \left(\frac{3c}{U} < f < \frac{1}{2}f_c\right) \tag{6.52}$$

$$\sigma_{\text{res}} \cong \frac{c^2}{Sf_c^2} \qquad \left(f_{11} < f < \frac{3c}{U}\right) \tag{6.53}$$

$$\sigma_{\text{res}} \cong \frac{4Sf^2}{c^2} \qquad (f < f_{11}) \tag{6.54}$$

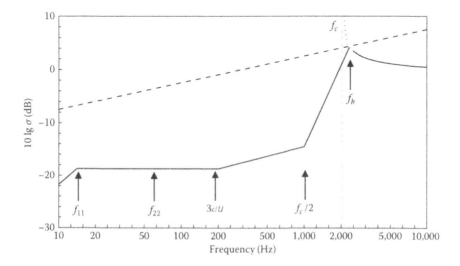

Figure 6.19 Radiation efficiency from resonant vibrations in a large thin plate ($f_{11} < f_c$). The maximum radiation from Equation 6.50 is shown as a dashed line.

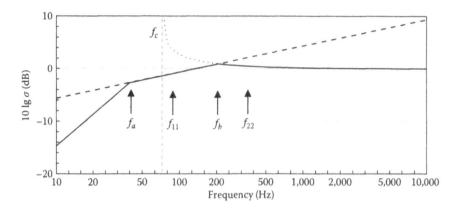

Figure 6.20 Radiation efficiency from resonant vibrations in a small thick plate ($f_{11} > f_c$). The maximum radiation from Equation 6.50 is shown as a dotted line.

The area is S and U is the perimeter. The range between ½ f_c and f_c is not covered by any formula, but a simple interpolation may be sufficient. The frequency calculated from $3c/U$ represents the approximate transition from mainly corner radiation at low frequencies to mainly edge radiation at somewhat higher frequencies. The frequencies that separated the range for Equations 6.49 through 6.54 are marked in Figure 6.19. In addition, the f_{22} frequency is shown, which is the lower frequency limit for application of statistical modal density (see Section 6.1).

As an example, a 6.5 mm glass pane with dimensions 1.4 m × 1.4 m is considered (Table 6.3). The first normal mode is at 15 Hz and the statistical modal density may be applied above 60 Hz.

Table 6.3 Examples of key frequencies for a 6.5 mm glass pane and a 250 mm brick wall

	Glass pane	Brick wall
h, mm	6.5	250
l_x, m	1.4	4
l_y, m	1.4	2.5
S, m²	2	10
f_{11}, Hz	15	91.5
f_{21}, Hz	37	168
f_{12}, Hz	37	289
f_{22}, Hz	60	366
f_a, Hz	89	43
$3c/U$, Hz	183	79
f_c, Hz	2000	72
f_b, Hz	2305	204

In contrast, Figure 6.20 shows the radiation efficiency in the case of a thick plate. The critical frequency is below the first normal mode of the plate, $f_c < f_{11}$. This implies that there are no resonant modes available around and below the critical frequency, so the regions of edge radiation and corner radiation have disappeared from the graph in this case. Another difference between Figures 6.19 and 6.20 is the frequency of maximum radiation. Whereas the maximum is close to the critical frequency of the thin plate, this is not the case for the thick plate. The maximum radiation is shifted to a somewhat higher frequency, f_b, which can be found as the frequency of the crossing of the two expressions, Equations 6.49 and 6.50:

$$f_b = f_c + \frac{5c}{U} \tag{6.55}$$

At very low frequencies, the crossing of Equations 6.50 and 6.54 is at the frequency f_a:

$$f_a = \sqrt[3]{\frac{\pi U}{64c}\left(f_{11}f_c\right)^2} \tag{6.56}$$

As an example of a thick plate, a 250 mm brick wall with dimensions 2.5 m×4.0 m is considered (Table 6.3). Here $5c/U = 5 \times 344/13 = 132\,\text{Hz}$, which means that although the critical frequency is around $f_c = 72\,\text{Hz}$, the frequency of maximum radiation is actually shifted more than one octave up, to 204 Hz. The course of the curve in Figure 6.20 is quite flat. Actually, the radiation efficiency is close to unity (0 dB) at all frequencies above 50 Hz.

6.7.2 Directivity of resonant radiation

The radiation pattern of resonant vibrations in a plate can be found for each of the normal modes by integration of the velocity distribution over the plate according to Rayleigh's method. The problem is reduced from three to two dimensions by assuming that the plane of radiation lies in the xz-plane. So, only modes related to the l_x dimension of the plate are considered. Taking only the most important contributions into account (the 'odd-1' mode), the approximate result is:

$$\sigma_{res,\theta} = \frac{k^2 S}{2\pi^3}\left(\text{sinc}X_1 + \text{sinc}X_2\right)^2 \tag{6.57}$$

where

$$X_1 = \frac{1}{2}kl_x\left(\sin\theta_r + c/c_B\right) \text{ and } X_2 = \frac{1}{2}kl_x\left(\sin\theta_r - c/c_B\right) \tag{6.58}$$

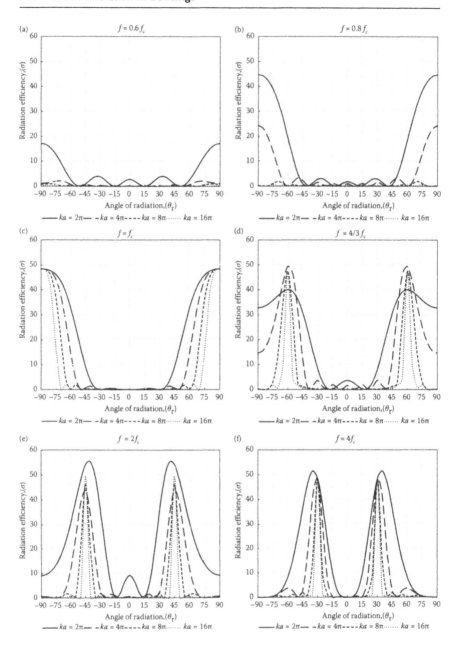

Figure 6.21 Examples of the radiation efficiency as a function of the angle of radiation calculated by Equation 6.57. (a) $f = 0.6\ f_c$, (b) $f = 0.8\ f_c$, (c) $f = f_c$, (d) $f = 1.33\ f_c$, (e) $f = 2\ f_c$, (f) $f = 4\ f_c$.

It is seen from the result (Equation 6.57) that the radiation has two dominating, symmetric lopes as long as $c_B > c$ $(f > f_c)$. The main direction of
radiation is:

$$\theta_r = \pm \arcsin(c/c_B) = \pm \arcsin\left(\sqrt{f_c/f}\right) \tag{6.59}$$

Some examples of calculated radiation patterns are shown in Figure 6.21. At
frequencies below the critical frequency, the radiation is weak and without
any specific directionality, but at higher frequencies, the radiation is highly
directional and symmetric around the normal of the plate. For increasing
plate dimensions compared to the wavelength (ka increasing), the directivity
of the radiation becomes sharper. At and just below the critical frequency, the
radiation is mainly in directions close to 90°, but covering a relatively broad
angle, especially for low ka values. Above the critical frequency, we have
radiation from supersonic bending waves, and the radiation is highly directive in accordance with Equation 6.59, especially for high ka values.

6.7.3 Combined forced and resonant radiation

A summary of the radiation properties of forced and resonant vibrations in
a plate is given in Table 6.4.

A sound wave incident on a plate generates a forced bending wave, and
when the vibrations are reflected from the boundaries, the resonant modes
of vibration are excited, mainly those with a natural frequency close to
the frequency of the incident sound. At high frequencies, where the modal
density in the plate is high, many modes will be excited simultaneously and
contribute to resonant radiation, but at low frequencies, the modal density
in the plate is low and the resonant radiation can only take place at the
natural frequencies of the plate.

Table 6.4 Comparison of forced and resonant vibrations

	Forced vibrations	Resonant vibrations
Excitation	Incident sound wave	Any excitation
Wave type	Propagating wave	Standing wave
Phase speed c_B	Depends on angle of incidence $c_B = c / \sin\theta \geq c$	Depends on frequency $c_B = c\sqrt{f / f_c}$
Direction of radiation	One direction $\theta_r \cong \arcsin(c / c_B)$	No direction for $f < f_c$ Two directions for $f \geq f_c$ $\theta_r \cong \pm\arcsin\left(\sqrt{f_c / f}\right)$
Strength of radiation	$\sigma < 1$ for $ka < 10$ $\sigma > 1$ for $ka > 10$	$\sigma \ll 1$ for $f < f_c$ $\sigma \cong 1$ for $f > f_c$

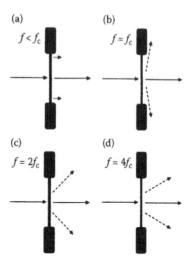

Figure 6.22 Sound radiation from a panel with excitation by a plane wave at normal
incidence. (a) Below the critical frequency. (b) At the critical frequency. (c)
One octave above the critical frequency. (d) Two octaves above the critical
frequency. Dotted line arrows indicate resonant radiation.

Examples of combined forced and resonant sound radiation are shown in
Figure 6.22. Below the critical frequency, the resonant vibrations can only
radiate from corners or edges, and the contribution is relatively weak, but
above the critical frequency, the resonant radiation is strong and may be
more important than the forced transmission.

6.8 THE REFRACTION EFFECT

The sound transmission through a surface of finite area includes diffraction
effects when the wavelength is not very small compared to the characteristic dimension of the surface. Diffraction effects include the scattering of
radiated sound in many directions and the refraction of transmitted sound
for incident sound near grazing incidence. Hence, the angle of radiation
may be different from the angle of incidence (Figure 6.23).

We return to Figure 6.10, which shows that near the grazing incidence,
i.e. θ between 75° and 90°, the radiation from a forced bending wave is limited and the maximum radiation depends on ka. The asymptotic value of
the radiation efficiency is $1/\cos \theta$ (see Equation 6.37). This means that the
actual value of the radiation efficiency can be converted into an equivalent
angle of radiation:

$$\theta_r = \arccos\left(\frac{1}{\sigma_\theta}\right) \tag{6.60}$$

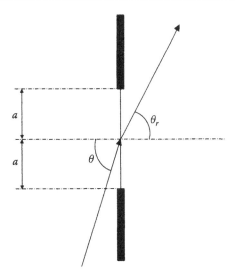

Figure 6.23 Angle of incidence θ and main direction of sound radiation θ_r, for forced vibrations in a finite plate with characteristic dimension a.

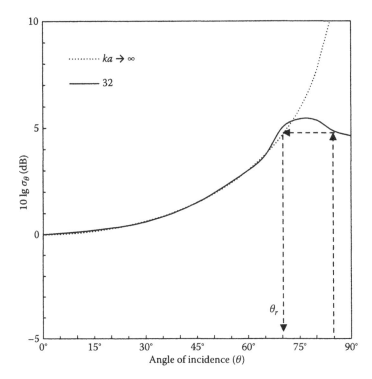

Figure 6.24 Equivalent angle of radiation from a square panel is 70° when the angle of incidence is 85° and $ka = 32$.

Figure 6.24 shows the principle of this conversion for the angle of incidence $\theta=85°$ and $ka=32$. The equivalent angle of radiation is found to be around $\theta_r=70°$.

The simulations shown in Figure 6.25 are four examples of radiation from resonant vibrations in a plate that is excited by an incident plane wave. The normal modes radiate sound equally to both sides of a plate, and the radiation is symmetric around the normal to the plate. In Figure 6.25a, the mode with modal number 1 is excited, and the radiation is nearly omnidirectional. With higher modal numbers, the radiation is more directive, but with a clear refraction effect (Figure 6.25b and c). In these simulations, the resonant radiation is strong and the radiation from the forced bending wave is not visible.

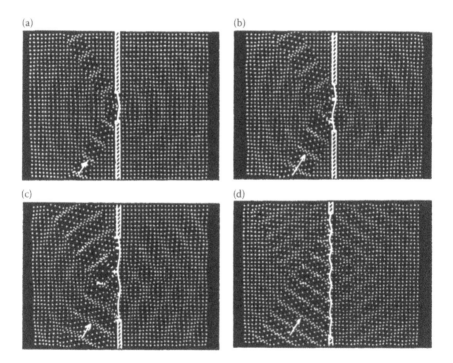

Figure 6.25 Computer simulation of sound radiation from a plate at the coincidence frequency. The excitation is a plane wave incident at angle 60°. (a) $ka=\pi$, (b) $ka=2\pi$, (c) $ka=4\pi$, (d) $ka=14\pi$. (Reproduced from Heckl, M., *J. Sound Vib.*, 77, 165–189, 1981. With permission from Elsevier.)

REFERENCES

K. Gösele (1953). Schallabstrahlung von Platten, die zu Biegeschwingungen angeregt sind (Sound radiation from plates that are excited to bending waves, in German). *Acustica* 3, 243–248.

M. Heckl (1981). The tenth Sir Richard Fairey memorial lecture: Sound transmission in buildings. *Journal of Sound and Vibration* 77, 165–189.

M. Heckl, M. Heckl (1976). Darstellung von abstrahlvorgängen im film. *Fortschritte der Akustik – DAGA 1976*, VDI-Verlag, Düsseldorf, 165–168.

J. Kristensen, J.H. Rindel (1989). *Bygningsakustik – Teori og praksis* (Building acoustics – Theory and practice, in Danish). SBI Anvisning 166. Danish Building Research Institute, Hørsholm, Denmark.

G. Maidanik (1962). Response of ribbed panels to reverberant acoustic fields. *Journal of the Acoustical Society of America* 34, 809–826. Erratum (1975). *Journal of the Acoustical Society of America* 57, 1552.

J.W.S. Rayleigh (1896). *The Theory of Sound*. Vol. 2. Reprint of 2nd edition by Dover Publications, New York, 1945.

J.H. Rindel (1975). *Transmission of traffic noise through windows. Influence of incident angle on sound insulation in theory and experiment*. Report No. 9, The Acoustics Laboratory, Technical University of Denmark, Lyngby.

J.H. Rindel (1993). Modelling the angle-dependent pressure reflection factor. *Applied Acoustics* 38, 223–234.

H. Sato (1973). On the mechanism of outdoor noise transmission through walls and windows. *Journal of the Acoustical Society of Japan* 29, 509–516. (In Japanese).

E.C. Sewell (1970). Transmission of reverberant sound through a single-leaf partition surrounded by an infinite rigid baffle. *Journal of Sound and Vibration* 12, 21–32.

H. Stenzel (1952). Die akustische Strahlung der rechteckiger kolbenmembran. *Acustica* 2, 263–281.

C.E. Wallace (1972). Radiation resistance of a rectangular panel. *Journal of the Acoustical Society of America* 51, 946–952.

Chapter 7

Statistical energy analysis, SEA

Sound fields in rooms and mechanical vibrations in plates are examples of systems with a high number of modes, the so-called resonant systems. In case of combinations of two or more such resonant systems, it can often be advantageous to take a statistical approach in order to analyse the propagation and distribution of the acoustic energy in the coupled resonant systems. The method of *statistical energy analysis* (SEA) was originally developed by Lyon and Maidanik (1962) and applied for sound insulation by Crocker and Price (1969) and others. The method has been particularly successful for the investigation of sound propagation through complicated solid structures like ships and buildings.

Even if SEA is a powerful tool, there are some general underlying assumptions that put restrictions and drawbacks on the methodology. Some of the underlying assumptions are as follows:

- Only resonant fields are of importance.
- The modal density should be high and smooth.
- There should be a more or less diffuse field.
- The coupling between different subsystems should be weak.

Some of these assumptions can be violated if one is careful, but it is important to be aware of these underlying facts. One consequence of the underlying assumptions is that structures that are periodic in nature – as in many lightweight and wooden building structures – are not well suited for SEA (these structures violate the first three points). However, for heavy and homogeneous traditional building structures, the underlying assumptions are well met.

7.1 ENERGY BALANCE IN A RESONANT SYSTEM

The total mechanical energy in a vibration system is the sum of the kinetic and the potential energies. In a plate with normal modes, the kinetic and the potential energies are on average equal. An explanation of this is that at resonance the kinetic and the potential energies are per definition equal, and the average response magnitude is mainly determined by the response at the resonances. Since it is common to describe mechanical vibrations by the velocity of the surface, i.e. the kinetic part of the energy, we calculate the total energy E as twice the kinetic energy, as in:

$$E = mS\langle \tilde{v}^2 \rangle \tag{7.1}$$

where m is the mass per unit area of the plate, S is the area of the plate and $\langle \tilde{v}^2 \rangle$ is the mean square of the velocity averaged over the area of the plate.

For a sound field in a room, the energy is commonly described by the sound pressure, i.e. the potential part of the energy. The total energy E in the room can be expressed as twice the potential energy, as follows:

$$E = \frac{\langle \tilde{p}^2 \rangle}{\rho c^2} V \tag{7.2}$$

where V is the volume of the room, $\langle \tilde{p}^2 \rangle$ is the mean square of the sound pressure averaged over the volume of the room, ρ is the density of air and c is the speed of sound in air.

The *loss factor* η of a resonant system is defined as the dissipated energy in a time period of vibration divided by 2π and the total mechanical energy:

$$\eta = \frac{P_d T_0}{2\pi E} = \frac{P_d}{\omega E} \tag{7.3}$$

where P_d is the dissipated power (energy per time unit), T_0 is the time period of the vibration, E is the current total energy in the system and $\omega = 2\pi f = 2\pi / T_0$ is the angular frequency.

In a room, the energy losses are usually described by the equivalent absorption area A or by the reverberation time T. If the volume is V and the sound field in the room can be assumed to be approximately diffuse, the energy density in the stationary sound field is $E/V = 4P_{in}/cA$, where P_{in} is the sound power emitted in the room. Energy balance at stationary conditions means that $P_{in} = P_d$ and, thus, the loss factor of a room is:

$$\eta = \frac{P_d}{\omega} \cdot \frac{cA}{4P_d V} = \frac{cA}{4\omega V} \tag{7.4}$$

The reverberation time is given by Sabine's equation $T = 55V/cA$, and the loss factor of a room can also be expressed as:

$$\eta = \frac{55}{4\omega T} \cong \frac{2.2}{fT} \tag{7.5}$$

The latter expression can be shown to be valid for the vibrations in a solid structure as well, and it is a well-established measuring method for the loss factor. The reverberation time is derived from the slope of the measured decay curve after the source is stopped. The frequency f is taken as the centre frequency of the frequency band used during the measurement.

7.2 COUPLING BETWEEN RESONANT SYSTEMS

Two resonant systems are said to be coupled if energy can be transferred from the normal modes in one system to the normal modes in other system. This requires two conditions to be fulfilled, a physical one and an acoustical one:

- The resonant systems must have a *physical connection* in order that the energy can be exchanged, e.g. a junction between two plates or a common surface between a plate and a room.
- The resonant systems must have normal modes with *natural frequencies sufficiently close to each other* in order that the energy can be exchanged. This may be fulfilled if each system has at least one natural frequency within the frequency band under consideration, e.g. a one-third octave band.

The *coupling loss factor* is introduced to describe the transfer of energy from one system to another. The definition is an analogue to that of the loss factor; the coupling loss factor η_{12} from system 1 to system 2 is:

$$\eta_{12} = \frac{P'_{12}}{\omega E_1} \tag{7.6}$$

where P'_{12} is the power (energy per time unit) transferred from system 1 to system 2, E_1 is the current total energy in system 1 and ω is the angular frequency.

Example. The sound power radiated from resonant modes in a plate (system 1) into a room (system 2) can be derived. The radiated sound power from one side of the plate with velocity of vibration v_1, area S_1 and radiation efficiency $\sigma_{1,\text{res}}$ is (Equation 6.24):

$$P'_{12} = \left\langle \tilde{v}_1^2 \right\rangle S_1 \rho c \sigma_1 \tag{7.7}$$

Insertion of this and the total energy (Equation 7.1) into the definition yields:

$$\eta_{12} = \frac{\rho c}{\omega m_1} \sigma_{1,\text{res}} \tag{7.8}$$

This shows that the energy transfer to the room is strongest at low frequencies and for low mass per unit area if the radiation efficiency is constant.

Two coupled resonant systems are considered (Figure 7.1). Each system may have a source that supplies the system with mechanical or acoustic energy, P_{1i} and P_{2i}, respectively. The dissipated energy is P_{1d} and P_{2d}, respectively.

The net energy transfer per unit time from system 1 to system 2 can be written as:

$$P_{12} = P'_{12} - P'_{21} = \omega\eta_{12}E_1 - \omega\eta_{21}E_2 \qquad (7.9)$$

However, the consideration of the energy balance between the two systems should be made on the basis of the normal modes that carry the energy. It is the energy per mode in each system that is important for the energy balance, as it is the modes that are the degree of freedom of the system. In system 1, the average energy per mode E_{m1} is:

$$E_{m1} = \frac{E_1}{\Delta N_1} = \frac{E_1}{\Delta f} \cdot \frac{\Delta f}{\Delta N_1} \qquad (7.10)$$

where Δf is the frequency band under consideration and ΔN_1 is the number of modes in system 1 within this frequency band. The ratio $\Delta N_1/\Delta f$ is called the *modal density* of system 1, i.e. the number of modes per Hz around the centre frequency f.

The energy balance equation for the two systems can now be expressed in terms of the average energy per mode:

$$P_{12} = \omega\Delta N_1\eta_{12}E_{m1} - \omega\Delta N_2\eta_{21}E_{m2} \qquad (7.11)$$

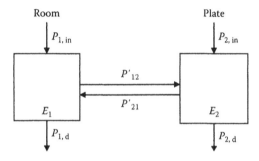

Figure 7.1 Block diagram to show the energy transfer between two coupled resonant systems.

The idea is that each normal mode in a resonant system can be considered a degree of freedom, and it is a general physical principle that energy transfer between coupled systems will always be in the direction from the system with higher energy per degree of freedom to the system with lower energy per degree of freedom. This means that in the case of energy balance, $E_{m1} = E_{m2}$, the net power flow P_{12} is zero and thus Equation 7.11 yields,

$$\eta_{21} = \frac{\Delta N_1}{\Delta N_2} \eta_{12} \qquad (7.12)$$

This is the very important *reciprocity relation* for coupled resonant systems. It says that it is possible to derive the coupling loss factor in one direction if the coupling loss factor is known in the opposite direction and if the modal densities of the two systems are also known. As an example, it is well known how a plate radiates sound into a room, and with the reciprocity relation, it will then be possible to estimate how much energy is transferred into a plate from the sound field in a room.

The energy balance equation can be expressed by only one coupling loss factor using Equation 7.12:

$$P_{12} = \omega \Delta N_1 \eta_{12} \left(E_{m1} - E_{m2} \right) \qquad (7.13)$$

This is sometimes called the *fundamental equation* of SEA. It clearly shows that the difference in average modal energy $(E_{m1} - E_{m2})$ determines the direction and amount of energy flow between the two systems. Equation 7.13 also shows a certain analogy to thermodynamics; the average energy per mode has the same role in acoustics as the temperature has in thermodynamics, and the heat is transferred from a body with a higher temperature to a body with a lower temperature.

It is emphasised that two statistical assumptions have been made:

- The energy in each system is assumed to be equally distributed on the number of normal modes ΔN_i available within the frequency band Δf with the centre frequency f.
- All pairs of normal modes in the two systems have the same coupling loss factor within the frequency band Δf with the centre frequency f.

These assumptions are the reason for calling the method statistical. None of the assumptions are strictly correct, but on average they may be the best available estimate. Later, it will be shown that the results from the SEA method may also be used without the statistical part in case of very low modal density in one of the systems. The modal densities of some typical systems are collected in Table 7.1.

Table 7.1 Statistical modal density in different one-, two- and three-dimensional systems

System	Modal density, $\Delta N_s/\Delta f$
Room, 3D modes only Volume V More accurate, see Equation 4.13	$\dfrac{4\pi}{c^3}Vf^2$
Narrow cavity, 2D modes only Area $S=l_x\,l_y$	$\dfrac{2\pi}{c^2}Sf$
Tube, both ends either closed or open, 1D modes only Length l_x	$\dfrac{2}{c}l_x$
Thin plate, bending waves Area S, critical frequency f_c More accurate, see Equation 6.9	$\dfrac{\pi}{c^2}Sf_c$
Thick plate, shear waves Area S, speed of shear waves c_s More accurate, see Equation 6.10	$\dfrac{2\pi}{c_s^2}Sf$
Beam, longitudinal waves Length l_x, speed of sound c_m	$\dfrac{2}{c_m}l_x$

7.3 RESONANT TRANSMISSION THROUGH A SINGLE WALL

A partition wall between two rooms can be considered a resonant system that is coupled to the two resonant systems representing the rooms. A sketch is shown in Figure 7.2 together with a SEA flow diagram. System 1 is the source room, system 2 is the wall and system 3 is the receiving room.

The forced sound transmission through the wall cannot be derived by SEA since the normal modes of the wall are not involved in this transmission.

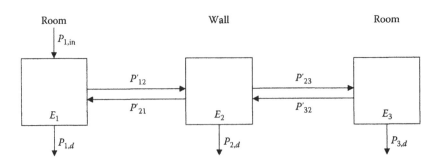

Figure 7.2 Block diagram to show the energy transfer between three coupled resonant systems.

So, only the resonant transmission is considered in the following. Later, it will be discussed how to combine the two different kinds of transmission.

With the usual assumption of diffuse sound fields in the two rooms, the transmission loss of the wall can be written in terms of the total energies in the rooms using Equation 7.2 in combination with Equation 5.5:

$$R = 10 \lg \frac{\langle \tilde{p}_1^2 \rangle S_2}{\langle \tilde{p}_3^2 \rangle A_3} = 10 \lg \frac{E_1 V_3 S_2}{E_3 V_1 A_3} \tag{7.14}$$

where S_2 is the area of the wall and A_3 is the absorption area of the receiving room.

The SEA method implies to set up the energy balance equations for each of the systems, i.e. three in this case:

$$P_{1,in} + P'_{21} = P_{1,d} + P'_{12} \tag{7.15}$$

$$P'_{12} + P'_{32} = P_{2,d} + P'_{21} + P'_{23} \tag{7.16}$$

$$P'_{23} = P_{3,d} + P'_{32} \tag{7.17}$$

System 1 is excited by a stationary sound source, and the equation for that system will only be used if it is relevant to analyse the results as a function of the power of the source. In the present situation, this is not the case; it is sufficient to find the ratio between the energy of systems 1 and 3, which is independent of the power of the source. So, it will be sufficient to continue with the equations from systems 2 and 3, i.e. the systems without a source. The loss factors (Equation 7.3) and the coupling loss factors (Equation 7.6) are inserted in Equations 7.16 and 7.17:

$$\eta_{12} E_1 + \eta_{32} E_3 = \eta_2 E_2 + \eta_{21} E_2 + \eta_{23} E_2 = \eta_{2,tot} E_2, \tag{7.18}$$

$$\eta_{23} E_2 = \eta_3 E_3 + \eta_{32} E_3 = \eta_{3,tot} E_3. \tag{7.19}$$

Here, the total loss factors have been introduced in systems 2 and 3. The loss factor of the wall is combined from internal losses in the solid material and boundary losses (both contained in η_2) and radiation losses to either side (these are the coupling loss factors from the wall to each of the rooms, η_{21} and η_{23}). The total loss factor is the one that can be measured experimentally since it is not possible to avoid the radiation losses. The total loss factor of the receiving room combines the absorption of sound from the wall (η_{32}) and the absorption from all other surfaces in the room ($\eta_{3,d}$). Again, the total loss factor is the one that can be measured experimentally.

It is possible to eliminate E_2 and derive the ratio E_1/E_3:

$$\frac{E_1}{E_3} = \frac{\eta_{2,\text{tot}}\eta_{3,\text{tot}}}{\eta_{12}\eta_{23}} - \frac{\eta_{32}}{\eta_{12}} \tag{7.20}$$

The last term can be neglected because it must be close to unity and the total ratio is expected to be $\gg 1$. In the first term, it is an advantage to change the coupling loss factor η_{12} with η_{21}, which is done by applying the reciprocity relation (Equation 7.12):

$$\frac{E_1}{E_3} \cong \frac{\Delta N_1}{\Delta N_2} \cdot \frac{\eta_{2,\text{tot}}\eta_{3,\text{tot}}}{\eta_{21}\eta_{23}}. \tag{7.21}$$

From Equations 7.4 and 7.8 and Table 7.1, we have

$$\eta_{3,\text{tot}} = \frac{cA_3}{4\omega V_3} \text{ and } \eta_{21} = \eta_{23} = \frac{\rho c\sigma_{2,\text{res}}}{\omega m_2} \text{ and } \frac{\Delta N_1}{\Delta f} \cong \frac{4\pi}{c^3}V_1 f^2$$

leading to

$$\frac{E_1}{E_3} \cong \frac{4\pi}{c^3}V_1 f^2 \cdot \frac{\Delta f}{\Delta N_2} \cdot \eta_{2,\text{tot}} \cdot \frac{cA_3}{4\omega V_3} \cdot \left(\frac{\omega m_2}{\rho c\sigma_{2,\text{res}}}\right)^2 \tag{7.22}$$

and the transmission loss for resonant transmission (Equation 7.14):

$$R_r = 10\lg\frac{E_1 V_3 S_2}{E_3 V_1 A_3} = 10\lg\left(\frac{\omega m_2}{2\rho c}\right)^2 - 10\lg\left(\frac{c^2\sigma_{2,\text{res}}^2}{2\eta_{2,\text{tot}}S_2 f} \cdot \frac{\Delta N_2}{\Delta f}\right) \tag{7.23}$$

In this result, the first term is recognised as the mass law at normal incidence R_0. In this form, Equation 7.23 is a generally valid expression for thin or thick walls, having either high or low modal density. The modal density can be inserted according to the type of wall (Table 7.1) (bending waves or shear waves). If a thin wall is assumed (only bending waves and high modal density), the resonant transmission loss is approximately:

$$R_r = R_0 - 10\lg\left(\frac{\pi\sigma_{2,\text{res}}^2 f_{c,2}}{2\eta_{2,\text{tot}}f}\right) \tag{7.24}$$

It should be noted that the radiation efficiency is squared, i.e. it appears twice. If we trace back in the derivation, it is seen that the radiation efficiency comes from the two coupling loss factors η_{21} and η_{23}. So, the first one represents the excitation of the plate from the source room (reciprocity), while the second one represents the radiation into the receiving room.

7.4 FORCED AND RESONANT TRANSMISSION THROUGH A SINGLE WALL

The forced part of the transmission cannot be handled with SEA. However, it will be more thoroughly studied in Chapter 8. The forced transmission can be added as an independent transmission path connecting source room and receiving room (Figure 7.3). This is equivalent of adding a direct field to a diffuse field in room acoustics.

The total transmission through the wall is found by adding the sound powers of forced and resonant transmission:

$$P_{tot} = P_{for} + P_{res} = \rho c S_2 \left\langle \tilde{v}_{2,for}^2 \right\rangle \sigma_{2,for} + \rho c S_2 \left\langle \tilde{v}_{2,res}^2 \right\rangle \sigma_{2,res} \qquad (7.25)$$

Here, the velocity of forced vibrations is $v_{2,for}$ and the radiation efficiency of the forced vibrations is $\sigma_{2,\,for}$ (Equation 6.40). This equation is a good basis for a sound transmission model; forced and resonant transmission is combined, and for both parts, the transmission process is divided into excitation (generating the vibrations in the panel) and radiation.

The transmission loss including both forced and resonant transmission is then:

$$R = 10 \lg \frac{P_{in}}{P_{tot}} = 10 \lg \frac{P_{in}}{P_{for} + P_{res}} = -10 \lg \left(10^{-0.1 R_f} + 10^{-0.1 R_r} \right) \qquad (7.26)$$

A simple approximation for the forced transmission loss R_f was derived in Chapter 5 (see Equation 5.37).

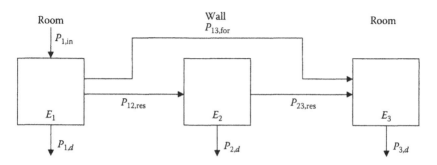

Figure 7.3 Block diagram to show the combination of forced and resonant transmission through a wall.

7.5 ACOUSTIC EXCITATION OF RESONANT VIBRATIONS IN A PLATE

7.5.1 Excitation from one side

The acoustic excitation of a plate by sound incident from one side can be derived from the previous example with one room on either side of the plate or it can be simplified to one room and one plate only. If the room is system 1, the plate is system 2, and if the room is excited by a stationary source, we only need to consider the energy balance equation for system 2:

$$P'_{12} = P_{2,d} + P'_{21} \tag{7.27}$$

In terms of loss factor and coupling loss factors, this leads to:

$$\eta_{12}E_1 = \eta_2 E_2 + \eta_{21}E_2 = \eta_{2,\text{tot}}E_2 \tag{7.28}$$

Here, it should be noted that in this particular case, the loss factor η_2 includes the radiation loss from the back of the plate, since η_{21} represents the radiation loss from the exposed side of the plate back to the source room. Compare Equation 7.28 with the three-system model in the previous section (Equation 7.18). In any case, it is more practical to introduce the total loss factor of the plate. Then, the energy ratio is:

$$\frac{E_2}{E_1} \cong \frac{\Delta N_2}{\Delta N_1} \cdot \frac{\eta_{21}}{\eta_{2,\text{tot}}} \tag{7.29}$$

If the plate is excited from one side as a wall that is part of the surfaces of a room, the coupling loss factor is as in Equation 7.8 except that the numbering is different here. Together with the modal density for room 1, this leads to the result:

$$\frac{\left\langle \tilde{v}_{2,\text{res}}^2 \right\rangle}{\left\langle \tilde{p}_1^2 \right\rangle} \cong \frac{c^2}{\left(\omega m_2 \right)^2} \cdot \frac{\sigma_{2,\text{res}}}{2f \eta_{2,\text{tot}} S_2} \cdot \frac{\Delta N_2}{\Delta f} \tag{7.30}$$

This is a general result for the acoustic excitation of the resonant modes in a plate from one side, as a function of the modal density of the plate. Assuming a thin plate with bending waves, this simplifies to:

$$\frac{\left\langle \tilde{v}_{2,\text{res}}^2 \right\rangle}{\left\langle \tilde{p}_1^2 \right\rangle} \cong \frac{\pi}{\left(\omega m_2 \right)^2} \cdot \frac{f_{c,2}\, \sigma_{2,\text{res}}}{2f \eta_{2,\text{tot}}} \tag{7.31}$$

7.5.2 Freely hung plate in a room

If instead the plate is freely hung in the room and is excited from both sides, the coupling loss factor must include the radiation to both sides, i.e. a factor of 2 must be added and instead of Equation 7.8, we get:

$$\eta_{21} = \frac{2\rho\, c\sigma_{2,\mathrm{res}}}{\omega\, m_2} \tag{7.32}$$

This leads to a factor of 2 in the results corresponding to Equations 7.30 and 7.31. However, this does not mean that the energy of the resonant modes in the plate is twice as high as in the case of excitation from one side. The radiation efficiency is not the same for a freely hung plate; it will be lower than for the plate in a wall because the free edges of the plate allow some degree of acoustic cancellation around the edges. At low frequencies, it will be considerably lower than for a plate surrounded by a baffle, but how much lower shall not be discussed here.

Assuming a freely hung thin plate with bending waves, we get instead of Equation 7.31:

$$\frac{\langle \tilde{v}_{2,\mathrm{res}}^2 \rangle}{\langle \tilde{p}_1^2 \rangle} \cong \frac{\pi}{\left(\omega m_2\right)^2} \cdot \frac{f_{c,2}\sigma_{2,\mathrm{res}}}{f\eta_{2,\mathrm{tot}}} \tag{7.33}$$

7.6 RESONANT TRANSMISSION THROUGH A SLAB WITH EXTENDED AREA

A more general example of sound transmission is shown in Figure 7.4. The source room (1) has the volume V_1 and the floor area S_1, while the receiving

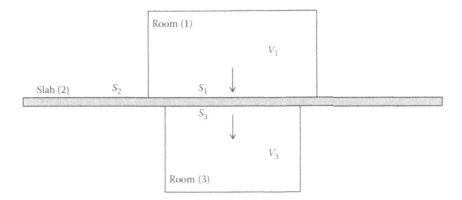

Figure 7.4 Case of a large slab separating a source room and a receiving room. Other rooms are not shown. All walls are assumed to be lightweight and the vibration energy equally distributed in the slab.

room (2) has the volume V_3 and the floor area S_3. The sound is transmitted through a large slab with area S_2 that is greater than both S_1 and S_3. We also assume that the walls are lightweight and, thus, the junction attenuation can be neglected. So, we have three coupled resonant systems as in Figure 7.3, but now the area of the slab is different from the excitation area S_1 and the radiation area S_3.

The steady-state energies in the three systems are:

$$E_1 = \frac{\langle \tilde{p}_1^2 \rangle}{\rho c^2} V_1 \tag{7.34}$$

$$E_2 = m_2 S_2 \langle \tilde{v}_2^2 \rangle \tag{7.35}$$

$$E_3 = \frac{\langle \tilde{p}_3^2 \rangle}{\rho c^2} V_3 \tag{7.36}$$

The coupling loss factor from plate to receiving room yields:

$$\eta_{23} = \frac{\rho c S_3 \sigma}{\omega m_2 S_2} \tag{7.37}$$

Resonant transmission, only, is assumed and σ is the radiation efficiency. The modal densities in source room and slab (with bending waves) are:

$$\frac{\Delta N_1}{\Delta f} \cong \frac{4\pi}{c^3} V_1 f^2 \tag{7.38}$$

$$\frac{\Delta N_2}{\Delta f} \cong \frac{\pi}{c^2} S_2 f_c \tag{7.39}$$

where f_c is the critical frequency of the slab. The reciprocity relation (Equation 7.12) leads to the coupling loss factor from source room to plate:

$$\eta_{12} = \frac{\Delta N_2}{\Delta N_1} \eta_{21} = \frac{\pi c^3 S_2 f_c}{4\pi c^2 V_1 f^2} \cdot \frac{\rho c S_1 \sigma}{\omega m_2 S_2} = \frac{\rho c^2 f_c}{4 V_1 f^2} \cdot \frac{S_1 \sigma}{\omega m_2} \tag{7.40}$$

In the same way as in Equation 7.22, we find the ratio of energy in source and receiver rooms:

$$\frac{E_1}{E_3} \cong \frac{2\eta_{2,\text{tot}} f}{\pi \sigma^2 f_c} \cdot \frac{V_1 A_3 S_2}{V_3 S_1 S_3} \cdot \left(\frac{\omega m_2}{2\rho c} \right)^2 \tag{7.41}$$

The transmission loss for resonant transmission yields:

$$R_r = 10\lg\frac{E_1 V_3 S_1}{E_3 V_1 A_3} = 10\lg\left(\frac{\omega m_2}{2\rho c}\right)^2 - 10\lg\left(\frac{\pi\sigma^2 f_c}{2\eta_{2,\text{tot}}f}\right) + 10\lg\left(\frac{S_2}{S_3}\right)$$

$$R_r = R_0 - 10\lg\left(\frac{\pi\sigma^2 f_c}{2\eta_{2,\text{tot}}f}\right) + 10\lg\left(\frac{S_2}{S_3}\right) \tag{7.42}$$

Comparing with Equation 7.24, we see that the extended area of the slab gives rise to an increased transmission loss. If the area of the slab is 10 times the area of the ceiling radiating into the receiving room, the transmission loss may be up to 10 dB higher than in the normal case where $S_2 = S_3$.

REFERENCES

M.J. Crocker, A.J. Price (1969). Sound transmission using statistical energy analysis. *Journal of Sound and Vibration* 9, 469–486.

R.H. Lyon, G. Maidanik (1962). Power flow between linearly coupled oscillators. *Journal of Acoustical Society of America* 34, 623–639.

Chapter 8

Airborne sound transmission through single constructions

The sound insulation of single constructions is rather complicated. One reason for this is that two transmission mechanisms, the forced and the resonant, are in play at the same time, as mentioned in Section 7.4. The resonant vibrations using statistical energy analysis (SEA) were discussed in Chapter 7. Thus, the forced vibrations remain and will be discussed in Section 8.1. Another reason for sound insulation being rather complicated is the large variation in building constructions covering a range from thin plates of steel, wood or glass to thick structures of brick or concrete. To this picture, orthotropic and sandwich panels with complicated structural waves are added. The aim of this chapter is to describe the physical principles of sound transmission and derive a general formula for the transmission loss that covers the range from thin to thick panels and includes effects due to the finite area of the panels. The influence of the angle of incidence is also analysed with special address to the enigma of grazing incidence; according to classical theory, the transmission loss approaches 0 dB (total transmission), but the incident sound power also approaches zero. Obviously, the classical theory is insufficient in the case of grazing incidence of sound.

8.1 SOUND INSULATION OF HOMOGENEOUS, ISOTROPIC PLATES AT RANDOM INCIDENCE

The physical behaviour of a single homogeneous isotropic plate is now to be studied. The vibrations generated in a plate by excitation of an incident sound field can, as already discussed, be divided into forced and resonant vibrations. In the present chapter, the forced transmission is analysed in some detail and combined with the results from SEA in order to establish more general models for the total transmission.

8.1.1 Total transmitted sound power

In Figure 8.1, the principle of sound transmission through a single panel is shown. The excitation of the panel gives rise to two kinds of vibrations: forced and resonant. So, at the same time, the panel exhibits forced and

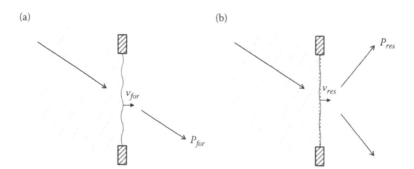

Figure 8.1 Principle of sound transmission through a panel. Excitation and radiation are different for forced transmission (a) and resonant transmission (b).

resonant waves with different velocities of vibration and different radiation efficiencies. When S is the area of the panel, the total transmitted power can be written as:

$$P_{tot} = P_{for} + P_{res} = \rho c S \left\langle \tilde{v}_{for}^2 \right\rangle \sigma_{for} + \rho c S \left\langle \tilde{v}_{res}^2 \right\rangle \sigma_{res} \qquad (8.1)$$

For each transmission part, the physical process can be divided into excitation and radiation. The excitation gives rise to the vibrations of the panel, and the velocity is controlled by the impedance (mass and stiffness) of the panel. The radiation is described by the radiation efficiency, which is different for forced and resonant radiation (see Chapter 6).

8.1.2 The mass law

The mass law expresses the airborne sound insulation of a plate in dependency of its mass per unit area. A very large uniform/homogeneous plate of mass m per unit area is examined as the resonant normal modes are disregarded. The plate is excited by a plane sound wave at normal incidence, so all parts of the plate will move in phase and the bending stiffness can be disregarded. Under these idealised conditions, the transmission loss is R_0, where only m and the frequency f are variable. Using 1 kg/m² and 100 Hz as reference values, we get from Equation 5.27:

$$R_0 = 10 \lg \left[1 + \left(\frac{\omega m}{2 \rho c} \right)^2 \right] \text{(dB)} \qquad (8.2)$$

$$R_0 \approx 20 \lg \left(\frac{m}{1\text{kg/m}^2} \right) + 20 \lg \left(\frac{f}{100\,\text{Hz}} \right) - 2\,\text{dB} \qquad (8.3)$$

This simple relation between transmission loss, mass per unit area and frequency is called the mass law, and is the basis of all the following considerations on airborne sound insulation. The mass law is characterised by a 6 dB increase every time m or f is doubled.

8.1.3 Wall impedance

The resistance of a wall or a plate against a sound wave is called the wall impedance Z_w or the transmission impedance. Z_w is defined as the ratio between pressure difference Δp across the plate and the velocity v of the vibrations of the plate (Equation 5.23):

$$Z_w = \frac{\Delta p}{v} \qquad (8.4)$$

When Δp and v are out of phase, the wall impedance is a complex number, which is normally the case. The wall impedance is determined from the bending wave equation (see Equation 1.24). Assuming a harmonic time dependency (with angular frequency $\omega = 2\pi f$) of the pressure exciting the plate, the bending wave equation is:

$$B\left(\frac{\partial^2}{\partial x^2} + \frac{\partial^2}{\partial y^2}\right)^2 v - m\omega^2 v = j\omega\Delta p \qquad (8.5)$$

where v is the vibration velocity, B is the bending stiffness, m is the mass per unit area and Δp is the applied force per unit area. Wave motion is in both x- and y-directions. Assuming an applied pressure from the surrounding air in the form:

$$p = \hat{p}e^{-jkr\sin\theta} \qquad (8.6)$$

where $k = \omega/c$ and θ is the angle of incidence and r is an arbitrary distance parameter, the bending wave equation will have the form:

$$B\left(\frac{\partial^2}{\partial x^2} + \frac{\partial^2}{\partial y^2}\right)^2 \hat{v}e^{-jkr\sin\theta} - m\omega^2\hat{v}e^{-jkr\sin\theta} = j\omega\Delta\hat{p}e^{-jkr\sin\theta} \qquad (8.7)$$

and the wall impedance can be written as:

$$Z_w = \frac{\Delta\hat{p}}{\hat{v}} = \frac{1}{j\omega}\left(B(k\sin\theta)^4 - m\omega^2\right) = Z_B + Z_m \qquad (8.8)$$

where the mass impedance is:

$$Z_m = j\omega m \qquad (8.9)$$

and the bending wave impedance is:

$$Z_B = \frac{1}{j\omega} B \left(\frac{\omega}{c} \sin\theta \right)^4 = -j\omega m \cdot \frac{c_B^4}{c^4} \cdot \sin^4\theta \qquad (8.10)$$

This would be sufficient for thin panels. However, in order to derive results applicable also to thick panels, it is necessary to consider shear waves in addition to bending waves. The shear wave impedance can be found from a wave equation for shear waves in a way similar to that for bending waves and this leads to:

$$Z_S = \frac{1}{j\omega} Gh \left(\frac{\omega}{c} \sin\theta \right)^2 = -j\omega m \cdot \frac{c_S}{c} \cdot \sin^2\theta \qquad (8.11)$$

where G is the shear modulus, h is the thickness of the panel and c_s is the speed of shear waves in the panel.

Like a simple mechanical system, the wall impedance contains three elements representing mass, stiffness and losses. The wall impedance can be described by an electrical analogy, as shown in Figure 8.2. The mass is represented by an inductor and the losses by a resistance $r \approx \eta\omega m$ (Equation 2.28). The inductor is $L=m$, and the capacitors are:

$$C_1 = \frac{c^4}{\omega^4 B}, \quad C_2 = \frac{c^2}{\omega^2 Gh}$$

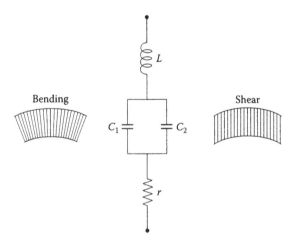

Figure 8.2 Electrical analogy of the wall impedance of a thick panel.

Bending stiffness and shear stiffness are considered to work in parallel, so the weaker one is always dominating. The wall impedance yields:

$$Z_w = j\omega L + \frac{1}{j\omega}\left(C_1 + C_2\right)^{-1} + r = Z_m + \frac{Z_B Z_S}{Z_B + Z_S} + r$$

$$= j\omega m\left(1 - \frac{(c_S/c)^2 (c_B/c)^4}{(c_S/c)^2 + (c_B/c)^4} \cdot \sin^4\theta\right) + \eta\omega m \qquad (8.12)$$

$$= j\omega m\left(1 - \sin^4\theta \cdot \left(\left(\frac{c}{c_S}\right)^2 + \left(\frac{c}{c_B}\right)^4\right)^{-1}\right) + \eta\omega m$$

In case of a material with $c_S \gg c$ like glass or steel, the wall impedance can be approximated by the more simple formula 5.35.

The imaginary part of the wall impedance (Equation 8.12) relative to the mass impedance is displayed as a function of the frequency in Figure 8.3a for grazing incidence and in Figure 8.3b for the angle of incidence 45°.

The wall impedance is zero when coincidence occurs. We notice that coincidence at grazing incidence is not exactly at the critical frequency, but slightly above f_c for $c_S/c = 2$. The minimum of the impedance is shifted to very high frequencies for $c_S/c \rightarrow 1$. If $c_S/c < 1$, which can be the case in sandwich panels, there is no possibility of coincidence, simply because the speed of the transverse waves is subsonic. At the angle of incidence 45°, the coincidence dip is shifted to higher frequencies by one octave for $c_S/c \gg 1$.

The wall impedance (Equation 8.12) at grazing incidence ($\theta = 90°$) can be used to derive the effective speed of free transverse vibrations, as shown by Kurtze and Watters (1959). A plane sound wave with variable speed c is assumed to propagate along the surface of the panel and excite a forced transverse motion. The ratio of the pressure to the transverse velocity is the wall impedance, which is a function of the speed of the exciting wave. The speed of free transverse waves in the panel $c_{B,eff}$ is found by the *minimum energy condition*, i.e. the impedance must be minimum that means $\text{Im}\{Z_w\} = 0$. This leads to an equation in c of the fourth degree:

$$c_S^2 c^4 + c_B^4 c^2 - c_B^4 c_S^2 = 0 \qquad (8.13)$$

The speed $c_{B,eff}$ equals the speed c of the exciting wave in this equation, and according to Rindel (1994), the solution is:

$$c_{B,eff} = \frac{c_B^2}{c_S}\sqrt{-\frac{1}{2} + \frac{1}{2}\sqrt{1 + 4\left(\frac{c_S}{c_B}\right)^4}} \cong \left(\frac{1}{c_B^3} + \frac{1}{c_S^3}\right)^{-1/3} \qquad (8.14)$$

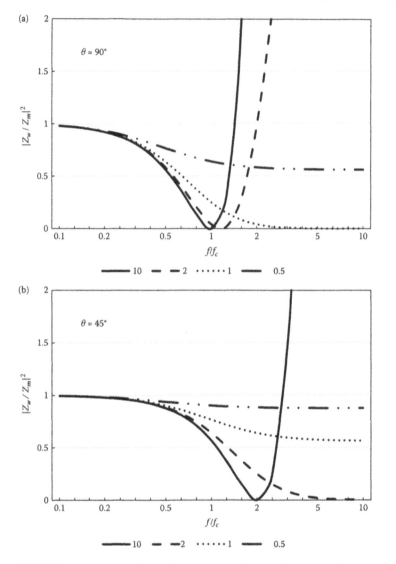

Figure 8.3 Wall impedance relative to the mass impedance as function of frequency for different values of c_s/c. (a) At the angle of incidence 90° and (b) at the angle of incidence 45°. Solid line: $c_s/c=10$, dashed line: $c_s/c=2$, dotted line: $c_s/c=1$, dashed–dotted line: $c_s/c=0.5$.

where the approximation is accurate within ±1% (see also the dispersion curve in Figure 1.4).

At low frequencies, the impedance of the plate is dominated by the resonance at the first normal mode. If the size of the panel is not extremely small, the cross-over frequency from bending waves to shear waves is $f_s \gg f_{11}$ and bending waves can be assumed at the first normal mode. Thus, the natural frequency of the first normal mode f_{11} and the critical frequency f_c are closely related, as seen from Equation 6.7 with $(m, n) = (1, 1)$:

$$f_{11} = \frac{c^2}{4f_c}\left[\left(\frac{1}{l_x}\right)^2 + \left(\frac{1}{l_y}\right)^2\right] \tag{8.15}$$

where l_x and l_y are the plate dimensions. Therefore, a high f_c means a low f_{11} and vice versa. Some examples are given in Table 8.1. It is noted that the thick plates (concrete and brick) have a much lower modal density than thin plates (porous concrete, glass or gypsum), even if the latter have smaller dimensions.

At frequencies below and around the natural frequency of the first structural mode, f_{11}, the stiffness comes not only from bending the plate but also from the connections to the surroundings at the boundaries. As an approximation, we assume the system to behave as a simple mechanical system with the natural frequency f_{11}.

$$Z_w = j\left(\omega m - \frac{k_d}{\omega}\right) + r$$

$$= j\omega m\left(1 - \left(\frac{f_{11}}{f}\right)^2\right) + \eta\omega m \tag{8.16}$$

Combining with the above result (Equation 8.12) and using Equation 5.32 to replace c_B with f_c, we get the wall impedance at the angle of sound incidence θ:

$$Z_w = j\omega m \cdot \left(1 - \left(\frac{f_{11}}{f}\right)^2\right)\left(1 - \sin^4\theta \cdot \left(\left(\frac{c}{c_S}\right)^2 + \left(\frac{f_c}{f}\right)^2\right)^{-1}\right) + \eta\omega m \tag{8.17}$$

In analogy with Equation 6.38, we can take into account the finite size of the panel by replacing the term $\sin^2\theta$ with the radiation efficiency:

$$\sin^2\theta = 1 - \cos^2\theta \rightarrow 1 - \sigma_{for,\theta}^{-2} \tag{8.18}$$

Table 8.1 Calculated critical frequency, first natural frequency and number of modes for examples of thin and thick plates

	Plate Dimensions (m)	f_c (Hz)	f_{11} (Hz)	ΔN in One-Third Octave	
				100 Hz	500 Hz
4 mm glass	1.2×1.3	3250	12	3	15
13 mm gypsum board	2.5×1.2	2615	10	5	24
75 mm porous concrete	2.5×4.0	707	9	4	22
180 mm concrete	2.5×4.0	106	62	0.7	3
250 mm brick	2.5×4.0	72	91	0.4	2

where $\sigma_{\text{for},\theta}$ is the radiation efficiency of forced waves (Equation 6.39). This can be used for relatively high frequencies where $\sigma_{\text{for},\theta} > 1$, but for low frequencies (small ka) coincidence is not possible, and $\sin^2\theta \to 0$.

For a diffuse sound field with random incidence, a good approximation valid for $ka > 3.84 \approx 4$ is obtained by the corresponding value for grazing incidence, $\theta = 90°$ in Equation 6.39:

$$\sin^2\theta \to 1 - \sigma_{\text{for},90}^{-2} \cong 1 - \frac{3.84}{ka} \tag{8.19}$$

For $ka < ca.$ 4, the plate is acoustically small and coincidence is not possible. The reason is that the dimensions of the plate are too small compared to the wavelength of the incident sound. So, instead of sound propagation along the plate, diffraction effects mean that the transmitted sound is refracted (see Figure 6.25; also Figure 6.10 for a graphical presentation of the radiation efficiency as function of ka and the angle of incidence).

8.1.4 The loss factor

There are three different contributions to the energy losses of vibrations in a plate: internal losses, boundary losses and radiation losses.

The internal loss is part of the energy that is transferred to heat and this is quite small for most materials and approximately frequency independent (see examples in Table 1.4). The loss factor due to boundary losses can be calculated as shown by (Craik 1981):

$$\eta_{\text{border}} = \frac{U \lambda_B \alpha}{\pi^2 S} = \frac{U c \alpha}{\pi^2 S} (f \cdot f_c)^{-1/2} \tag{8.20}$$

where α is the average structural absorption coefficient of vibrations along the boundary, which has a total length of U.

Typical values of the structural absorption coefficient are within $0 < \alpha \leq 0.3$, and it depends both on the type of connection to the surrounding structures and on the thickness of the plate relative to the thickness (or mass) of the connecting structures. In Figure 8.4, theoretical values of the structural absorption coefficient for junctions where all plates are made of the same material but with different thickness are shown.

The radiation loss factor was dealt with in Chapter 7 and is closely related to the resonant radiation efficiency (Equation 7.8). Since the resonant modes of the plate radiate to both sides, the radiation loss factor is twice the coupling loss factor from plate to room:

$$\eta_{\text{rad}} = \frac{2\rho c}{\omega m} \cdot \sigma_{\text{res}} \tag{8.21}$$

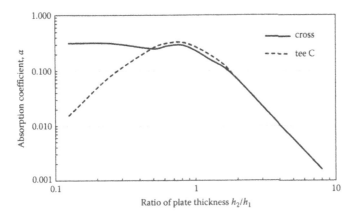

Figure 8.4 Absorption at the boundary where all plates are connected rigidly and made of the same material but with different thickness. The absorption relates to the plate with thickness h_1, and h_2 is the thickness of the crossing plate. (Adapted from Craik, R.J.M., *Appl. Acoust.*, 14, 347–359, 1981.)

Below the critical frequency, the radiation loss factor is small, but it can be quite significant around the critical frequency. The total loss factor is:

$$\eta_{tot} = \eta_{int} + \eta_{border} + \eta_{rad} = \eta_{int} + \frac{Uc\alpha}{\pi^2 S}(f \cdot f_c)^{-1/2} + \frac{2\rho c}{\omega m} \cdot \sigma_{res} \tag{8.22}$$

When measuring the loss factor of a plate, it is the total loss factor that is measured, as it is not possible to eliminate the radiation losses.

Examples of measured loss factors are shown in Figure 8.5. The wall elements were 2.54 m high and 1.80 m wide and were connected in a T-junction to a 180 mm clinker concrete slab. The loss factor generally increases towards the lower frequencies. One of the walls was measured with an elastic connection to the slab, and in this case the loss factor was clearly higher than with the rigid connection, which may be due to viscous losses in the elastic connection.

8.1.5 Forced transmission

Let the sound pressure of the incident, reflected and transmitted sound be denoted as p_i, p_r and p_t, respectively. Then, $\Delta p = p_i + p_r - p_t \cong 2p_i$ and the squared sound pressure in the source room in some distance from the wall is $\tilde{p}_1^2 \cong \tilde{p}_i^2 + \tilde{p}_r^2 \cong 2\tilde{p}_i^2$ because we assume nearly total reflection from the panel, i.e. $p_i \approx p_r$. The relation between velocity of the forced vibrations of the plate v_{for} and the sound pressure p_1 in the source room yields:

$$\langle \tilde{v}_{for}^2 \rangle \cong \frac{2\langle \tilde{p}_1^2 \rangle}{|Z_w|^2} \tag{8.23}$$

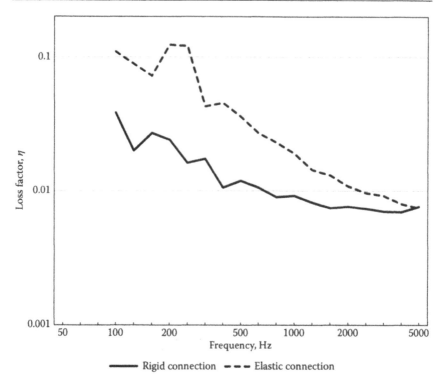

Figure 8.5 Measured loss factor as a function of frequency for two walls, which were the lower part of T-junction to a slab of 180 mm clinker concrete (1700 kg/m³). Full line: 150 mm porous concrete wall (660 kg/m³) with rigid connection and dotted line: 100 mm porous concrete wall (660 kg/m³) with elastic connection. (Adapted from Rindel, J.H., *Appl. Acoust.*, 41, 97–111, 1994.)

Forced transmission means the sound transmission due to transverse waves, which are directly created by the sound pressure of incident sound wave. The radiated (transmitted) sound power is:

$$P_{\text{for}} = \rho c S \langle \tilde{v}^2_{\text{for}} \rangle \sigma_{\text{for}} \tag{8.24}$$

where σ_{for} is the radiation efficiency for forced bending waves (Equation 6.40). From Equation 5.5 and the above Equations 8.23 and 8.24, we get the transmission loss for forced transmission:

$$R_f = 10 \lg \frac{\langle \tilde{p}^2_1 \rangle S}{4 \rho c P_{\text{for}}} = 10 \lg \frac{|Z_w|^2 \cdot \langle \tilde{v}^2_{\text{for}} \rangle}{8 \rho c \cdot P_{\text{for}}}$$

$$= 20 \lg \left| \frac{Z_w}{2 \rho c} \right| - 10 \lg (2 \sigma_{\text{for}}) \, (\text{dB}) \tag{8.25}$$

Considering a diffuse sound field in the source room, the wall impedance (Equation 8.17) is inserted with $\theta = 90°$ because this angle of incidence corresponds to the coincidence dip at random incidence. The forced transmission for random incidence yields:

$$R_f = R_0 + 10\lg\left(\left(1-\left(\frac{f_{11}}{f}\right)^2\right)^2 \cdot \left(1-\left(\left(\frac{c}{c_S}\right)^2 + \left(\frac{f_c}{f}\right)^2\right)^{-1}\right)^2 + \eta_{tot}^2\right)$$

$$-10\lg(2\sigma_{for})\,(dB) \tag{8.26}$$

The loss factor η_{tot} determines how sharp the dip is, either at coincidence or at the first structural mode.

For thin plates $(f_{11} \ll f_c)$ and at frequencies well below the critical frequency, the forced transmission (Equation 8.26) can be approximated by the transmission loss:

$$R_f \cong R_0 - 10\lg(2\sigma_{for}) \cong R_0 - 5 \text{ (dB)} \left(f_{11} < f < \frac{1}{2}f_c\right) \tag{8.27}$$

Here, the last approximation is valid within ±2 dB for surfaces above $10\,\text{m}^2$ and frequencies above 100 Hz. However, for small areas and low frequencies, the area has a significant influence on the transmission loss (Figure 8.6). The reason for this *area effect* of sound transmission is the same as applied to a loudspeaker; the diameter of the loudspeaker membrane determines how well low frequencies can be radiated. The low-frequency limit is determined by the ratio between wavelength and diameter.

At frequencies above the critical frequency, the bending stiffness starts to dominate the wall impedance and the transmission loss increases rapidly with frequency:

$$R_f \cong R_0 + 40\lg\ (f/f_c) - 5 \text{ (dB)}\ (f > f_c) \tag{8.28}$$

However, in this frequency range, the forced transmission has no practical importance because the resonant transmission dominates over the forced transmission above the critical frequency, as shown in the following.

At very low frequencies, $(f < f_{11})$ the transmission loss is controlled by the stiffness of the plate, and the transmission loss increases rapidly towards low frequencies by 12 dB per octave. The approximate result from Equation 8.26 is:

$$R_f \cong R_0 - 40\lg\ (f/f_{11}) - 10\lg(2\sigma_{for})\,(dB)\ (f < f_{11}) \tag{8.29}$$

We shall return to this subject in Section 8.3.

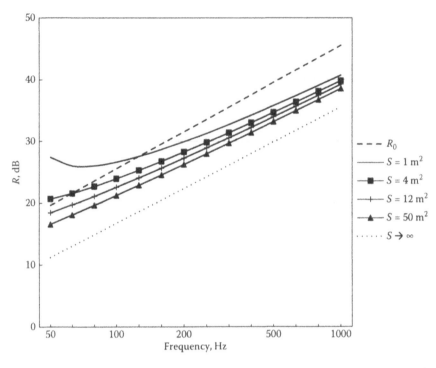

Figure 8.6 Influence of area on the transmission loss of a square panel with $m = 25$ kg/m². (Adapted from Sato, H., *J. Acoust. Soc. Jpn.* 29, 509–516, 1973.)

8.1.6 Resonant transmission

As explained earlier, a plate that is excited by airborne sound will exhibit forced vibrations and resonant vibrations, if the plate has any normal modes in the frequency band of excitation. The resonant transmission cannot be described in the same way as the forced transmission by the wall impedance method, but instead the SEA can be applied (see Chapter 7).

The transmission loss for resonant transmission alone R_r has been found in Equation 7.23:

$$R_r = R_0 - 10 \lg \left(\frac{c^2 \sigma_{res}^2}{2 \eta_{tot} Sf} \cdot \frac{\Delta N}{\Delta f} \right) (\text{dB}) \tag{8.30}$$

Here, the frequency dependency of the radiation efficiency σ_{res} is very important (see Figure 6.18). For many constructions, especially thin plates, the resonant transmission has only minor importance at frequencies below the critical frequency, but it is the dominating transmission above the critical frequency where $\sigma_{res} \approx 1$.

In the frequency range where bending waves can be assumed, the modal density can be approximated by Equation 6.9 and the resonant transmission loss is:

$$R_r \cong R_0 - 10\lg\left(\frac{\pi f_c \sigma_{res}^2}{2\eta_{tot}f}\right)(dB) \quad \left(4f_{11} < f < \frac{1}{2}f_s\right) \tag{8.31}$$

The cross-over frequency from bending to shear waves is f_s (Equation 1.29). At frequencies above $f_s/2$, the wall is too thick for bending waves and the modal density for shear waves (Equation 6.10) must be used instead, leading to:

$$R \cong R_0 - 10\lg\left(\frac{\pi c^2 \sigma_{res}^2}{\eta_{tot}c_s^2}\right)(dB) \quad \left(f > \frac{1}{2}f_s\right) \tag{8.32}$$

A slight change in the slope with frequency can be noted at $f_s/2$. While R_0 has a slope of 6 dB per octave, the resonant transmission increases with 9 dB per octave below $f_s/2$ and continues with 6 dB per octave above $f_s/2$.

As pointed out in Chapter 6, the statistical approximations for modal density in a plate should not be applied at frequencies below $4 f_{11}$.

8.1.7 Total transmission: a calculation model

The transmission loss of a single construction can be described by adding the forced and the resonant transmissions, as in Equation 7.26:

$$R = -10\lg\left(10^{-0.1R_f} + 10^{-0.1R_r}\right) \tag{8.33}$$

Using this with Equation 8.26 for forced transmission and Equation 8.30 for resonant transmission, this makes a general calculation model for the random incidence transmission loss of homogeneous and isotropic constructions.

In addition to the area of plate, angle of incidence, speed of sound and frequency, the main parameters are

R_0: mass law (Equation 8.2)
f_{11}: first natural frequency (Equation 8.15)
f_c: critical frequency (Equation 1.28)
c_S: speed of shear waves (Equation 1.26)

$\dfrac{\Delta N}{\Delta f}$: modal density in the plate (Equation 6.9) bending or (Equation 6.10) shear

η_{tot}: total loss factor (Equation 8.22)

σ_{for}: radiation efficiency of forced vibrations at random incidence
(Equation 6.40)

σ_{res}: radiation efficiency of resonant vibrations at random incidence
(Equation 6.51)

We shall discuss this result for thin and thick plates, which behaves quite differently from an acoustical point of view. An acoustically thin plate has $f_{11} \ll f_c$, whereas an *acoustically thick plate* can be defined as one having $f_c \le f_{22} = 4\,f_{11}$.

The critical frequency plays an important role for the transmission loss of thin plates, and there is usually a local dip in the transmission loss curve at or slightly above the critical frequency. Below the critical frequency, the radiation from resonant vibration is weak and the forced transmission will normally be the dominating part of the transmission. However, above the critical frequency, the resonant vibrations radiate efficiently and dominate over the forced transmission. This behaviour is shown graphically in Figure 8.7. Between f_{22} and ½ f_c, the sound transmission is controlled by the mass per unit area following the mass law, i.e. with a slope of 6 dB per octave. It is noted that the increase of R by 6 dB per doubling of the mass (the mass law) is only valid within a limited frequency range. Increasing the thickness of the plate causes a lower critical frequency and a higher modal resonance frequency, narrowing the mass law region.

Figures 8.8 and 8.9 show two examples of application of the calculation model for an acoustically thin panel, namely, a 10 mm glass pane, but with different dimensions. In the calculations, the internal loss factor is 0.01 and

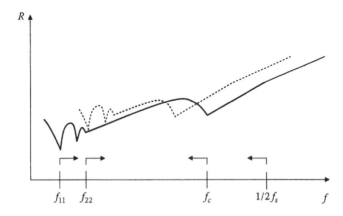

Figure 8.7 Principle of sound transmission loss for an acoustically thin single construction as a function of frequency. The critical frequency is high, $f_c > f_{22}$. Increasing the thickness of the plate causes a shift from the solid curve to the dotted curve and the key frequencies move up and down as indicated by the arrows, narrowing the mass law region.

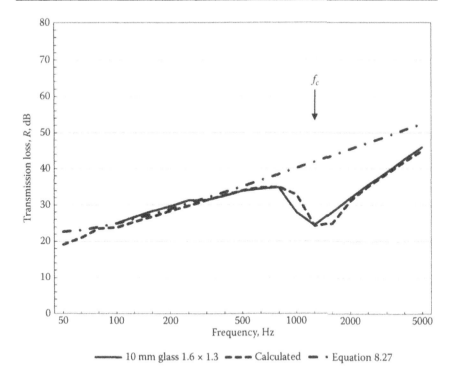

Figure 8.8 Measured and calculated transmission loss of a 10 mm glass pane (25 kg/m²). Dimensions 1.6 m × 1.3 m. The solid line is measured, the dashed line is calculation with the complete model (Equation 8.33) and the dashed-dotted line is calculation with a simplified model (Equation 8.27). (Adapted from Cops, A., Myncke, H., *The Sound Insulation of Glass and Windows – Theory and Practice*. IV Jornadas G.A.L.A., Cordoba, 1971.)

absorption coefficient at the edges is 0.2. The coincidence dip at the critical frequency (ca. 1300 Hz) is clearly seen in the large glass pane (Figure 8.8). Here, the first structural modes are below $f_{22} = 90$ Hz and thus of no importance. Below the critical frequency, the transmission loss is controlled by forced transmission, and the agreement with the simplified equation 8.27 is good.

However, several changes can be observed in the case of a small pane (Figure 8.9). Here, $f_{11} = 90$ Hz and $f_{22} = 360$ Hz and the low frequencies are strongly influenced by the structural modes. The coincidence dip is shallow and shifted to a somewhat higher frequency and the loss factor is higher due to increased influence of boundary losses. The simplified model does not work in this case, but the agreement with the complete model (Equation 8.33) is very good. The area effect shown in Figure 8.6 is not so clear in reality; it is to some extent overruled by dips at the lower structural modes.

An example of a wall with a somewhat lower critical frequency is a 100 mm thick porous concrete wall. The measured transmission loss is

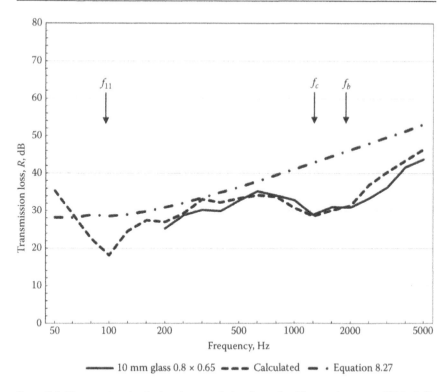

──── 10 mm glass 0.8 × 0.65   ▄ ▄ ▄ Calculated   ▄ • Equation 8.27

Figure 8.9 Measured and calculated transmission loss of a 10 mm glass pane (25 kg/m²). Dimensions 0.8 m×0.65 m. The solid line is measured, the dashed line is calculation with the complete model (Equation 8.33) and the dashed-dotted line is calculation with a simplified model (Equation 8.27). (Adapted from Cops, A., Myncke, H., *The Sound Insulation of Glass and Windows – Theory and Practice*. IV Jornadas G.A.L.A., Cordoba, 1971.)

compared with the calculated one, using Equation 8.33 in Figure 8.10. In the calculations, the internal loss factor is 0.001 and absorption coefficient at the edges is 0.1. The critical frequency is calculated to 374 Hz, but the coincidence dip is located at a somewhat higher frequency, $f_b=508$ Hz (Equation 6.55). The first structural modes are below $f_{22}=67$ Hz and thus of no importance.

8.2 SOUND INSULATION OF ACOUSTICALLY THICK PLATES

We define an acoustically thick plate as one, with a critical frequency f_c lower than the f_{22} structural mode. In this case, the critical frequency has no practical importance and we have a general picture of the transmission loss, which is quite different from the previous one (Figure 8.11). Since the critical frequency is very low, the resonant radiation is efficient ($\sigma_{res}\approx 1$) and

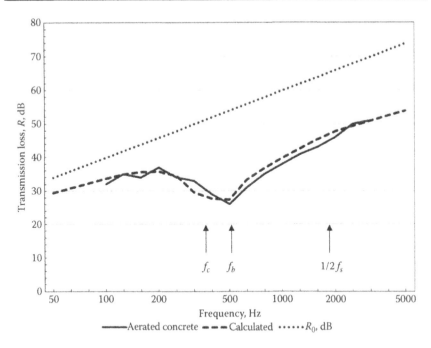

Figure 8.10 Measured and calculated transmission loss of a 100 mm thick wall of porous concrete. Dimensions are 3.7 m×2.69 m. Mass per unit area is 130 kg/m². (Data from Lydteknisk Institut, *Bygningsakustiske laboratoriemålinger*, Building Acoustic Laboratory Measurements, in Danish, Lyngby, Denmark, LL 871/68.)

the resonant transmission will dominate in most of the frequency range. The transmission loss is controlled by the bending stiffness of the plate, gradually taken over by the shear stiffness above the crossover frequency ½ f_s. The transmission loss is well approximated by the simplified equations

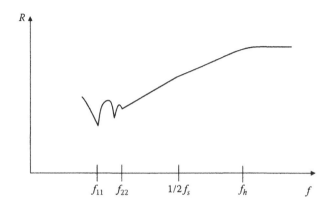

Figure 8.11 Principle of sound transmission loss for an acoustically thick single construction as a function of frequency. The critical frequency is low, $f_c \leq f_{22}$.

8.31 and 8.32 above the frequency $f_{22}=4\ f_{11}$. The slope of the transmission loss curve changes from 9 dB per octave below $\frac{1}{2}\ f_s$ to 6 dB per octave above $\frac{1}{2}\ f_s$ (Figure 8.11). At very high frequencies, the longitudinal waves cause an upper limit of the transmission loss, although this region if very seldom seen in practice. The cross-over frequency is f_h as previously defined (Equation 5.21).

At the lower frequencies, the sound transmission of thick plates is dominated by single structural modes. In the example of a 180 mm concrete wall, the first structural mode is at 106 Hz, and already below 400 Hz the modal density is so low that the equations derived from SEA cannot be used directly. A statistical approach is not valid when the modal density is low i.e. below f_{22}.

However, we may still use the general Equation 8.33 but with great caution. With low modal density, the SEA can be replaced by a *modal energy analysis*. This implies that the first structural modes are calculated one by one, and then the actual mode count ΔN in each frequency band is observed. If one-third octave bands are used, $\Delta f=0.23\ f$ and the actual number of modes ΔN are inserted in Equation 8.30. In some frequency bands $\Delta N=0$, which means that no resonant transmission is present in that particular frequency band; only forced transmission remains. Other bands may contain just one mode, $\Delta N=1$, e.g. the one-third octave band where f_{11} is located. Since the first structural modes are much spread (there are normally only two modes between f_{11} and f_{22} that are two octaves apart), the transmission loss may exhibit great variations and pronounced dips between f_{11} and f_{22}.

The example in Figure 8.12 is a 185 mm hollow concrete slab with critical frequency around 88 Hz and the first natural frequencies are within the important frequency range ($f_{11}\approx67$ Hz and $f_{22}\approx268$ Hz). In the calculations, the internal loss factor is 0.01 and absorption coefficient at the edges is 0.05. Above 250 Hz, the statistical modal density can be used and, thus, the transmission loss can be estimated from Equations 8.31 and 8.32. However, below 250 Hz there are very few resonant modes, so the one-third octave bands 63 Hz, 160 Hz and 200 Hz each contain just one mode ($\Delta N=1$) while the one-third octave bands 80 Hz, 100 Hz and 125 Hz have none ($\Delta N=0$). Insertion in the general Equation 8.33 and using the statistical modal density above 200 Hz leads to the result shown in Figure 8.12. The dip at 160 Hz might look like a coincidence dip, but it is not. It is actually the opposite, namely, the consequence of relatively high sound insulation in the lower frequency range from 80 Hz to 125 Hz. In this frequency range, the slab has no modes and, thus, the resonant transmission is missing and the sound insulation at these frequencies is relatively high. At frequencies above 160 Hz, the slope is close to 6 dB per octave and the transition to shear waves is around 840 Hz (calculated $f_s=1680$ Hz). The critical frequency is also marked in the figure, but it is typical for an acoustically thick construction that there is no dip at the critical frequency.

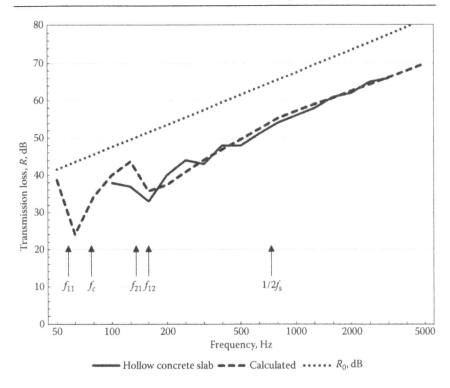

Figure 8.12 Measured and calculated transmission loss for a 185 mm hollow concrete slab (density 1700 kg/m³). Dimensions are 3.37 m×2.99 m. Mass per unit area is 315 kg/m². (Data from Lydteknisk Institut, *Bygningsakustiske laboratoriemålinger*, Building Acoustic Laboratory Measurements, in Danish, Lyngby, Denmark, LL 709/74.)

Another example is a 240 mm clinker concrete wall with transmission loss as shown in Figure 8.13. In the calculations, the internal loss factor is 0.01 and absorption coefficient at the edges is 0.3. Again, it is remarkable that there is no dip at the critical frequency (91 Hz), but some small dips at the low structural modes ($f_{11}=69$ Hz, $f_{21}=140$ Hz, $f_{12}=204$ Hz). The speed of shear waves is assumed to be around 1100 m/s, and this means that the curve at frequencies above ca. 500 Hz is controlled by shear waves, which yields the slope of 6 dB per octave.

8.3 SOUND INSULATION AT VERY LOW FREQUENCIES

At very low frequencies, e.g. under 100 Hz, the sound insulation is a complicated interaction between the structure and the connected rooms. Particularly, the dimensions of the receiving room have been shown to be important (Gibbs and Maluski 2004). A pronounced dip in the transmission loss curve can be expected in the one-third octave band that contains

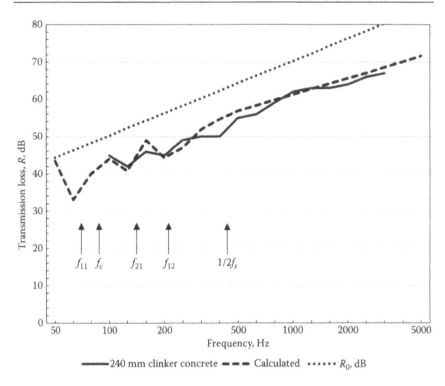

Figure 8.13 Measured and calculated transmission loss for a 240 mm clinker concrete wall (density 1800 kg/m³). Dimensions are 3.70 m×2.69 m. Mass per unit area is 430 kg/m². (Data from Lydteknisk Institut, *Bygningsakustiske laboratoriemålinger*, Building Acoustic Laboratory Measurements, in Danish, Lyngby, Denmark, LL 1156/77.)

the frequency of the first axial room mode perpendicular to the partition wall (or floor). If the room dimension perpendicular to the surface is called L_x, the dip will be at the frequency $c/(2L_x)$. Gibbs and Maluski (2004) have found this dip to be on average

- 5 dB for lightweight constructions
- 10 dB for heavyweight walls/floor greater than 10 m²
- 20 dB for heavyweight walls/floor less than 10 m²

Heavyweight constructions are normally acoustically thick plates ($f_c < f_{22}$) and at frequencies below the critical frequency, the transmission loss is controlled by forced transmission. Thus, the transmission loss below f_{11} can be approximated by the simple Equation 8.29. This means that theoretically the transmission loss below f_{11} increases towards lower frequencies with 12 dB per octave. The plate may behave like a stiff membrane, and the stiffness comes from the fixations at the edges of the plate.

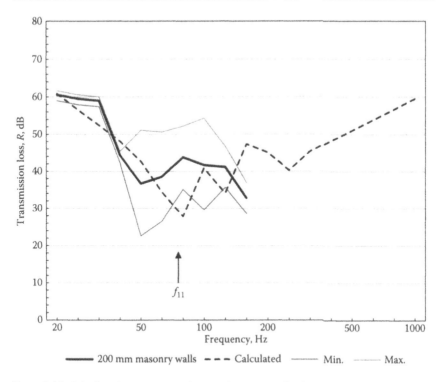

Figure 8.14 Calculated transmission loss and average of a large number of measured sound level difference between equal rooms of various length. Heavyweight walls (200 mm masonry walls, 385 kg/m²). The dashed line is calculation with the complete model (Equation 8.33). (Adapted from Gibbs, B.M., Maluski, S., *Build. Acoust.*, 11, 61–78, 2004.)

The results from the survey conducted by Gibbs and Maluski (2004) included ~100 room pairs and showed a clear difference in the low-frequency level difference for lightweight and heavyweight constructions. In the frequency range from 20 Hz to 50 Hz, the lightweight walls had average sound insulation around 20 dB to 25 dB decreasing towards lower frequencies, whereas the heavyweight walls had average sound insulation around 40 dB to 60 dB increasing towards lower frequencies (Figures 8.14 and 8.15).

The explanation behind this finding must be that the frequency of the first structural mode in lightweight walls is very low ($f_{11}<20\,Hz$), whereas for heavyweight walls, this frequency is relatively high ($f_{11}>50\,Hz$). It is quite remarkable how different the heavyweight walls and the lightweight walls behave at the low frequencies. The measured data are averaged over many different room dimensions and, thus, the above-mentioned dip due to room depth has been averaged out. For the heavyweight walls, the structural modes have a strong influence and cause great variations in the frequency range from 50 Hz to 125 Hz, but below 50 Hz the spread of the measured data is surprisingly small.

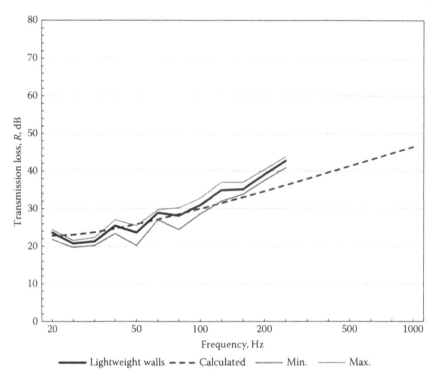

Figure 8.15 Calculated transmission loss and average of a large number of measured sound level difference between equal rooms of various length. Lightweight walls (approximated by 58 kg/m²). The dashed line is calculation with the complete model (Equation 8.33). (Adapted from Gibbs, B.M., Maluski, S., *Build. Acoust.,* 11, 61–78, 2004.)

8.4 SOUND INSULATION OF SUBDIVIDED STRUCTURES

Some structures like drywalls and windows are often subdivided into smaller areas, and this may influence the sound insulation in various ways:

- The frequency of the first structural mode f_{11} is increased when the plate dimensions get smaller (Equation 8.15).
- The modal density is decreased (Equation 6.9).
- The boundary losses are increased (Equation 8.20).
- The resonant radiation efficiency below and around the critical frequency is increased (Equations 6.45, 6.50 and 6.52, 6.53).

This means that the resonant transmission is changed, and most significantly above the critical frequency, where the transmission loss is controlled by the modal density. The dimensions, circumference and area of the subdivided sections should be used in all the above-mentioned equations.

However, the forced transmission is only influenced to a minor extent, since the excitation of the forced vibrations is correlated for the whole area, and the radiation is always from the entire area. So, in the calculations, σ_{for} in Equation 8.26 and S in Equation 8.30 refer to the total area of the construction, not the area of each subdivision.

As a calculated example we consider an 8 mm glass window with dimensions 2 m×2 m. If subdivided into a 2×2 grid or a 4×4 grid by a perfectly rigid structure, the transmission loss is changed as seen in Figure 8.16. The first structural mode is increased from far below 50 Hz (9 Hz) for the large pane to around 146 Hz for the small panes measuring 0.5 m×0.5 m. At frequencies above the critical frequency, the transmission loss is increased due to the decreased modal density in the smaller panes (see Equation 8.30). This high-frequency effect of subdivided surfaces has been found experimentally (e.g. Figure 7.10 in Lewis 1971) but was previously explained by an increase of the boundary loss factor.

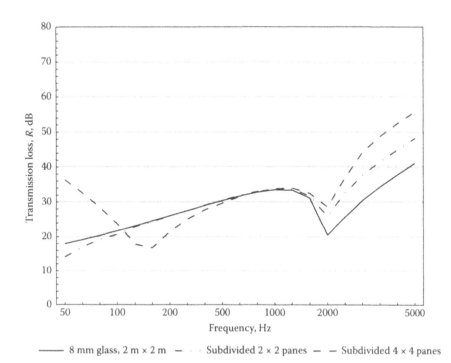

Figure 8.16 Calculated transmission loss of 8 mm glass with dimensions 2 m×2 m, either as a large pane or subdivided into 4 or 16 square panes.

8.5 SOUND INSULATION AT OBLIQUE INCIDENCE

8.5.1 Definition of external transmission loss

In the case of windows and façades in a building, there is no source room, and the incident sound power is not well defined. For instance, the incident power of a plane wave at grazing incidence is normally assumed to be zero, but nevertheless some sound will be transmitted. A solution to this problem is to introduce the so-called *external transmission loss* R_E. The usual definition of transmission loss is based on the transmitted sound power in relation to the incident sound power. However, for this purpose, we define the external transmission loss directly by Equation 5.5:

$$R_E = 10 \lg\left(\frac{\langle p_1^2 \rangle}{\langle p_2^2 \rangle} \cdot \frac{S}{A}\right) = L_1 - L_2 + 10 \lg \frac{S}{A} \quad (\text{dB}) \tag{8.34}$$

Figure 8.17 shows the parameters in Equation 8.34. The angle of incidence does not go into the definition of external transmission loss.

Note that $\langle p_1^2 \rangle$ (and thus L_1) is not really a spatial average since this is the sound pressure in an open space. Instead, L_1 is the sound pressure level close to the wall including the energy reflection but excluding the interference fluctuations.

It is also noted that R_E is in reality the same as the apparent sound reduction index R'_S as defined in ISO 16283-3 when the source is traffic noise [road, rail or air traffic (see Equation 12.10)]. The only difference being that in the ISO standard L_1 is measured as the sound pressure level at the façade −3 dB. The sound pressure level at a reflecting surface is known to be ~3 dB higher than the average sound pressure level a few meters in front of the surface (see Figures 3.7 and 3.8).

An advantage of the definition (Equation 8.34) is that information about the direction of the incident sound is not necessary. It may have any angle of incidence, or it may be a more complicated sound field, like sound from line

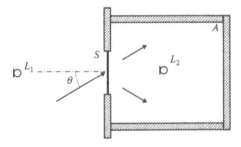

Figure 8.17 External transmission loss of a window with area S defined from the sound pressure levels L_1 and L_2 and the absorption area A in the receiving room. The angle of incidence is θ.

(e.g. a road). In the special case of a diffuse sound field on the source side, the external transmission loss equals the normal transmission loss, $R_E = R$.

8.5.2 Transmission from outside to inside

First, a point source with spherical radiation located outside in the distance r from the window is considered. It is assumed that r is much greater than the dimensions of the sound-transmitting surface. The squared sound pressure outside the window p_1^2 is taken as twice the value in a free field because of the reflection from the façade. In a real situation, this is approximately the sound pressure that can be measured in a distance of $2\,\mathrm{m}$ in front of the façade.

$$p_1^2 = \frac{2P_a}{4\pi r^2}\rho c \tag{8.35}$$

Thus, the external transmission loss can be expressed in terms of the sound power of the outside point source and the sound pressure in the receiving room:

$$R_E = 10\lg\left(\frac{P_a\rho c}{2\pi r^2 \langle p_2^2 \rangle} \cdot \frac{S}{A}\right)(\mathrm{dB}) \tag{8.36}$$

The *reciprocity principle* states that in linear sound fields, the sound pressure in one point due to an omnidirectional point source in another point remains the same if the source and receiver are switched. Figure 8.18 shows the reciprocity principle applied to the case of sound transmission through a window. The room with absorption area A can be either the source or the receiving room. The relation between the sound power of the source P_a and the sound pressure p_2 remains the same when the source and receiver switch positions.

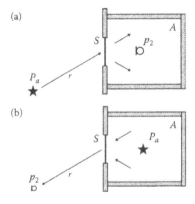

Figure 8.18 Reciprocity principle applied to the sound transmission through a window in a façade. (a) From outside to inside and (b) from inside to outside.

8.5.3 Transmission from inside to outside

With transmission from inside to outside, the source is in the room with the absorption area A. The source still has the sound power P_a and assuming a diffuse sound field, it produces the average sound pressure in the source room:

$$p_1^2 = \frac{4P_a}{A}\rho c \tag{8.37}$$

So, P_a in Equation 8.36 can be replaced by p_1 and the external transmission loss can be written as:

$$R_E = 10\lg\left(\frac{P_a\rho c}{2\pi r^2 \langle p_2^2\rangle}\cdot\frac{S}{A}\right) = 10\lg\left(\frac{\langle p_1^2\rangle}{\langle p_2^2\rangle}\cdot\frac{S}{8\pi r^2}\right)(\mathrm{dB}) \tag{8.38}$$

This gives the following formula for calculation of the sound pressure level L_2 in the outside receiver position when the indoor sound pressure level is L_1:

$$L_2 = L_1 - R_E + 10\lg\frac{S}{8\pi r^2}(\mathrm{dB}) \tag{8.39}$$

8.5.4 External transmission loss for a single construction

For the forced transmission, it is necessary to go back to the wall impedance as a function of the angle of incidence (Equation 8.17) and insert in Equation 8.25. The radiation efficiency of the forced waves depends on the angle of incidence $\sigma_{\mathrm{for},\,\theta}$, and it can be found in Equation 6.39. The forced external transmission loss yields:

$$R_{E,f} = R_0 + 10\lg\left(\left(1-\left(\frac{f_{11}}{f}\right)^2\right)^2\cdot\left(1-\sin^4\theta\cdot\left(\left(\frac{c}{c_S}\right)^2+\left(\frac{f_c}{f}\right)^2\right)^{-1}\right)^2+\eta_{\mathrm{tot}}^2\right)$$

$$-10\lg\left(2\sigma_{\mathrm{for},\theta}\right)(\mathrm{dB}) \tag{8.40}$$

The resonant transmission was derived by SEA in Equation 7.22. The result contains twice the radiation efficiency for the resonant modes, one representing the excitation of the vibrations in plate and one representing the sound radiation to the receiving room. In the case of the external transmission loss, these two radiation efficiencies must be treated separately because the sound field is very different on either side. With transmission from the outside to the inside, the first radiation efficiency is a function of the angle of incidence $\sigma_{res,\theta}$ (Equation 6.57) and the other one represents the radiation into a room $\sigma_{res,d}$ (Equation 6.51). Thus, the resonant external transmission loss at the angle of incidence θ can be calculated from the following formula:

$$R_{E,r} = R_0 - 10 \lg\left(\frac{c^2 \sigma_{res,\theta} \sigma_{res,d}}{2\eta_{tot} Sf} \cdot \frac{\Delta N}{\Delta f}\right) (dB)$$ (8.41)

Finally, the combined, total transmission loss can be calculated using Equation 8.33.

Examples of measured and calculated external transmission losses are shown in Figure 8.19. The measurements were made in an anechoic room on a window with 3 mm glass in scale model 1:4. This corresponds to 12 mm glass in full scale (Rindel 1975). Dimensions of the window correspond to $1.68\,m \times 1.20\,m = 2.0\,m^2$ in full scale. The radiation efficiency $\sigma_{res,\theta}$ has strong fluctuations with frequency: so, for these calculations, the radiation efficiency was averaged over ten frequency steps within each one-third octave. It is noted that the calculation model works quite well, even in the extreme case of grazing incidence (see Figure 8.19d).

8.6 ORTHOTROPIC PLATES

Many building materials are orthotropic, i.e. they have different characteristics in different directions as opposed to isotropic plates. In many wooden plate materials, the bending stiffness is larger in the direction parallel with the veins than perpendicular to the veins. Another example of an orthotropic plate is a corrugated metal plate where a trapeze or waving profile involves a very large difference on the bending stiffness in the two orthogonal directions.

As shown by Heckl (1960), it is possible to insert a geometric average value B_{xy} of the two bending stiffnesses B_x and B_y as a good approximation in many of the results already known from the isotropic plates:

$$B_{xy} = \sqrt{B_x B_y}$$ (8.42)

For sound insulation, the important difference between isotropic and orthotropic plates is that the latter has two different critical frequencies corresponding to sound propagation in the direction of the largest or the smallest bending stiffness, respectively. If the two directions are called x and y, we have as in Equation 1.28:

$$f_{cx} = \frac{c^2}{2\pi}\sqrt{\frac{m}{B_x}} = \frac{K_{cx}}{h}$$ (8.43)

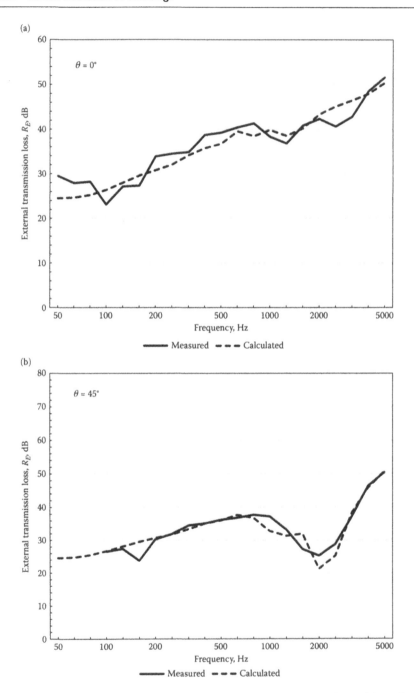

Figure 8.19 Measured and calculated external transmission loss of a scale model window made from 12 mm thick glass. (a) Angle of incidence 0°, (b) angle of incidence 45°

(Continued)

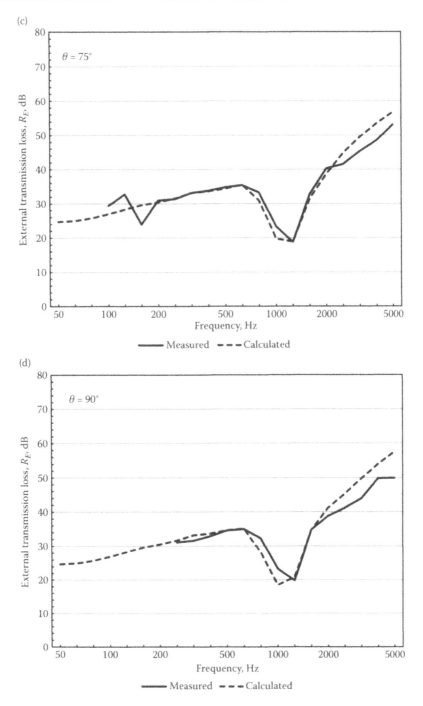

Figure 8.19 (c) angle of incidence 75° and (d) angle of incidence 90°. Dotted line: measured; full line: calculated.

$$f_{cy} = \frac{c^2}{2\pi} \sqrt{\frac{m}{B_y}} = \frac{K_{cy}}{h} \tag{8.44}$$

where f_{cx} is the critical frequency in the direction with the largest bending stiffness, i.e. $f_{cx} < f_{cy}$, h is the plate thickness and K_{cx} and K_{cy} are material constants. As a result of Equation 8.42, the critical frequency can in many relations be replaced by:

$$f_c \to \sqrt{f_{cx} f_{cy}} \tag{8.45}$$

However, this should be done with caution. For the natural frequencies of the structural modes, Equation 6.7 changes to:

$$f_{mn} = \frac{c^2}{4} \left(\frac{m^2}{f_{cx} l_x^2} + \frac{n^2}{f_{cy} l_y^2} \right) \tag{8.46}$$

The modal density is changed from Equation 6.9 to:

$$\frac{\Delta N}{\Delta f} \cong \frac{\pi \cdot S}{c^2} \sqrt{f_{cx} f_{cy}} \tag{8.47}$$

The two different critical frequencies involve the possibility of two dips in the sound insulation and a relatively low sound insulation in the frequency range between the dips. According to Heckl (1960), the radiation efficiency in the frequency range $f_{cx} < f < f_{cy}$ is determined by

$$\sigma_{res} \cong \frac{1}{\pi^2} \sqrt{\frac{f_{cx}}{f_{cy}}} \left(\ln \frac{4f}{f_{cx}} \right)^2 \tag{8.48}$$

Below the lower critical frequency ($f < f_{cx}$), the transmission loss can be calculated from the forced transmission (Equation 8.26) modified by introducing two critical frequencies instead of only one and simplified by neglecting shear waves and the frequencies below f_{11}:

$$R \cong R_f = R_0 + 10 \lg \left(\left| 1 - \left(\frac{f}{f_{cx}} \right)^2 \right| \left| 1 - \left(\frac{f}{f_{cy}} \right)^2 \right| + \eta_{tot}^2 \right) - 10 \lg (2\sigma_{for}) \tag{8.49}$$

Above the lower critical frequency f_{cx}, the resonant transmission will normally be dominant, and the transmission loss is determined by inserting Equations 8.47 and 8.48 in Equation 8.30:

$$R \cong R_r = R_0 - 10 \lg \left[\frac{1}{2\pi^3 \eta_{tot}} \cdot \frac{f_{cx}}{f} \sqrt{\frac{f_{cx}}{f_{cy}}} \left(\ln \frac{4f}{f_{cx}} \right)^4 \right] (f_{cx} < f < f_{cy}) \tag{8.50}$$

In some cases, the forced and the resonant transmission are both important in this frequency range, and Equation 8.33 is used to find the total transmission loss.

In the upper frequency range above the higher critical frequency, Equation 8.30 can be applied with the modal density (Equation 8.47) and simplified with the approximation $\sigma_{res} \approx 1$:

$$R \cong R_r = R_0 - 10\lg \frac{\pi\sqrt{f_{cx}f_{cy}}}{2\eta_{tot}f} \quad (f > f_{cy}) \tag{8.51}$$

For some orthotropic plates, the interval between f_{cx} and f_{cy} is so small that only one coincidence dip appears but the dip may cover a larger frequency range than with isotropic plates. Other orthotropic plates may have a very large interval between the critical frequencies; some corrugated plates can have critical frequencies from below 200 Hz to above 20 kHz.

Figure 8.20 shows an example of measured and calculated transmission loss of a corrugated steel plate. With reference to the sketch in Figure 8.20, the dimensions of the corrugation are $a = 120$ mm, $b = 120$ mm, $c = 45$ mm, $d = 30$ mm and $h = 0.7$ mm. The plate has a mass of 6.3 kg/m² and assumed internal loss factor of 0.001. The critical frequencies used for

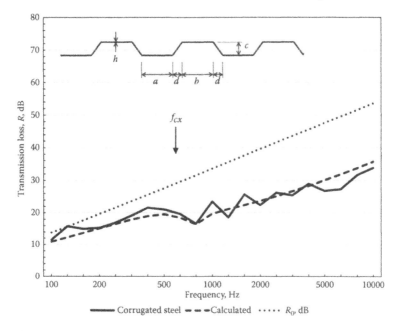

Figure 8.20 Example of measured and calculated transmission loss of a 0.7 mm corrugated steel plate with a profile as shown in the sketch. The critical frequencies are $f_{cx} = 750$ Hz and $f_{cy} = 18,500$ Hz. (Data from Cederfeldt, L. Sound insulation of corrugated plates – a summary of laboratory measurements. Report no. 55, Division of Building Technology, Technical University of Lund, Sweden.)

the calculations are 750 Hz and 18,500 Hz, respectively. Below the lower critical frequency, the transmission is forced (Equation 8.48) and above that frequency, the transmission is resonant (Equation 8.49). It is typical for a corrugated plate that the transmission loss is far below the mass law.

8.7 SANDWICH PLATES

A sandwich construction is a plate composed of two parallel, thin laminas or face plates united with a lightweight core material. Such constructions can combine a high static stiffness with low weight, especially useful in airplanes and ships. The principle is also known from building materials, e.g. the so-called zero-energy elements where the face plates are plywood plates and the core is mineral wool placed edgeways with the fibre at right angles to the plates. Other core materials could be plastic foam or honeycomb core, which resembles honeycombs in a beehive but is made by metal foil, paper or plastic. A very common building material that slightly behaves as a sandwich plate is the gypsum (or plaster) board, covered on each side of a sheet of paper. The paper is very thin and has ignorable bending stiffness, but the in-plane tension stiffness is higher than for the gypsum core, forcing extra shear motion on the gypsum core.

Figure 8.21 shows a sandwich plate and some of the wave types that may appear in the sound transmission through sandwich plates. The dilatation wave (Figure 8.21e) will be considered later, but is disregarded at this point, i.e. the core material is assumed to act as an incompressible material that simply keeps the two face plates at a constant distance from each other. The face plates are assumed to be identical.

8.7.1 Speed of transverse waves

The mass per unit area of the sandwich plate is:

$$m = m_c + 2m_f = \rho_c h_c + 2\rho_f h_f \tag{8.52}$$

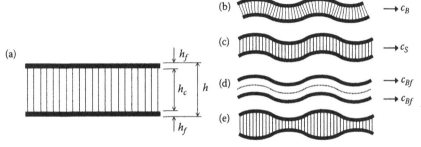

Figure 8.21 (a) Section of sandwich panel, (b) bending wave, (c) shear wave, (d) bending waves in the face plates and (e) dilatation resonance in elastic core.

Here, index c refers to the core material and index f refers to the face plates.

According to Ver and Holmer (1971, p. 313), the bending stiffness of the sandwich plate is:

$$B = \frac{1}{2} E_f h_f \left(h_f + h_c \right)^2 \tag{8.53}$$

The bending wave is shown in Figure 8.21b. This has the speed of propagation c_B determined by Equation 1.25:

$$c_B = \sqrt{2\pi f} \; \sqrt[4]{\frac{B}{m}} \tag{8.54}$$

One peculiarity of sandwich plates is that the bending stiffness is very high even if it is a lightweight construction and, thus, the speed of the bending wave is extremely high. Shear wave motion plays an important role in sandwich plates. In the shear waves, there is no compression or dilatation of the face plates, and the movements in the core material are perpendicular to the direction of propagation (Figure 8.21c). The propagation speed of the shear wave is:

$$c_S = \sqrt{\frac{Gh}{m}} \tag{8.55}$$

where G is the shear modulus, h is the total thickness of the plate and m is the total mass per unit area. The core material is enclosed such that it cannot extend to the sides. The shear modulus is determined by the modulus of elasticity E_c of the core material and the Poisson's ratio μ:

$$G = \frac{E_c}{2(1+\mu)} \tag{8.56}$$

At high frequencies (short wavelengths), the shear wave is influenced by the bending stiffness of the face plates. So, if we divide the sandwich plate into two parts along the dotted line in Figure 8.21d, the shear wave actually becomes a bending wave in each of the face plates. The bending stiffness of the face plate is (Equation 1.22):

$$B_f = \frac{E_f h_f^3}{12(1-\mu^2)} \tag{8.57}$$

The propagation speed c_{Bf} is determined by Equation 1.25, but with half the mass of the core material added to the mass of the face plate:

$$c_{Bf} = \sqrt{2\pi f} \; \sqrt[4]{\frac{B_f}{m_f + m_c/2}} = \sqrt{2\pi f} \; \sqrt[4]{\frac{2B_f}{m}} \tag{8.58}$$

The described waves (Figure 8.21b through d) will not appear separately as pure bending or pure shear waves, but there is a gradual transition as a function of the frequency. An electrical model of the wall impedance of a sandwich panel is seen in Figure 8.22. The bending and shear stiffnesses are in parallel as for a thick panel, but in addition the bending stiffness of the face plates adds to the shear stiffness.

The inductor is $L=m$, and the capacitors are:

$$C_1 = \frac{c^4}{\omega^4 B}, \quad C_2 = \frac{c^2}{\omega^2 Gh}, \quad C_3 = \frac{c^4}{\omega^4 B_f}$$

The wall impedance is then calculated from:

$$Z_w = j\omega L + \frac{1}{j\omega}\left(C_1 + \frac{1}{1/C_2 + 2/C_3}\right)^{-1} + r$$

$$= j\omega m\left(1 - \frac{c_B^4}{c^4} \cdot \frac{c_{Bf}^4 + c^2 c_S^2}{c_B^4 + c_{Bf}^4 + c^2 c_S^2}\right) + \eta\omega m$$

(8.59)

The possibility of compression or dilatation of the core material is not taken into account at this point, but the core is assumed to keep a constant distance between the face plates. The resulting free transverse wave propagates with the effective bending wave speed $c_{B,\mathrm{eff}}$, which can be found as the speed c by setting $\mathrm{Im}\{Z_w\}=0$. Using that $c_{Bf}\ll c_B$, this leads to an equation in c of the sixth degree (Kurtze and Watters 1959):

$$\frac{c_S^2}{c_B^4}c^6 + c^4 - c_S^2 c^2 - c_{Bf}^4 = 0$$

(8.60)

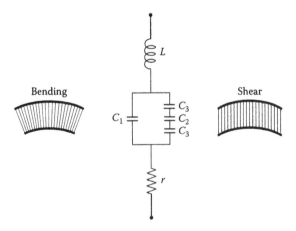

Figure 8.22 Electrical analogy of the wall impedance of a sandwich panel.

The solution of this equation is not simple, but a good approximation can be derived (Rindel 1988). The effective bending wave speed has an S-shaped function of frequency, as shown in Figure 8.23. It is obvious that $c_B \gg c_{Bf}$ and at the crossing point marked 1 in the figure $c_{Bf} \ll c_S$ and thus:

$$c_S = c_B \Rightarrow c_{B,\text{eff}} = c_s \sqrt{\frac{1}{2}\left(\sqrt{5}-1\right)} \cong c_S \cdot 0.7862 \cong c_S \cdot 2^{-1/3}$$

At the crossing point marked 2 in Figure 8.23, $c_B \gg c_s$ and thus:

$$c_S = c_{Bf} \Rightarrow c_{B,\text{eff}} = c_s \sqrt{\frac{1}{2}\left(\sqrt{5}+1\right)} \cong c_S \cdot 1.2720 \cong c_S \cdot 2^{1/3}$$

Based on these observations, the approximate solution to Equation 8.60 becomes:

$$c_{B,\text{eff}} \cong \left(\frac{1}{c_B^3} + \frac{1}{c_S^3 + c_{Bf}^3}\right)^{-1/3} \tag{8.61}$$

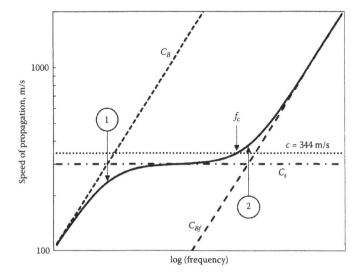

Figure 8.23 Dispersion curve for the effective bending wave speed in a sandwich plate. Here, the shear wave speed is assumed to be less than the speed of sound in the air. The critical frequency is where the effective bending wave speed equals the speed of sound in the air. The crossing points 1 and 2 are explained in the text.

The influence of $c_{B,\text{eff}}$ is discussed on the basis of the dispersion curve (propagation speed of transverse waves as function of frequency, see Figure 8.23). It is noted that $c_{B,\text{eff}}$ increases with the frequency in a relatively complicated way. At low frequencies, $c_{B,\text{eff}} \rightarrow c_B$ and the asymptotic slope is $c_B \sim \sqrt{f}$. In a mid-frequency range, the vibrations have the character of shear waves and $c_{B,\text{eff}}$ only increases slowly with the frequency. At high frequencies, the slope of the curve increases again as $c_{B,\text{eff}} \rightarrow c_{Bf}$ because the bending stiffness in the face plates adds to the total stiffness and becomes determining.

The critical frequency f_c can be found by the definition, i.e. where $c_{B,\text{eff}} = c$ (Figure 8.23). f_c will be somewhere in the interval between the low critical frequency corresponding to the plane bending and the high critical frequency for the face plates alone. The crucial thing for the location of f_c can be seen to be the ratio of the speed of shear waves to the speed of sound in air, i.e. c_S/c. The best construction in terms of sound insulation will be achieved when $c_S < c$ as this gives a high critical frequency and thus the lowest possible contribution from the resonant transmission. The forced transmission will dominate and the transmission loss will follow Equation 8.27. However, a core material with very low shear modulus G will be needed and in practice G will often be so large that $c_S \approx c$ or $c_S > c$. Increasing the shear wave speed to exceed c has the consequence that the critical frequency changes drastically from very high to a low frequency. So, instead of forced transmission, the resonant transmission can be expected to dominate the transmission loss.

8.7.2 Resonant transmission

The dispersion function for the transverse waves also has consequences for the natural frequencies of the structural modes. Assuming simply supported edges, the natural frequencies f_{mn} are determined as in Equation 6.6, but now using the wavelength $\lambda_{B,\text{eff}} = c_{B,\text{eff}}/f$:

$$\left(\frac{2}{\lambda_{B,\text{eff}}}\right)^2 = \left(\frac{2f_{mn}}{c_{B,\text{eff}}}\right)^2 = \left(\frac{m}{l_x}\right)^2 + \left(\frac{n}{l_y}\right)^2 \tag{8.62}$$

However, the solution for the natural frequencies is not simple because of the dispersion function of $c_{B,\text{eff}}$.

Figure 8.24 shows a geometrical representation of Equation 8.62, each dot represents a structural mode and the frequency is represented by the radius $2/\lambda_{B,\text{eff}}$. The statistical number of modes below a certain frequency f is the number of dots within the quarter circle with the radius corresponding to that frequency:

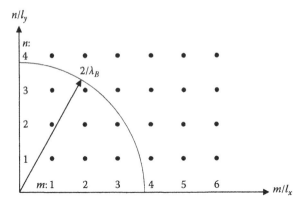

Figure 8.24 Geometrical representation of Equation 8.62, each dot representing a structural mode.

$$N = \frac{1/4\pi \cdot \left(2 / \lambda_{B,\text{eff}}\right)^2}{\left(1/l_x\right)\cdot\left(1/l_y\right)} = \pi l_x l_y \frac{f^2}{c_{B,\text{eff}}^2} \tag{8.63}$$

With the area $S=l_x\cdot l_y$, the modal density is found by differentiation with frequency:

$$\frac{\Delta N}{\Delta f} = 2\pi S \frac{f}{c_{B,\text{eff}}^2}\left[1 - \frac{f}{c_{B,\text{eff}}}\cdot\frac{\mathrm{d}\, c_{B,\text{eff}}}{\mathrm{d}f}\right] \tag{8.64}$$

For pure bending waves, we have $c_B \sim \sqrt{f}$, while the speed is constant for shear waves.

The resonant transmission depends on the modal density and by insertion in Equation 8.30, we get:

$$R_r = R_0 - 10\lg\left(\frac{c^2\sigma_{\text{res}}^2}{2\eta S f}\cdot\frac{\Delta N}{\Delta f}\right)$$

$$= R_0 - 10\lg\left(\frac{\pi\sigma_{\text{res}}^2}{2\eta_{\text{tot}}}\cdot\left(\frac{c}{c_{B,\text{eff}}}\right)^2\right) - \Delta R_c(\text{dB}) \ (f > f_c) \tag{8.65}$$

where the correction term ΔR_c is determined from the slope of the dispersion curve:

$$\Delta R_c = 10\lg\left(2\left[1 - \frac{f}{c_{B,\text{eff}}}\cdot\frac{\mathrm{d}\, c_{B,\text{eff}}}{\mathrm{d}f}\right]\right)(\text{dB}) \tag{8.66}$$

This term takes values between 0 dB and 3 dB, the former referring to pure bending waves and the latter referring to pure shear waves.

The sound radiation from the resonant modes depends on the speed of the transverse wave relative to the speed of sound in air (see Equation 6.49). This leads to the radiation efficiency above the critical frequency:

$$\sigma_{\text{res}} \cong \left(1 - \left(\frac{c}{c_{B,\text{eff}}}\right)^2\right)^{-1/2} \quad (f > f_c) \tag{8.67}$$

This is close to unity, $\sigma_{\text{res}} \approx 1$ for $c_{B,\text{eff}} \gg c$. Insertion in Equation 8.65 gives the following result for the transmission loss above the critical frequency:

$$R_r \cong R_0 + 10\lg\frac{2\eta_{\text{tot}}}{\pi} + 10\lg\left(\left(\frac{c_{B,\text{eff}}}{c}\right)^2 - 1\right) - \Delta R_c(\text{dB}) \ (f > f_c) \tag{8.68}$$

8.7.3 Forced transmission

The transmission loss for forced transmission through a sandwich panel can be calculated as for a homogeneous panel (Equation 8.25). The coincidence dip around the critical frequency can be quite pronounced if $c_S \gg c$. However, some sandwich constructions have $c_S < c$, and this means that the coincidence dip is more shallow or disappears completely.

Inserting the wall impedance Equation 8.59 in Equation 8.25 yields the transmission loss for forced transmission:

$$R_f = R_0 + 10\lg\left(\left(1 - \left(\frac{f_{11}}{f}\right)^2\right)^2 \cdot \left(1 - \frac{c_B^4}{c^4} \cdot \frac{c_{Bf}^4 + c^2 c_S^2}{c_B^4 + c_{Bf}^4 + c^2 c_S^2}\right)^2 + \eta_{\text{tot}}^2\right)$$

$$-10\lg(2\sigma_{\text{for}})(\text{dB}) \tag{8.69}$$

An example of measured and calculated transmission loss for a sandwich panel with $c_S/c > 1$ is shown in Figure 8.25. With a stiff core material and a low critical frequency ($f_c = 170\,\text{Hz}$), the sound transmission is resonant in most of the frequency range, and in general the sound insulation is low. The calculated forced and resonant transmission losses are shown separately.

8.7.4 Dilatational resonance

So far, the core material has been assumed to be incompressible. However, for porous core materials like foam and mineral wool, the core is compressible and dilatational resonances may occur (Figure 8.21e). The first

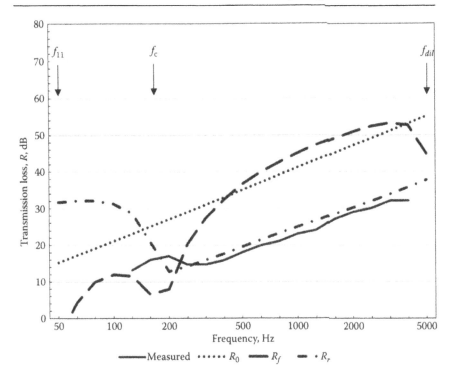

Figure 8.25 Example of measured and calculated transmission loss of a sandwich panel made from two sheets of 6.4 mm plywood and a 76 mm core of honeycomb plastic. Panel dimensions 3.0 m×2.4 m, $G=95$ MN/m^2, $c_s/c=2.18$, $f_c=170$ Hz and $f_{11}=50$ Hz. The dilatational resonance is above 5000 Hz. (Measured data from Jones, R.E., *Control Eng.*, 16, 90–105, 1981.)

dilatational resonance is a mass-spring-mass resonance in which the faces act as masses connected by a spring. Higher-order dilatational resonances are also possible, in which a standing wave in the core connects the masses of the face plates. The first dilatational resonance frequency can be calculated by Ver and Holmer (1971, p. 314):

$$f_{\text{dil}} = \frac{1}{2\pi} \sqrt{\frac{4E_{ct}}{h_c\left(2m_f + m_c\,/\,3\right)}} \tag{8.70}$$

where E_{ct} is the compression modulus of the core material given by:

$$E_{ct} = \frac{E_c}{3\left(1-2\mu\right)} \tag{8.71}$$

E_c is Young's modulus of the core material and μ is the Poisson ratio. It is seen that μ has a major influence on the dilatation resonance frequency, which increases towards infinity if μ approaches 0.5. Rubber is an example of a material with such a high Poisson ratio and rubber makes a very hard and stiff spring if the material cannot expand sidewise. Normally, f_{dil} cannot be made sufficiently low to achieve a positive effect of compressible core material. Therefore, this is in contrast to double constructions, which can often have a low resonance frequency and very high sound insulation above the resonance frequency.

Since the dilatational resonance behaves very much like the mass-spring-mass resonance of a double wall, the transmission loss for forced transmission (Equation 8.69) can be extended with an additional term like (Equation 5.52):

$$R_{f,\text{dil}} = R_f + 10\lg\left(\left(1 - \left(\frac{f}{f_{\text{dil}}}\right)^2\right)^2 + \eta_{\text{tot}}^2\right)(\text{dB}) \qquad (8.72)$$

In summary, sandwich panels can be divided into two types. One type has a core material that provides a shear wave speed less than the speed of sound in air. Then, the critical frequency is high, and the transmission loss is controlled by the forced transmission and the result can be close to the mass law. However, the dilatational resonance may spoil the sound insulation. The other type has a core with a high shear modulus and the shear wave speed exceeds the speed of sound in air. The critical frequency is low and the transmission loss is controlled by resonant transmission. The mass law makes an upper limit for the obtainable sound insulation of sandwich plates and even small alterations in the core material may lead to major deteriorations of the sound insulation.

Another example of the transmission loss for a sandwich plate with a relatively stiff but compressible core of mineral wool is given in Figure 8.26. Here, $c_S/c < 1$ meaning that the free transverse vibrations propagate at subsonic speed and, thus, the resonant transmission is negligible. In such constructions, one or more dilatation resonances may occur as the masses of the face plates oscillate with the core material as a spring (Figure 8.21e). The fundamental dilatational resonance is clearly seen in the example in Figure 8.26, butthere is also a dip at a higher-order dilatational resonance. Although not included in the current calculation model, this second dip can be associated with the thickness of the core h_c being equal to half a wavelength of longitudinal waves in the core material, the

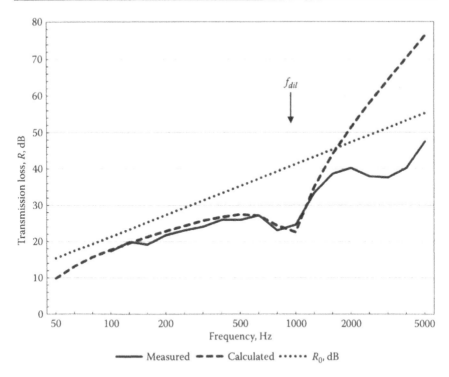

Figure 8.26 Example of measured and calculated transmission loss of a sandwich panel made from two sheets of 0.6 mm steel and a 50 mm core of mineral wool with fibres perpendicular to the surface. Panel dimensions 3.7 m × 2.69 m, $G=1.67$ MN/m², $c_s/c=0.22$, $f_c=20{,}500$ Hz and $f_{\parallel}=34$ Hz. The dilatational resonance is around 950 Hz.

speed of longitudinal waves being calculated from Equation 1.20 for an extended solid medium. This example has a very high critical frequency, and the forced transmission is dominating over the resonant transmission. The sound insulation is better than in the previous example, but it is obvious that it is difficult to obtain good sound insulation with a sandwich panel.

Examples of data for materials applicable as core material in sandwich panels are collected in Table 8.2. Attempts to optimise the sound insulation of sandwich panels should include a core material that can give a low speed of shear waves and a high critical frequency, like this example, but with a Young's modulus high enough to avoid problems with the dilatational resonance.

Table 8.2 Examples of typical data for core materials in sandwich panel

Material	Density, ρ_m (kg/m^3)	Young Modulus, E (10^6 N/m^2)	Shear Modulus, G (10^6 N/m^2)
Urethane foam	33–72	7–19	2.1–3.7
Polystyrene foam	33–42	9–26	8.5
Honeycomb, aluminium	45–123	482–368	–
Honeycomb, plastic	64–149	248–1360	–
Plastic foam	30	6.2	2.1
Glass wool, fibres ≠ surface	120	0.34–0.48	–
Glass wool, fibres ⊥ surface	120	<6.5	9
Glass wool, fibres ⊥ surface	80	<1.9	3.6
Glass wool, fibres ⊥ surface	45	<1.2	1.3

Source: Bodlund, K. *Luftlydisolering. En sammanstälning av tillämplig teori*, (Airborne sound insulation. A collection of applicable theory, in Swedish). Rapport R60:1980, Statens råd för Byggnadsforskning, Stockholm, 1980.

REFERENCES

K. Bodlund (1980). *Luftlydisolering. En sammanstälning av tillämplig*, (Airborne sound insulation. A collection of applicable theory, in Swedish). Rapport R60:1980, Statens råd för Byggnadsforskning, Stockholm.

L. Cederfeldt (1974). *Sound insulation of corrugated plates – A summary of laboratory measurements*. Report No. 55, Division of Building Technology, Technical University of Lund, Sweden.

R.J.M. Craik (1981). Damping of building structures. *Applied Acoustics* 14, 347–359.

A. Cops and H. Myncke (1971). *The Sound Insulation of Glass and Windows – Theory and Practice*. IV Jornadas G.A.L.A., Cordoba, September 1975.

B.M. Gibbs and S. Maluski (2004). Airborne sound level difference between dwellings at low frequencies. *Building Acoustics* 11, 61–78.

M. Heckl (1960). Untersuchungen an orthotropen Platten. *Acustica* 10, 109–115.

ISO 16283-3 (2016). *Acoustics – Field measurements of sound insulation in buildings and of building elements – Part 3: Façade sound insulation*. International Organization for Standardization, Geneva, Switzerland.

R.E. Jones (1981). Field sound insulation of load-bearing sandwich panels for housing. *Noise Control Engineering* 16, 90–105.

G. Kurtze and B.G. Watters (1959). New wall design for high transmission loss or high damping. *Journal of Acoustical Society of America* 31, 739–748.

P.T. Lewis (1971). Real windows. Chapter seven in *Building Acoustics* (Eds. T. Smith, P.E. O'Sullival, B. Oakes, R.B. Conn). British Acoustical Society Special, vol. 2, Oriel Press, London.

Lydteknisk Institut. *Bygningsakustiske laboratoriemålinger*, Building Acoustic Laboratory Measurements, in Danish, Lyngby, Denmark.

J.H. Rindel (1975). *Transmission of traffic noise through windows. Influence of incident angle on sound insulation in theory and experiment.* Report No. 9, The Acoustics Laboratory, Technical University of Denmark, Lyngby.

J.H. Rindel (1988). Sound transmission through sandwich panels. *Proceedings of NAM-88*, Tampere, Finland, 55–58.

J.H. Rindel (1994). Dispersion and absorption of structure-borne sound in acoustically thick plates. *Applied Acoustics* 41, 97–111.

H. Sato (1973). On the mechanism of outdoor noise transmission through walls and windows. *Journal of the Acoustical Society of Japan* 29, 509–516. (In Japanese).

I.L. Ver and C.I. Holmer (1971). Interaction of sound waves with solid structures. Chapter 11 in *Noise and Vibration Control* (Ed. L.L. Beranek). McGraw-Hill, New York.

Chapter 9

Airborne sound transmission through double constructions

The theoretical basis of sound insulation of double constructions is described in this chapter. However, the theoretical model is combined with semi-empirical models, which allows relatively simple mathematical formulae. This may be a basis for prediction of the sound insulation of a number of building components, e.g. double-glazed windows and double wall constructions with or without mechanical connections. The difficulties in calculating the sound insulation of double constructions are especially related to mechanical connections including boundary connections.

9.1 TRANSMISSION VIA THE CAVITY

A double construction can consist of two parallel plates separated by a cavity d, which contains an elastic material with the density ρ_d, and the speed of sound c_d. The cavity will often contain air (with or without mineral wool), but special window constructions may use other gases in the cavity. As shown in Figure 9.1, the sound transmission will be partly through the cavity and partly via possible mechanical connections, the so-called sound bridges. The latter will be dealt with in the following paragraphs; here only transmission via the cavity is dealt with. The two plates are characterized by their mass per unit area m_1 and m_2, their critical frequencies f_{c1} and f_{c2} and the transmission losses R_1 and R_2 for each of the two plates.

9.1.1 Normal incidence of sound

If the boundary conditions of a plate are neglected, a plane sound wave at normal incidence will not create bending waves in the plate, but the plate will move with all parts in phase. From the equations of movement (Newton's second law) and from the conditions of continuity, the result in Equation 5.43 was derived. Assuming the gas in the cavity to be different from the atmospheric air outside, and wall impedances to be determined from the mass per unit area, the following expression for the transmission loss can be obtained:

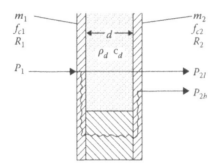

Figure 9.1 Sound transmission through a double construction can be divided into one contribution via the cavity and another contribution via connections (sound bridges). (Reproduced from Kristensen, J., Rindel, J.H., *Bygningsakustik – Teori og praksis.* (Building acoustics – Theory and practice, in Danish). SBI-Anvisning 166. Danish Building Research Institute, Hørsholm, Denmark, 1989. With permission.)

$$R = 10\lg\left(\left[\cos(kd) - n\left(K_1 + K_2\right)\sin(kd)\right]^2 \right.$$

$$\left. + \left[\frac{1}{2}(n+1/n)\sin(kd) + \left(K_1 + K_2\right)\cos(kd) - 2nK_1K_2\sin(kd)\right]^2 + \eta_0^2 \right) \tag{9.1}$$

where

$k = 2\pi f/c_d$ is the angular wave number

d is the depth of the cavity

η_0 is the loss factor, which is only significant at the resonance

$n = \dfrac{\rho c}{\rho_d c_d}$ is the ratio of impedance of the air outside the wall and the air in the cavity

$K_1 = \dfrac{\omega m_1}{2\rho c}$ is the normalized wall impedance for plate 1

$K_2 = \dfrac{\omega m_2}{2\rho c}$ is the normalized wall impedance for plate 2

Normally, there is no need to include all terms in Equation 9.1. For most double constructions, only two terms (and the loss factor) are of importance:

$$R \cong 10\lg\left(\left(\left(K_1 + K_2\right)\cos(kd) - 2nK_1K_2\sin(kd)\right)^2 + \eta_0^2\right) \tag{9.2}$$

It is seen from this that R will have a minimum if the two important terms become numerically equal, which gives the following equation:

$$\tan(kd) = \tan\frac{\omega d}{c_d} = \frac{m_1 + m_2}{m_1 m_2} \cdot \frac{\rho_d c_d}{\omega} \tag{9.3}$$

9.1.2 The resonance frequency

At frequencies where the wavelength is large compared to the depth of the cavity, i.e. $kd \ll 1$, the tangent function equals the argument, and from Equation 9.3 we find the resonance frequency:

$$f_0 = \frac{1}{2\pi}\sqrt{\frac{\rho_d c_d^2}{d}\left(\frac{1}{m_1} + \frac{1}{m_2}\right)} \tag{9.4}$$

A mass-spring-mass resonance is caused by the spring effect of the gas in the cavity. At the resonance frequency, a change occurs in the character of the sound transmission compared to the single wall, as it is at frequencies above the resonance frequency that double constructions are also acoustically double.

Figure 9.2 shows how to determine the resonance frequency for a symmetric double construction with air in the cavity.

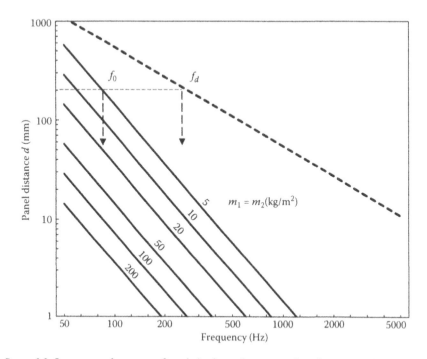

Figure 9.2 Resonance frequency f_0 and the knee frequency f_d as functions of the cavity depth for a symmetrical double construction with air in the cavity. The knee frequency involves only the cavity depth, whereas the resonance frequency also depends on the mass per unit area of the two plates. The indicated example is for $d=200$ mm and $m_1=m_2=5$ kg/m², which yields $f_0=85$ Hz and $f_d=275$ Hz.

9.1.3 Standing waves in the cavity

At higher frequencies where the wavelength is not large compared to the depth of the cavity, the spring effect of the air is no longer active. This happens at a knee frequency f_d corresponding to $k_d = 1$ or $d \approx \lambda/6$:

$$f_d = \frac{c_d}{2\pi d} \tag{9.5}$$

Above this frequency, there are natural frequencies that are solutions to Equation 9.3 as $\tan(kd) \approx 0$. These frequencies represent standing waves in the cavity orthogonally on the two plates, and in principle this gives a number of dips in the frequency curve for the transmission loss. In practise, however, these dips cannot be seen as the sound incidence not only occurs orthogonally but also at diffuse sound incidence. For $f > f_d$, the maximum transmission loss is achieved by the insertion of $\cos(kd) \approx 0$ and $\sin(kd) \approx 1$ when calculating the transmission loss. These are good approximations when the cavity is attenuated with absorbing material.

9.1.4 Calculation for random incidence

The previously given expressions for the transmission loss of double constructions Equations 9.1 and 9.2 give useful results by inserting the transmission losses R_1 and R_2 of the two plates; or even better the measured transmission losses at random incidence. In this case, the equivalent wall impedances for the two plates can be taken as:

$$K_1 = 10^{R_1/20} \quad \text{and} \quad K_2 = 10^{R_2/20} \tag{9.6}$$

For estimated calculations, it can be fair to simplify Equation 9.2 depending on the different frequency ranges. This can be done by means of Equations 9.5 and 9.6 and using that $\sin(kd) \approx kd$ for $f < f_d$ and $\sin(kd) \approx 1$ for $f > f_d$.

$$R \cong \begin{cases} R_{(1+2)} = 20\lg(K_1 + K_2) & (f < f_0) \\ R_1 + R_2 + 20\lg\left(2n\dfrac{f}{f_d}\right) & (f_0 < f \le f_d) \\ R_1 + R_2 + 20\lg(2n) & (f > f_d) \end{cases} \tag{9.7}$$

The corresponding simplified frequency course is seen in Figure 9.3. In the case of thin plates and at frequencies below the critical frequency ($f < f_c$), a simplified mass law may be used for an estimate of R_1, R_2 and R_{1+2}. For the latter, the simplified estimate comes from Equations 8.3 and 8.27:

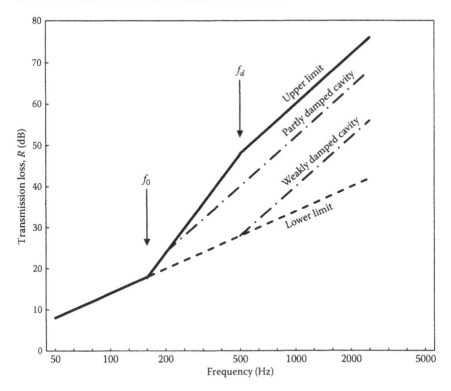

Figure 9.3 Simplified frequency course of the transmission loss of double constructions with more or less sound absorption in the cavity, but without rigid connections. The three main sections correspond to Equation 9.7. The upper limit for the sound insulation that can be achieved in practice is indicated. The lower limit is also indicated. The two dashed-dotted lines show different examples of a partly attenuated cavity.

$$R_{(1+2)} \cong 20\lg\left(\frac{m_1 + m_2}{1 \text{ kg m}^{-2}}\right) + 20\lg\left(\frac{f}{100 \text{ Hz}}\right) - 7 \text{ dB} \tag{9.8}$$

However, if R_1 and R_2 are known from measurements or a more reliable prediction than the simple mass law, the application of the equivalent wall impedances (Equation 9.6) should be preferred, instead of Equation 9.8.

It is implied that there are some absorbing material in the cavity and, thus, standing waves are efficiently attenuated. As sound bridges are not included at this stage, this is the maximum achievable sound insulation for an ideal double construction.

It should be emphasized that normally $n=1$, even if the cavity contains mineral wool, as long as this does not contribute to the spring stiffness of the cavity. The only case where $n \neq 1$ is when the cavity is filled with a special gas as sometimes used in windows for improved thermal insulation. If the cavity contains atmospheric air, $n=1$, and we get:

$$R \cong \begin{cases} R_1 + R_2 + 20\lg\left(\dfrac{f}{f_d}\right) + 6\text{ dB} & (f_0 < f < f_d) \\[2ex] R_1 + R_2 + 6\text{ dB} & (f \geq f_d) \end{cases} \tag{9.9}$$

9.1.5 Weakly attenuated cavity

Double constructions, in which the cavity is not efficiently attenuated with sound absorbing material, will give a sound insulation between the upper and lower bounds, as shown in Figure 9.3. With weak attenuation, a higher sound pressure is built up in the cavity leading to an increased sound transmission. In window constructions, a sound absorbing treatment of the window frame – a so-called *frame-absorber* along the perimeter of the cavity – is sometimes used. The transmission loss for such a construction can be estimated by:

$$R \cong R_1 + R_2 + 10\lg\left(\alpha\,\frac{d \cdot U}{S}\right) \tag{9.10}$$

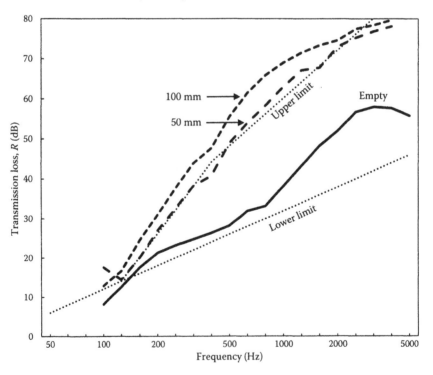

Figure 9.4 Measured transmission losses of a double wall with 6.4 mm and 3.2 mm hardboard plates and a cavity of 160 mm without sound bridges. Curve 1: No absorption in the cavity, curves 2 and 3: attenuated with 50 mm glass wool and 100 mm glass wool, respectively. Upper and lower limits are indicated. (Adapted from Sharp, B.H., *Noise Control Eng.*, 11, 53–63, 1978.)

where

    α is the sound absorption coefficient of the boundary in the cavity
    d is the depth of the cavity
    U is the perimeter of the boundary
    S is the area of the plates

In a cavity wall without frame-absorber, the absorption coefficient α will be very small. However, if the depth of the cavity is less than approximately 100 mm, some absorption will occur because of the friction in the viscous boundary layer along the internal surface of the plates. From empirically derived values (Brekke, 1980), the following equivalent absorption coefficient of an empty cavity can be used:

$$\alpha \cong \begin{cases} 0.5 & \text{for} \quad d \le 20 \text{ mm} \\[2mm] \dfrac{10 \text{ mm}}{d} & \text{for} \quad d > 20 \text{ mm} \end{cases} \tag{9.11}$$

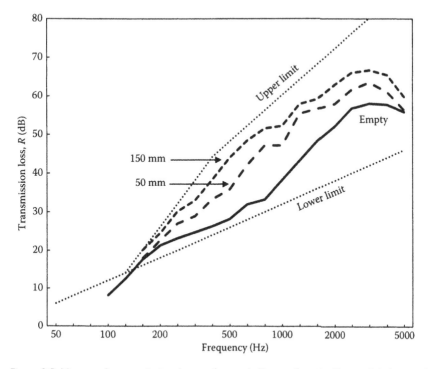

Figure 9.5 Measured transmission losses for a similar wall as in Figure 9.4, but with frame-absorber of glass wool (48 kg/m³). The thickness of the frame-absorber is either 150 mm or 50 mm. (Adapted from Sharp, B.H., *Noise Control Eng.*, 11, 53–63, 1978.)

The use of Equation 9.10 is restricted to the range between lower and upper limit as indicated in Figure 9.3. In Figure 9.4, some examples of measured transmission losses can be seen with and without sound absorbing material in the cavity. Corresponding results with a frame-absorber can be seen in Figure 9.5.

9.2 TRANSMISSION VIA SOUND BRIDGES

A sound bridge means a mechanical connection by which vibration energy is transferred between two separated plates in a double construction. As illustrated in Figure 9.1, partly a sound power P_{2b} is radiated due to the transmission via sound bridges and partly a sound power P_{2l} is radiated due to the transmission via the cavity. The sum of the radiated sound powers goes into the transmission loss of the construction.

$$R = 10\lg\left(\frac{P_1}{P_{2\ell} + P_{2b}}\right) = -10\lg\left[10^{-R_\ell/10} + 10^{-R_b/10}\right] \tag{9.12}$$

The transmission loss R_l refers to the corresponding construction without sound bridges and R_b is the transmission loss when only the transmission via sound bridges is considered. The latter can be expressed by the transmission loss $R_{(1+2)}$ (Equation 9.7) for a single construction with the plates 1 and 2 close together plus a correction term ΔR_m that depends on the type of the sound bridge.

$$R_b = R_{(1+2)} + \Delta R_m \tag{9.13}$$

In the following, it is shown how ΔR_m can be determined for different types of sound bridges. For the time being, they are presumed completely stiff and, thus, the velocities are locally the same on each side of the sound bridge, i.e. $v_{1,0} = v_{2,0}$ (Figure 9.6). It is further presumed that the mass of the sound bridges is insignificant compared to the total mass of the construction.

As shown by Sharp (1978), the relationship between the average velocity in plate 1 and the local velocity at the sound bridge is determined by the impedance of the two plates (point or line impedances) Z_1 and Z_2:

$$\frac{\langle \tilde{v}_1^2 \rangle}{\tilde{v}_{2,0}^2} = \left|\frac{Z_1 + Z_2}{Z_1}\right|^2 \tag{9.14}$$

This expresses that there is a reduced movement at a sound bridge because of the mutual effect of the plates.

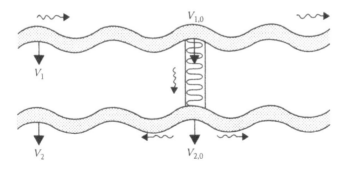

Figure 9.6 Transmission of bending waves from plate I to plate 2 through a sound bridge. (Reproduced from Kristensen, J., Rindel, J.H., *Bygningsakustik – Teori og praksis*. (Building acoustics – Theory and practice, in Danish). SBI-Anvisning 166. Danish Building Research Institute, Hørsholm, Denmark, 1989. With permission.)

If the average velocity in plate 1 is compared to that in the equivalent single construction (where plates 1 and 2 are assumed close together), a fair approximation should be:

$$\frac{\langle v_{1+2}^2 \rangle}{\langle \tilde{v}_1^2 \rangle} = \left(\frac{m_1}{m_1 + m_2} \right)^2 \tag{9.15}$$

As the radiation efficiency for the equivalent single construction is presumed to be $\sigma_{1+2} \approx 1$, the following expression for ΔR_m is obtained:

$$\begin{aligned}
\Delta R_m &= 10\lg \frac{\langle \tilde{v}_{1+2}^2 \rangle \rho c S}{P_{2b}} \\
&= 10\lg \left(\frac{\tilde{v}_{2,0}^2 \rho c S}{P_{2b}} \right) + 20\lg \left(\frac{m_1}{(m_1 + m_2)} \cdot \left| \frac{Z_1 + Z_2}{Z_1} \right| \right)
\end{aligned} \tag{9.16}$$

Further development of this expression depends on the type of the sound bridge (see the following).

9.2.1 Point connections

The point excitation of a plate results in both a near-field radiation and a resonant radiation, which is caused by the natural vibrations in plate 2. For simplicity, the resonant radiation is disregarded here, but it will be dealt with later. The sound power radiated from a number of equal excitation points N_p is (Cremer and Heckl, 1967, p. 471):

$$P_{2b} \cong N_p P_{2n} = \rho c v_{2,0}^2 \frac{8c^2}{\pi^3 f_{c2}^2} N_p \tag{9.17}$$

The *point impedance* for a plate is (Equation 2.13)

$$Z_o = \frac{F}{v_o} = 8\sqrt{mB} = \frac{4c^2 m}{\pi f_c}$$

(9.18)

Applying this for plates 1 and 2 (and inserted in Equation 9.16) yields:

$$\Delta R_{m,p} \cong 10 \lg\left(\frac{S\pi^3 f_{c2}^2}{N_p 8c^2}\right) + 20 \lg\left(\frac{m_1(m_1/f_{c1} + m_2/f_{c2})}{(m_1 + m_2)(m_1/f_{c1})}\right)$$

(9.19)

Here, the last term can be combined with f_{c2} from the first term to yield an auxiliary quantity f_{cp}, which is a combination of the critical frequencies of the two plates applying particularly for point connections:

$$f_{cp} = \frac{m_1 f_{c2} + m_2 f_{c1}}{m_1 + m_2}$$

(9.20)

and thus:

$$\Delta R_{m,p} \cong 10 \lg\left(\frac{S\pi^3 f_{cp}^2}{N_p 8c^2}\right)$$

(9.21)

The principal course of the transmission loss as a function of frequency is seen in Figure 9.7 for double constructions with sound bridges.

A high value of ΔR_m can be obtained with a high critical frequency and with the lowest possible number of connections per unit area. If the sound bridges are placed in a quadratic pattern with a centre-to-centre distance $e = \sqrt{S/N_p}$, Equation 9.21 can be further simplified:

$$\Delta R_{m,p} \cong 20 \lg(e \cdot f_{cp}) - 45 \text{ dB}$$

(9.22)

From Equation 9.20, it is seen that for asymmetric double constructions, the critical frequency of the lighter of the two plates is decisive for the size of f_{cp}. The direction of the transmission is of no importance. The simple formula Equation 9.22 is illustrated in Figure 9.8. The advantage of choosing a plate with a high critical frequency is obvious, if it is otherwise acceptable to use plates with a low bending stiffness.

It is characteristic that ΔR_m is not dependent on the frequency and the transmission loss will, thus, go parallel with the transmission loss for the corresponding single construction, i.e. typically with a slope of 6 dB/octave. $R_{(1+2)}$ is the lower bound of the transmission loss; thus:

$$\Delta R_{m,p} \geq 0 \text{ dB}$$

(9.23)

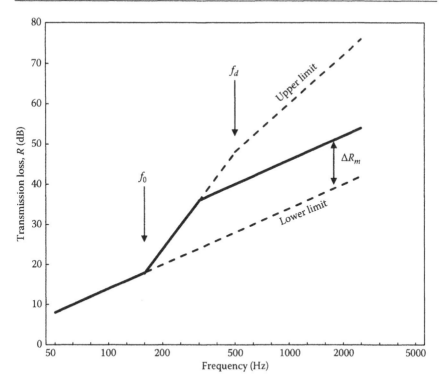

Figure 9.7 Transmission loss for a double construction with rigid sound bridges.

In the above expression for point connections, it is an important assumption that the frequency is below the critical frequency for the radiating plate. Otherwise, the resonant radiation will result in $\Delta R_{m,p} \approx 0\,\mathrm{dB}$.

9.2.2 Line connections

The sound transmission via line connections can be discussed analogously to point connections. Instead of the point impedance, the line impedance is used for each of the plates where the expression for a beam with excitation in a point can be found (Cremer and Heckl, 1967, p. 273):

$$Z_\ell = 2(1+j)mcL_\ell\sqrt{\frac{f}{f_c}} \tag{9.24}$$

where L_l is the length of the line connection. As shown in the work of Cremer and Heckl (1967, p. 481), the radiated sound power is:

$$P_{2b} \cong P_{2n} = \rho c\tilde{v}_{2,0}^2\,\frac{2c}{\pi f_{c2}}\,L_\ell \tag{9.25}$$

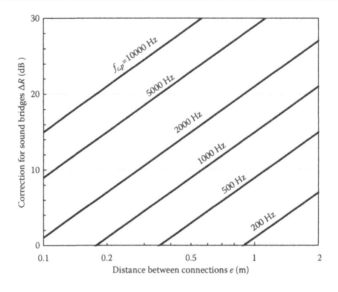

Figure 9.8 Influence of rigid point connections on the transmission loss. The connections are assumed to have a square pattern with the centre distance *e* between the points. The equivalent critical frequency f_{cp} can normally be approximated by the critical frequency of the thinner of the plates in the double construction.

disregarding the contribution from the resonant radiation. By inserting in Equation 9.16 and by means of the auxiliary quantity

$$f_{c\ell} = \left[\frac{m_1\sqrt{f_{c2}} + m_2\sqrt{f_{c1}}}{m_1 + m_2} \right]^2 \tag{9.26}$$

the following result applies for the sound transmission through a stiff line connection:

$$\Delta R_{m,\ell} \cong 10\lg\left(\frac{S}{L_t} \frac{\pi f_{c\ell}}{2c} \right) \tag{9.27}$$

The equivalent critical frequency $f_{c\ell}$ can normally be approximated by the critical frequency of the thinner of the plates in the double construction. If the sound bridges are located as parallel laths with a centre-to-centre distance $b = S/L_l$, the simple formula is:

$$\Delta R_{m,\ell} \cong 10\lg\left(b f_{c\ell} \right) - 23 \text{ dB} \tag{9.28}$$

When comparing this with the result for point connections (Equation 9.22), a slightly lower dependency of the critical frequency (3 dB/octave instead of 6 dB/octave) can be observed. Thus, it is more difficult to achieve high values of ΔR_m with line connections than it is with point connections, as is evident from Figures 9.8 and 9.9.

9.2.3 Boundary connections

A special variant of linear sound bridges are connections along the boundary of a double construction. Compared to the line connections, the line impedance for the boundaries is smaller by a factor of 4 (Cremer and Heckl, 1967, p. 273).

$$Z_r = \frac{1}{2}(1+j)mcL_r\sqrt{\frac{f}{f_c}} \tag{9.29}$$

where L_r is the length of the boundary connection. From a simplified consideration, the radiated sound power becomes one-half of that from a line excitation at the middle of the plate. As in the previously described cases, the resonant sound radiation is disregarded so far. The sound power radiated from the boundary excitation is:

$$P_{2b} \cong P_{2n} = \rho c \tilde{v}_{2,0}^2 \frac{c}{\pi f_{c2}} L_r \tag{9.30}$$

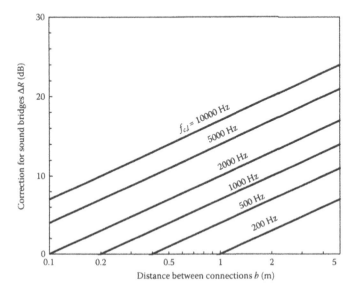

Figure 9.9 Influence of rigid line connections on the transmission loss. The connections are assumed to have the centre distance b between the lines. The equivalent critical frequency f_{cl} can normally be approximated by the critical frequency of the thinner of the plates in the double construction.

For a stiff boundary connection:

$$\Delta R_{m,r} \cong 10 \lg\left(\frac{S}{L_r}\frac{\pi f_{c\ell}}{c}\right) \tag{9.31}$$

where $f_{c\ell}$ is the previously introduced auxiliary quantity (Equation 9.26). By comparison with Equation 9.27 we see that the result for $\Delta R_{m,r}$ differs by 3 dB from an ordinary line connection of the same length. The ΔRm,r correction is displayed in Figure 9.10 for the case of boundary connection all around the circumference.

9.2.4 Resonant sound radiation

The power that is provided to plate 2 partly causes a near-field radiation from the affected area with the power P_{2n}, and partly causes the build-up of normal modes in the plate, and thus to a resonant sound radiation with the power P_{2r}. As the velocity in the excitation point is $v_{2,0}$, and the velocity of the normal mode vibrations is v_2, an energy balance equation can be set up for plate 2. The input power is determined from the impedance by

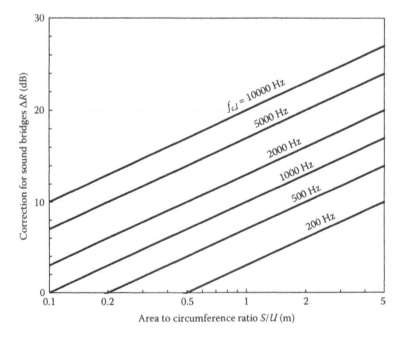

Figure 9.10 Influence of rigid boundary connections on the transmission loss. The abscissa is the ratio between the plate area S and the total length of the boundary connection U. The equivalent critical frequency $f_{c\ell}$ can normally be approximated by the critical frequency of the thinner of the plates in the double construction.

Equation 2.57 and the dissipated power is expressed through the loss factor by Equation 7.3:

$$\tilde{v}_{2,0}^2 \text{Re}\{Z_2\} = 2\pi f m_2 S \langle \tilde{v}_2^2 \rangle \eta_2 \tag{9.32}$$

where η_2 is the total loss factor for plate 2. The resonant radiated sound power is:

$$P_{2r} = \langle \tilde{v}_{2,0}^2 \rangle \rho c S \sigma_2 = \frac{\tilde{v}_{2,0}^2 \text{Re}\{Z_2\} \rho c \sigma_2}{2\pi f m_2 \eta_2} \tag{9.33}$$

The near-field radiation depends on the type of excitation; please refer to the overview given in Table 9.1. For a point connection, the near-field radiation (cf. Equation 9.17) is:

$$P_{2n} = \rho c \tilde{v}_{2,0}^2 \frac{8c^2}{\pi^3 f_{c2}^2} \tag{9.34}$$

The totally radiated sound power for a single point connection is:

$$P_{2b} = P_{2n} + P_{2r} = P_{2n}\kappa_p \tag{9.35}$$

where κ_p is inserted as a factor for the resonant radiation. For point connections, this becomes:

$$\kappa_p = 1 + \frac{\pi \sigma_2 f_{c2}}{4\eta_2 f} \tag{9.36}$$

Table 9.1 Some quantities related to the excitation of a plate by point, line or boundary connections

	Point	Line	Boundary
Type of excitation			
Impedance Z_0	$\dfrac{4c^2 m}{\pi f_c}$	$2(1+j)mcL_\ell \sqrt{\dfrac{f}{f_c}}$	$\frac{1}{2}(1+j)mcL_r \sqrt{\dfrac{f}{f_c}}$
Admittance Y_0	$\dfrac{\pi f_c}{4c^2 m}$	$\dfrac{1-j}{4mcL_\ell}\sqrt{\dfrac{f_c}{f}}$	$\dfrac{1-j}{mcL_r}\sqrt{\dfrac{f_c}{f}}$
Near-field radiation P_{2n}	$\rho c v_0^2 \dfrac{8c^2}{\pi^3 f_c^2} \cdot N_p$	$\rho c v_0^2 \dfrac{2c}{\pi f_c} \cdot L_\ell$	$\rho c v_0^2 \dfrac{c}{\pi f_c} \cdot L_r$
Resonant radiation factor κ	$1 + \dfrac{\pi \sigma f_c}{4\eta f}$	$1 + \dfrac{\sigma}{2\eta}\sqrt{\dfrac{f_c}{f}}$	$1 + \dfrac{\sigma}{4\eta}\sqrt{\dfrac{f_c}{f}}$

Similarly, for line and boundary connections:

$$\kappa_\ell = 1 + \frac{\sigma_2}{2\eta_2}\sqrt{\frac{f_{c2}}{f}} \tag{9.37}$$

$$\kappa_r = 1 + \frac{\sigma_2}{4\eta_2}\sqrt{\frac{f_{c2}}{f}} \tag{9.38}$$

The significance of the resonant sound radiation will be that ΔR_m is reduced when the frequency approaches f_{c2} concurrently with an increasing value of the radiation efficiency. For $f > f_{cr}$ the radiation efficiency is $\sigma_2 \approx 1$ and $\kappa \gg 1$, which means that ΔR_m becomes infinitesimal.

9.2.5 Elastic connections

Two plates connected with a sound bridge as shown in Figure 9.6 are considered. The mechanical connection can be of point or line shape, and it is presumed to be elastic. The sound bridge is characterized by the mechanical impedance Z_b:

$$Z_b = \frac{F}{v_{1,0} - v_{2,0}} \cong \frac{k_d}{j\omega} \tag{9.39}$$

Here, F is the force whereby the sound bridge affects the plates, $v_{1,0}$ and $v_{2,0}$ are the velocities of the two plates at the point of the sound bridge. Z_b will primarily depend on the dynamic stiffness k_d of the sound bridge, whereas the internal losses can often be disregarded. However, the latter does not apply to sound bridges of viscous elastic materials.

A bending wave propagating in plate 1 with a velocity amplitude v_1 will get a slightly reduced velocity $v_{1,0}$ opposite the sound bridge:

$$v_{1,0} = v_1 - Y_1 F \tag{9.40}$$

where Y_1 is the mechanical admittance of the plate, i.e. the reciprocal of the impedance Z_1.

In plate 2, the velocity opposite the sound bridge will be determined by the admittance Y_2 of the plate.

$$v_{2,0} = Y_2 F \tag{9.41}$$

A new bending wave will consequently be generated in plate 2.

As shown by Ver (1971), the power P_{12} transferred from plate 1 to plate 2 can be determined:

$$P_{12} = \langle v_1^2 \rangle \frac{|Z_b|^2 Y_2}{1 + 2\,\mathrm{Re}\{Z_b\}(Y_1 + Y_2) + |Z_b|^2 (Y_1 + Y_2)^2} \tag{9.42}$$

For a completely stiff sound bridge, the result is:

$$P_{12} = \langle v_1^2 \rangle \frac{Y_2}{(Y_1 + Y_2)^2} \quad \text{for} \quad |Z_b| \to \infty \tag{9.43}$$

The effect of an elastic sound bridge will, thus, be described by a *coupling factor* γ, which is the ratio between the transferred power via an elastic sound bridge and a totally stiff sound bridge, respectively.

$$\gamma = \frac{|Z_b|^2 (Y_1 + Y_2)^2}{1 + 2\operatorname{Re}\{Z_b\}(Y_1 + Y_2) + |Z_b|^2 (Y_1 + Y_2)^2} \tag{9.44}$$

$$\gamma \cong \left(1 + \left(\frac{2\pi f}{k_d (Y_1 + Y_2)}\right)^2\right)^{-1} \tag{9.45}$$

In the latter formula, the internal losses of the sound bridge are disregarded, as Equation 9.39 has been inserted.

9.2.6 Coupling factor

In the following, it is shown how the coupling factor γ can be determined for three types of elastic sound bridges, point connections, line connections and boundary connections. Indices p, l and r are used, respectively.

The calculations are based on Equation 9.45, and the dynamic stiffness of the sound bridge is considered, but the internal losses are not. By inserting the admittances as given in Table 9.1, it is found that the coupling factors can be written as:

$$\gamma_p = \left(1 + \left(f/f_{kp}\right)^2\right)^{-1} \tag{9.46}$$

$$\gamma_\ell = \left(1 + \left(f/f_{k\ell}\right)^3\right)^{-1} \tag{9.47}$$

$$\gamma_r = \left(1 + \left(f/f_{kr}\right)^3\right)^{-1} \tag{9.48}$$

Here, some knee frequencies are inserted, below which the sound bridges behave as stiff connections:

$$f_{kp} = k_{dp} \frac{f_{cp}}{8c^2 m_m} \tag{9.49}$$

$$f_{k\ell} = \frac{1}{2}\left(k'_{d\ell}\frac{\sqrt{f_{c\ell}}}{2\pi cm_m}\right)^{2/3} \tag{9.50}$$

$$f_{kr} = \frac{1}{2}\left(k'_{dr}\frac{2\sqrt{f_{c\ell}}}{\pi cm_m}\right)^{2/3} \tag{9.51}$$

The dynamic stiffness of the point connection is k_{dp} and k'_{dl} and k'_{dr} are the dynamic stiffness's per unit length for line connection and boundary connection, respectively. In these expressions, some auxiliary terms are used, f_{cp} (Equation 9.20) and f_{cl} (Equation 9.26) and an equivalent mass per unit area m_m, which is determined by the mass per unit area of each of the two plates:

$$m_m = \frac{m_1 m_2}{m_1 + m_2} \tag{9.52}$$

9.2.7 Summary on sound bridges

The coupling factors and factors for resonant sound radiation that has been introduced above should be used with the expressions for radiated sound power from the near field for rigid sound bridges. Contributions from different types of sound bridges can be combined in a single expression that includes point connections, line connections and boundary connections:

$$\Delta R_m = -10\lg\left[\frac{8c^2}{S\pi^3 f_{cp}^2}N_p\gamma_p\kappa_p + \frac{2c}{S\pi f_{c\ell}}\left(L_\ell\gamma_\ell\kappa_\ell + \tfrac{1}{2}L_r\gamma_r\kappa_r\right)\right] \tag{9.53}$$

When this quantity has been determined, the transmission loss is found from Equations 9.13 and 9.12. Examples of measured transmission loss for light double walls with different types of sound bridges are shown in Figure 9.11.

Heavy double constructions are especially sensitive to sound bridges, as the critical frequency typically is rather low, which leads to a significant resonant sound radiation. Figures 9.12 and 9.13 show examples of the importance of sound bridges due to boundary coupling and point connections for double walls of clinker concrete.

Finally, Figure 9.14 shows an example of the use of secondary wall, i.e. a double construction made by a massive wall and a light plate, which reduces the sound radiation because of a high critical frequency. It is seen that there is a good improvement of sound insulation even with rigid connections, but the result is much better without sound bridges.

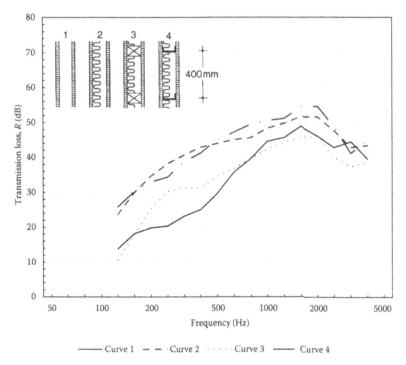

Figure 9.11 Measured transmission loss for double walls of 13 mm gypsum board with 106 mm cavity depth. Wall 1: No attenuation in the cavity, no sound bridges; wall 2: 50 mm of mineral wool in the cavity, no sound bridges; wall 3: rigid line connections of wooden laths, $b=400$ mm; wall 4: elastic line connections of steel laths, $b=400$ mm. (Adapted from Northwood, T.D., Transmission loss of plasterboard walls. Build. Res. Note No. 66. National Research Council of Canada, Ottawa, ON, 1970.)

9.3 SOUND INSULATION OF DRYWALLS

9.3.1 Influence of stud distance

In drywalls, e.g. made with gypsum boards, the stud distance can have a major influence on the low-frequency sound insulation (Rindel and Hoffmeyer, 1991). The studs have the function of stiffening the wall and, thus, the plate cladding exhibits normal modes in the subareas between the studs. Examples of measurement results from a laboratory test are shown in Figure 9.15. The test opening was 2700 mm high and 3700 mm wide. Thus, the standard stud distance of 600 mm could not be used throughout. In the case where the first and the last sections were 350 mm, the measured transmission loss was remarkably low in the range from 125 Hz to 160 Hz. If instead the first stud distance is either 100 mm or 700 mm, the dip at low frequencies is avoided and the overall result is much better. A difference of 18 dB is found at 160 Hz.

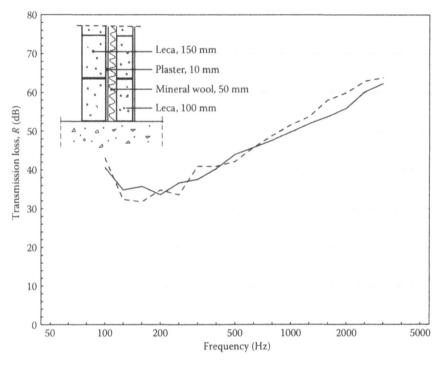

Figure 9.12 Measured transmission loss for a single wall and a double wall with a rigid boundary connection. Solid line: Single 150 mm clinker concrete wall with plaster; dashed line: double wall, 150 mm and 100 mm clinker concrete on common concrete foundation. (Adapted from Vigran, T.E., *Lydisolasjon i bygninger.* (Sound insulation in buildings, in Norwegian). Tapir, Trondheim, Norway, 1979.)

The results can be explained by the normal modes of the gypsum board in the sections between the studs. The boundary condition also plays a role here, and it may be something between simply supported and clamped depending on the construction details. The natural frequency of the fundamental mode $(m, n) = (1, 1)$ can be estimated from Equation 6.7 with an extra boundary condition factor γ:

$$f_{11} = \gamma \frac{c^2}{4f_c}\left(\left(\frac{1}{l_x}\right)^2 + \left(\frac{1}{l_y}\right)^2\right)$$

(9.53)

Here, c is the speed of sound in air, f_c is the critical frequency and the dimensions of the plate section are l_x and l_y. The factor γ is unity for supported plates but for a clamped boundary, it is a complicated function of dimensions and mode number, as shown by Timmel (1991). Typical values

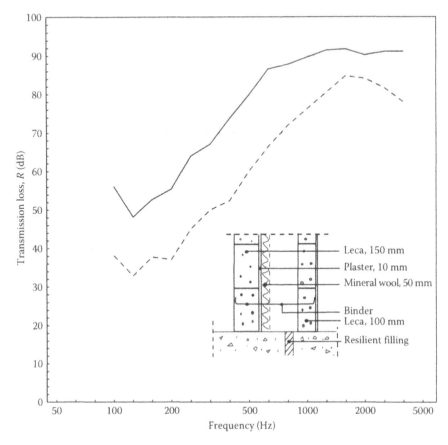

Figure 9.13 Measured transmission loss for a double wall of 150 mm and 100 mm clinker concrete. Solid line: No boundary connections, R_w = 74 dB, dashed line: With point connections, steel wire ties, e = 500 mm, R_w = 56 dB. (Adapted from Vigran, T.E., Lydisolasjon i bygninger. (Sound insulation in buildings, in Norwegian). Tapir, Trondheim, Norway, 1979.)

are in the range from 1.3 to 2, with the larger values for the fundamental mode (1, 1).

For a gypsum board with critical frequency 2500 Hz and height 2700 mm, the natural frequency f_{11} of the fundamental mode is calculated for some examples of stud distances in Table 9.2. For the case of partly clamped boundaries, the f_{11} is very low at cc 600 mm and very high at cc 100 mm, but at cc 350 mm, the frequency is exactly in the range from 125 Hz to 160 Hz where the severe dip in Figure 9.15 (solid line) is located. It is concluded that stud distances around 300 mm to 400 mm should be avoided in this type of wall.

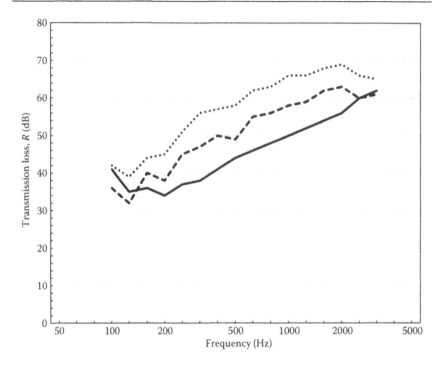

Figure 9.14 Measured transmission loss for single and double walls. Solid line: 150 mm clinker concrete wall with 10 mm plaster, R_w=48dB. Dashed line: Clinker wall and a light secondary wall of 13 mm gypsum board with rigid wooden line connections and 50 mm cavity with mineral wool, R_w=54dB. Dotted line: The same, but secondary wall on steel lath only fixed in top and bottom, 75 mm cavity with mineral wool, R_w=61 dB. (Adapted from Homb, A. et al., Lydisolerende konstruksjoner. (Sound insulation properties of windows, in Norwegian). NBI-anvisning 28. Oslo, Norway, 1983, p. 61c.)

Table 9.2 Calculated natural frequency of fundamental mode (1, 1) for sections of 13 mm gypsum board with height 2700 mm and different stud distances

Stud distance l_x (mm)	100	350	600
Simply supported, $\gamma=1$	1185 Hz	98 Hz	34 Hz
Partly clamped, $\gamma=1.41$	1671 Hz	138 Hz	49 Hz
Clamped, $\gamma=2$	2370 Hz	196 Hz	69 Hz

Note: Three different boundary conditions are assumed.

9.3.2 Examples

Dry walls can be designed within a wide range of sound insulation depending on the number of gypsum board layers, the cavity depth and whether the studs are common to both sides or separated. Table 9.3 gives some examples

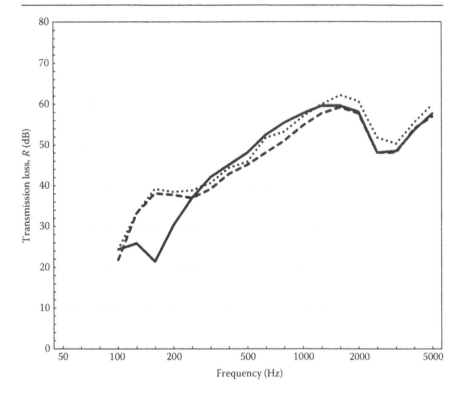

Figure 9.15 Measured transmission loss of a drywall with different stud distances at the border region. Solid line: Stud distances 350 mm/5 × 600 mm/350 mm; dashed line: stud distances 100 mm/6 × 600 mm; dotted line: Stud distances 700 mm/5 × 600 mm. (Adapted from Rindel, J.H., Hoffmeyer, D., *Proceedings of Inter-Noise '91*, Sydney, Australia, pp. 279–282, 1991.)

and the measured R_w values spread from 38 dB to 70 dB. It is important to note that the field performance in buildings is typically 3 dB to 10 dB lower than when the same wall is measured in a laboratory. The low-frequency performance is generally rather weak for lightweight double walls and, thus, the spectrum adaptation term $C_{50-5000}$ for these walls is typically −4 dB to −6 dB. The spectrum adaptation also applies to the apparent sound reduction index R'_w.

9.4 SOUND INSULATION OF WINDOWS

9.4.1 Gas filling in windows

Windows are often made as units with two or three layers of glass – especially in countries with a cold climate – for the reason of thermal insulation. Sometimes, a special gas is used in the sealed cavity of the

Table 9.3 Expected transmission loss of drywalls with 13 mm gypsum boards, in the laboratory (R_w and $C_{50-5000}$) and in the field (R'_w)

Type	Wall thickness (mm)	Number of plate layers	Cavity depth (mm)	Mineral wool thickness (mm)	R_w (dB)	$C_{50-5000}$ (dB)	R'_w (dB)
Common studs	100	1+1	75	0	38–41	−1	~35
	100	1+1	75	45	42–44	−3	~40
	125	2+2	75	45	48–50	−4	~46
	150	2+2	100	45	51–53	−4	~47
	175	2+2	125	90	56–58	−6	~52
Separate studs	150	2+2	100	100	56–58	−5	~51
	200	2+2	150	100	61–63	−6	~54
	275	1+1	250	250	60	−6	~55
	425	1+1	400	400	70	−4	~60

Source: Byggforsk 524.325, *Lydisolasjonsegenskaper til lette innervegger.* (Sound insulation properties of lightweight internal walls, in Norwegian). Byggforskserien, SINTEF Building and Infrastructure, Oslo, Norway, 2000.

window unit as this can improve the thermal insulation. Examples of measured sound insulation with different gasses in the cavity are shown in Figure 9.16. The windows have one 8 mm and one 4 mm glass with a distance of 12 mm. The gases are sulphur hexafluoride (SF_6), argon (Ar) and krypton (Kr). A window with normal air was also measured for comparison. The results show little difference between the transmission loss using argon, krypton, or normal air. However, using SF_6, the transmission loss changes significantly, which must be due to a high density (6.17 kg/m^3) and a low speed of sound (134 m/s). The mass-spring-mass resonance frequency is lowered and the dip is more pronounced, followed by a very steep increase above the resonance frequency. The other gases are also heavier than air and have a lower speed of sound, but the differences from normal air are smaller.

9.4.2 Examples

In Table 9.4, some examples of measured sound insulation of windows with two and three layers of glass are given. The constructions are explained by glass thickness in mm and distance between panes in mm. Laminated glass is denoted with a slash (/), so, for example, 4/1.14/4 means two panes of 4 mm glass with an interlayer of 1.14 mm.

For evaluation of the sound insulation against road traffic noise, it is necessary to include the spectrum adaptation term C_{tr}. The data used for the Table 9.4 did not include frequencies below 100 Hz, but if available, the

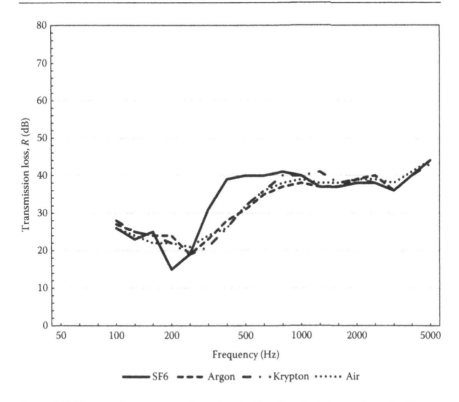

Figure 9.16 Measured transmission loss of sealed 8 – 12 – 4 window units with different gasses in the cavity. (Adapted from Rasmussen, B., *Optimering af vinduers lydisolation. Karm/rammekonstruktioner.* (Optimising the sound insulation of windows. Frame/sash constructions, in Danish). Danish Acoustical Institute, Denmark, 1990, p. A35.)

extended frequency range down to 50 Hz should be used for evaluation of the performance of the windows in practice.

Some rough rules of thumb can be extracted from the examples in Table 9.4:

- Asymmetry is an advantage. Better sound insulation is obtained in two-layer units if the glass panes have different thickness.
- Better sound insulation is obtained if laminated glass is used, at least in one of the panes.
- Three-layer units do not give better sound insulation than two layer units with the same total thickness. Although asymmetry in the two cavities is an advantage, the middle glass pane does not contribute noticeably to the sound insulation.

Table 9.4 Examples of two- and three-layer windows, total thickness, expected weighted sound reduction index (R_w) and practical sound insulation against road traffic noise $(R_w + C_{tr})$

Type	Total thickness (mm)	Construction (mm)	R_w (dB)	$R_w + C_{tr}$ (dB)
Two-layer units	20	4 – 12 – 4	34	30
	22	6 – 12 – 4	36	33
	24	8 – 12 – 4	37	34
	26	10 – 12 – 4	38	34
Two-layer units with laminated glass	25	4 – 12 – 4/1.14/4	37	33
	35	6 – 20 – 4/0.76/4	41	36
	37	8 – 20 – 4/0.76/4	43	39
	41	6/0.38/6 – 20 – 4/0.38/4	43	40
Three-layer units	24	4 – 6 – 4 – 6 – 4	33	28
	30	4 – 9 – 4 – 9 – 4	34	29
	32	6 – 12 – 4 – 6 – 4	39	33
	28	8 – 6 – 4 – 6 – 4	38	34
	48	8 – 20 – 4 – 12 – 4	40	35
Three-layer units with laminated glass	46	5 – 16 – 4 – 12 – 4/1.14/4	38	35
	51	8 – 18 – 4 – 12 – 4/0.2/4	43	39
	54	6/0.2/6 – 20 – 5 – 9 – 4/0.2/4	41	40
Single glass and two-layer unit in separate sashes	67	5 – 42 – (4 – 12 – 4)	38	31
	89	5 – 63 – (4 – 9 – 8)	40	35
	111	6 – 85 – (4 – 12 – 4)	43	40
	111	5 – 80 – (4 – 15 – 3/0.76/3)	45	38

Source: Byggforsk 533.109. Lydisolerende egenskaper for vinduer. (Sound insulation properties of windows, in Norwegian). Byggforskserien, SINTEF Building and Infrastructure, Oslo, Norway, 2013.

REFERENCES

A. Brekke (1980). *Beregningsmetoder for reduksjonstall til enkelt-, dobbelt- og trippelkonstruksjoner.* (Calculation methods for reduction index of single-, double- and tripple constructions, in Norwegian). Report STF 44 A 80021, ELAB, NTH, Trondheim, Norway.

Byggforsk 524.325 (2000). *Lydisolasjonsegenskaper til lette innervegger.* (Sound insulation properties of lightweight internal walls, in Norwegian). Byggforskserien, SINTEF Building and Infrastructure, Oslo, Norway.

Byggforsk 533.109 (2013). *Lydisolerende egenskaper for vinduer.* (Sound insulation properties of windows, in Norwegian). Byggforskserien, SINTEF Building and Infrastructure, Oslo, Norway.

L. Cremer and H. Heckl (1967). *Körpershall.* (*Structure-Borne Sound,* Translated by E.E. Ungar 1973). Springer, Berlin, Germany.

A. Homb, S. Hveem, S. Strøm (1983). *Lydisolerende konstruksjoner.* (Sound insulating constructions, in Norwegian). NBI-anvisning 28. Oslo, Norway.

J. Kristensen and J.H. Rindel (1989). *Bygningsakustik – Teorti og praksis.* (Building acoustics – Theory and practice, in Danish). SBI Anvisning 166. Danish Building Research Institute, Aalborg University Copenhagen, Hørsholm, Denmark.

T.D. Northwood (1970). *Transmission loss of plasterboard walls. Build. Res.* Note No. 66. National Research Council of Canada, Ottawa, ON.

B. Rasmussen (1990). *Optimering af vinduers lydisolation. Karm/rammekonstruktioner.* (Optimising the sound insulation of windows. Frame/sash constructions, in Danish). Danish Acoustical Institute, Lyngby, Denmark.

J.H. Rindel and D. Hoffmeyer (1991). Influence of stud distance on sound insulation of gypsum board walls. *Proceedings of Inter-Noise '91,* 2–4 December 1991, Sydney, Australia, pp. 279–282.

B.H. Sharp (1978). Prediction methods for the sound transmission of building elements. *Noise Control Engineering* 11, 53–63.

R. Timmel (1991). Der Abstrahlgrad rechteckiger, dünner, homogener Platten in der unendlich grossen Schallwand. (The radiation efficiency of rectangular, thin, homogeneous plates in the infinitely large acoustic wall, in German). *Acustica* 73, 1–11.

I.L. Ver (1971). Impact noise insulation of composite floors. *Journal of the Acoustical Society of America* 50, 1043–1050.

T.E. Vigran (1979). *Lydisolasjon i bygninger.* (Sound insulation in buildings, in Norwegian). Tapir, Trondheim, Norway.

Chapter 10

Impact sound insulation

The impact sound pressure levels that can be obtained with homogeneous massive plates can be described theoretically by the sound radiation from point excitation of plates. However, supplementary attenuation of the impact sound pressure level is normally necessary. This can be done by soft resilient floor coverings or by elastically layered floating floors, which can be considered double constructions. A suspended ceiling can also be relevant in some cases.

The impact sound is the most important cause of complaints from the dwellers in multi-storey houses, and the low frequencies below 100 Hz call for special attention in the acoustic design.

10.1 THE IMPACT SOUND PRESSURE LEVEL

The impact sound pressure level L_n is a measurement for the sound transmission into a room when the floor in another room is mechanically excited by a standardized tapping machine:

$$L_n = L_2 + 10 \lg \frac{A_2}{A_0} \, (\text{dB}) \tag{10.1}$$

where L_2 is the mean sound pressure level in dB re 20 µPa in the receiving room. A_2 is the equivalent sound absorption area of the receiving room, and $A_0 = 10 \, \text{m}^2$ is a standardized reference area. For theoretical calculations, Equation 10.1 can be rewritten by assuming the sound field in the receiving room to be diffuse:

$$L_n = 10 \lg \frac{\langle \tilde{p}_2^2 \rangle A_2}{p_0^2 A_0} = 10 \lg \frac{4 \rho c P_2}{p_0^2 A_0} \, (\text{dB}) \tag{10.2}$$

where $\langle \tilde{p}_2^2 \rangle$ is the mean sound pressure square in the receiving room, and P_2 is the sound power radiated into the receiving room. The formula in Equation 10.1 is the basis of measuring the impact sound pressure level in the laboratory as well as in buildings. However, the symbol L'_n is used in buildings in order to indicate that the result may be influenced by flanking transmission.

10.2 MASSIVE FLOORS

10.2.1 Point excitation of a plate

The tapping machine is standardized in Annex E of ISO 10140–5 and used for measuring impact sound pressure levels. The tapping machine has five identical steel hammers, the mass of each is $m_h = 500\,g$. The hammers supply a power corresponding to a free fall of height $h = 40\,mm$ and they deliver ten impacts per second, i.e. the tapping frequency is $f_{tap} = 10\,Hz$.

Under the assumption of a very heavy and stiff floor like a concrete slab, the duration of the force impulse is very small and the mean-square-force in a frequency band $\Delta f \gg 10\,Hz$ can be shown to (e.g. Beranek and Ver, 1992, pp. 328–329) be:

$$\tilde{F}^2 = 4f_{tap}m_h^2 g h \Delta f \cong 3.9\Delta f \ (N^2) \tag{10.3}$$

The approximation comes by inserting the data from the tapping machine and the acceleration of gravity, $g = 9.81\,m/s^2$. For less stiff structures like wooden floors, the relation becomes more complicated and (Equation 10.3) does not hold. For one-third octave bands, the relative bandwidth is 0.23 and the mean-square-force of the tapping machine on a heavy and stiff floor is:

$$\tilde{F}^2 \cong 0.91f \ (N^2) \tag{10.4}$$

where f is the centre frequency in Hz.

The relationship between the force F in a point and the velocity v_0 in the same point is called the point impedance, Z_0. For an infinite, homogeneous plate that is excited to bending waves, i.e. the thickness less than approximately one-sixth of the bending wave length, the point impedance is (Equation 2.13):

$$Z_0 = \frac{F}{v_0} \cong 8\sqrt{mB} = \frac{4c^2m}{\pi f_c} \tag{10.5}$$

The point impedance is in general a complex quantity, but in this case it takes a real value, i.e. the force and the velocity are in phase. For a finite plate, the point impedance varies from one point to another. However, it can be shown that the mean value of the point impedance is equal to that of the infinite plate. In a given point, the deviation from the mean value depends on the position of the point relative to the modal pattern of the plate.

A very important relationship exists between the modal density $\Delta N/\Delta f$ of a finite plate and the point impedance of the equivalent infinite plate. If m is the mass per unit area of the plate and S is the area, the real part of the point admittance (the reciprocal of the impedance) (see Cremer et al., 1988) is:

$$\mathrm{Re}\left\{\frac{1}{Z_0}\right\} = \frac{1}{4mS} \cdot \frac{\Delta N}{\Delta f} \tag{10.6}$$

The modal density of a homogeneous, isotropic plate depends on the thickness. If the plate is thin compared to the wavelength, bending waves can be assumed, but if the plate is thick, compared to the "wavelength" shear waves will occur instead. The approximate modal density is:

$$\frac{\Delta N}{\Delta f} \cong \begin{cases} \dfrac{\pi}{c^2}Sf_c & (f < f_s/2) \quad \text{(bending)} \\[2mm] \dfrac{2\pi}{c_s^2}Sf & (f > f_s/2) \quad \text{(shear)} \end{cases} \tag{10.7}$$

where f_c is the critical frequency, $f_s=f_c\,(c_s/c)^2$ is the crossover frequency for shear waves, and c_s is the speed of shear waves (see Equation 1.30). Both the general results in terms of the modal density as well as the bending wave approximation will be stated in the following.

If φ is the phase angle between force and velocity in the excitation point, the power input to the plate by the point force is (Equation 2.55):

$$P_i = \tfrac{1}{2}|F|\cdot|v_0|\cdot\cos\varphi = \tilde{v}_0^2\,\mathrm{Re}\{Z_0\} = \tilde{F}^2\,\mathrm{Re}\left\{\frac{1}{Z_0}\right\}$$

$$= \frac{\tilde{F}^2}{4mS}\cdot\frac{\Delta N}{\Delta f} \cong \frac{\tilde{F}^2\pi f_c}{4c^2 m} \tag{10.8}$$

In steady state, the power P_d that is lost by internal losses, boundary losses and radiation losses is expressed by the total loss factor η (using Equations 7.1 and 7.3):

$$P_d = 2\pi fmS\langle\tilde{v}_r^2\rangle\eta \tag{10.9}$$

At steady state, we have energy balance, $P_i=P_d$. Thus, the relationship between the average velocity and the input force can be determined:

$$\frac{\tilde{F}^2}{\langle\tilde{v}_r^2\rangle} = 8\pi fm^2S^2\eta\,\frac{\Delta f}{\Delta N} \cong 8Sm^2c^2\eta\,\frac{f}{f_c} \tag{10.10}$$

The sound power P_2 radiated into the receiving room depends on the radiation efficiency σ_{res} and the average velocity (see Equation 6.24). From this and Equation 10.10, the radiated sound power can be found as a function of the input force F:

$$P_2 = \frac{\tilde{F}^2 \rho c \sigma_{res}}{8\pi S m^2 \eta f} \cdot \frac{\Delta N}{\Delta f} \cong \frac{\tilde{F}^2 \rho \sigma_{res} f_c}{8 m^2 c \eta f} \tag{10.11}$$

The sound power radiated from the near field around the excitation point may also be included using the point impedance (Equation 10.5) together with Equation 9.17, which leads to:

$$P_2 = \frac{\tilde{F}^2 \rho}{2\pi m^2 c} + \frac{\tilde{F}^2 \rho c \sigma_{res}}{8\pi S m^2 \eta f} \cdot \frac{\Delta N}{\Delta f} \tag{10.12}$$

The first term is due to the near field radiation and the second term is the resonant contribution from the natural modes in the plate.

10.2.2 Impact sound pressure level for massive floors

For massive floors, the impact of sound pressure level can be determined from Equation 10.2 by inserting the radiated power (Equation 10.12):

$$L_n = 10 \lg \left[\frac{4\rho^2 \tilde{F}^2}{p_0^2 A_0 m^2} \left(\frac{1}{2\pi} + \frac{c^2 \sigma_{res}}{8\pi f S \eta} \cdot \frac{\Delta N}{\Delta f} \right) \right] \tag{10.13}$$

Insert the numerical constants and the force of the standardized tapping machine per one-third octave, and the impact sound pressure level per one-third octave is:

$$L_n \cong 82 - 20 \lg m + 10 \lg \left(\frac{4}{\pi} f + \frac{c^2 \sigma_{res}}{\pi S \eta} \cdot \frac{\Delta N}{\Delta f} \right) \tag{10.14}$$

where m shall have the unit kg/m^2. The last term represents the near field radiation plus the resonant radiation from the natural modes. When the loss factor η is small, the contribution of the near field is often insignificant compared to the resonant radiation. For massive floors with a low critical frequency, the radiation efficiency is $\sigma_{res} \approx 1$ because $f > fc$. Assuming bending waves, an approximation for the impact sound pressure level is:

$$L_n \cong 82 - 10 \lg \left(m^2 \frac{\eta}{f_c} \right) \quad (f_c < f < f_s / 2) \tag{10.15}$$

Above the frequency $f_s/2$, the vibrations in the plate propagate as shear waves and the approximation changes to:

$$L_n \cong 82 - 10\lg\left(m^2\,\frac{\eta}{f_c}\right) + 10\lg\left(\frac{2f}{f_s}\right) \quad (f > f_s/2) \tag{10.16}$$

If the frequency dependency of the loss factor is disregarded, these approximations show that the impact sound pressure level for a massive floor is independent of the frequency in the bending wave range, but increases with frequency at higher frequencies where the shear waves take over.

The influence of the plate thickness is seen from the fact that m is proportional to the thickness and f_c and f_s are inversely proportional to the thickness. The impact sound pressure level L_n decreases 9 dB per doubling of the plate thickness in the bending wave range, but decreases 12 dB per doubling of the plate thickness in the shear wave range. For a rough estimate, the relationship is shown for concrete and lightweight concrete in Figure 10.1. However, it is not realistic to use so thick plates that the impact sound pressure level is acceptable for dwellings without further attenuation of the impact sound.

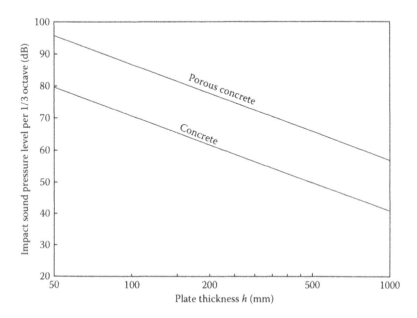

Figure 10.1 Approximate impact sound pressure level in one-third octave band for massive floor of concrete (2300 kg/m³) or of porous concrete (600 kg/m³) assuming bending waves, a radiation factor $\sigma = 1$, and a loss factor $\eta = 0.05$.

10.3 RESILIENT FLOOR COVERINGS

With a carpet or some other thin flooring, e.g. linoleum with soft underlay, a reduction of the hammer force of the floor is obtained by decreasing the speed gradually from maximum at the surface to zero when a certain pressure is reached in the flooring, after which the hammer returns to the surface level. The course corresponds to half a cycle for a mass-spring system where the spring constant is determined by the modulus of elasticity of the flooring, its thickness h_t and the actuated area S_h (Figure 10.2).

If the mass of the hammer is called m_h, the resonance frequency of the system can be determined by:

$$f_0 = \frac{1}{2\pi} \sqrt{\frac{S_h E_t}{m_h h_t}} \tag{10.17}$$

For the standardized tapping machine, $m_h = 0.500\,\text{kg}$ and $S_h = 700$ mm². The diagram in Figure 10.3 shows a relationship between the resonance frequency f_0 and the elastic properties of the flooring. For many elastic materials, the dynamic modulus of elasticity is about double that determined by a static load.

As shown by Ver (1971), if the resonance frequency f_0 is known, and $f > f_0$ the attenuation of the impact sound pressure level can be determined approximately for $f > f_0$:

$$\Delta L = L_{n,\text{without}} - L_{n,\text{with}} \cong 40 \lg \frac{f}{f_0} \tag{10.18}$$

For $f \leq f_0$, the attenuation is $\Delta L \cong 0\,\text{dB}$. The idealized frequency course of the impact sound attenuation is illustrated in Figure 10.4. For a good design, f_0 should be as low as possible and preferably under 90 Hz. In Figure 10.5, some examples of measured impact sound attenuation of thin flooring can be seen.

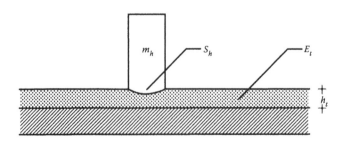

Figure 10.2 Resilient floor covering with thickness h_t and Young's modulus E_t. The hammer of the tapping machine has the mass $m_h = 500\,\text{g}$ and the area $S_h = 700\,\text{mm}^2$. (Reproduced from Kristensen, J., Rindel, J. H., *Bygningsakustik – Teori og praksis*, SBI-Anvisning 166 (In Danish), Danish Building Research Institute, Aalborg University Copenhagen, 1989. With permission.)

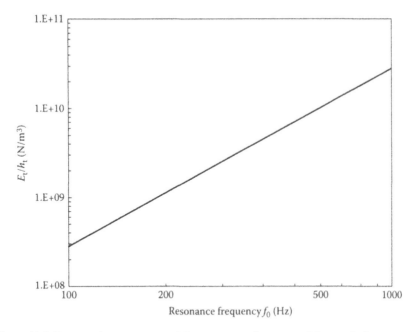

Figure 10.3 Diagram for estimation of the resonance frequency f_0 for a soft floor covering with thickness h_t and Young's modulus E_t.

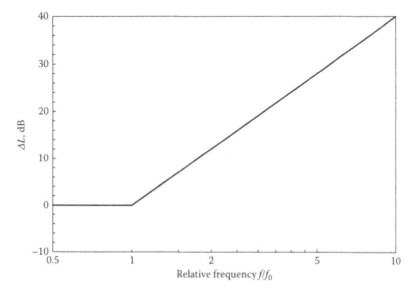

Figure 10.4 Idealized frequency course of the impact sound attenuation of a soft floor covering.

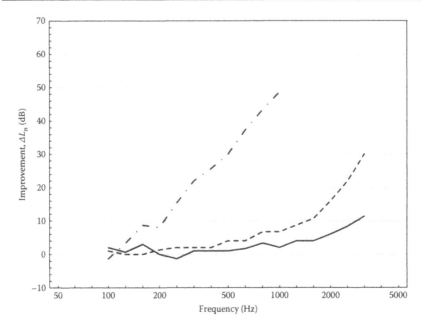

Figure 10.5 Examples of impact sound attenuation for floor coverings. Solid line: Thin linoleum. Dashed line: Vinyl on a layer of felt. Dashed–dotted line: Thick and soft carpet.

10.4 THE RELATIONSHIP BETWEEN R AND L_n

As regards massive floors, there exists a relationship between the airborne sound insulation and the impact sound pressure level. This relationship is surprisingly simple when the sound transmission is dominated by resonant transmission, i.e. for frequencies above the critical frequency ($f > f_c$). If soft flooring is assumed and only resonant radiation is considered (the near field radiation is neglected), we have from Equation 10.14:

$$L_n \cong 82 - 20\lg m + 10\lg\left(\frac{c^2\sigma_{res}}{\pi S\eta} \cdot \frac{\Delta N}{\Delta f}\right) - \Delta L \tag{10.19}$$

Here ΔL is the impact sound attenuation of the flooring. The transmission loss R is irrespective of the flooring given by (see Equation 7.23):

$$R \cong R_r = 10\lg\left(\frac{\omega m}{2\rho c}\right)^2 - 10\lg\left(\frac{c^2\sigma_{res}^2}{2f S\eta} \cdot \frac{\Delta N}{\Delta f}\right)$$

$$= 10\lg\left(m^2 f^3\right) - 10\lg\left(\frac{c^2\sigma_{res}^2}{\pi S\eta} \cdot \frac{\Delta N}{\Delta f}\right) - 44 \tag{10.20}$$

Adding R and L_n yields:

$$R + L_n = 38 + 30 \lg f - 10 \lg \sigma_{res} - \Delta L \qquad (10.21)$$

The radiation efficiency is approximately $\sigma_{res} \cong 1$ since we have assumed $f > f_c$. Thus, for floors with a hard surface, the relationship is as simple as:

$$R + L_n = 38 + 30 \lg f \qquad (10.22)$$

With a soft flooring or a floating floor that is locally reacting, the relationship changes above the resonance frequency f_0 of the flooring:

$$R + L_n = 38 + 30 \lg f_0 - 10 \lg \frac{f}{f_0} \quad (f > f_0) \qquad (10.23)$$

The relationship is illustrated in Figure 10.6 when applied to typical measurement results for heavy slabs in Danish domestic houses. The lowest structural mode f_{11} is clearly seen at a frequency around 80 Hz. However, the impact sound pressure mirror the course at frequencies below f_{11}, and the sum shown in Figure 10.6b is close to a straight line below f_0 with the slope 9 dB per octave. With soft flooring or a floating floor, the curve has a knee at f_0 and continues towards higher frequencies with the negative slope −3 dB per octave. In the example, f_0 is around 250 Hz.

This simple relationship can be utilized for control of measuring results as will be demonstrated in Chapter 11. If the measured airborne and impact sound insulation added together fails to follow the predicted curve, this is an indication of flanking transmission or leakage, i.e. some transmission paths that only affect the airborne sound transmission and not the impact sound transmission. It is emphasized that below the critical frequency of the floor, the relationship between R and L_n cannot be expected to be equally simple.

10.5 FLOATING FLOORS

A special type of double construction is a floating floor where the elasticity and construction of the inevitable mechanical connections are of importance for the sound insulating characteristics. As illustrated in Figure 10.7, some airborne flanking transmission paths are not affected by a floating floor, whereas the impact sound is effectively attenuated by a floating floor. Thus, floating floors tend to improve the impact sound insulation more than the airborne sound insulation.

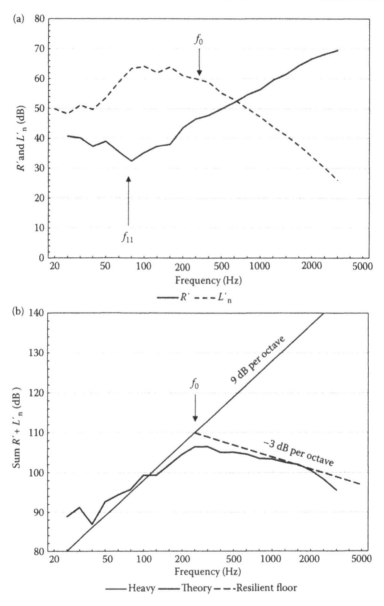

Figure 10.6 (a) Measured airborne and impact sound insulation of typical heavy floor
constructions. Average of 10 measurements. The frequency of the lowest
structural mode f_{11} and the resonance frequency of the flooring f_0 are indicated. (Below 50 Hz the results are not corrected for reverberation time.)
(b) The sum $R' + L'_n$ for a massive floor. Thick line: The sum of the measured
results in a. Thin straight line: Idealized frequency course for a hard floor.
Dashed line: Floor with impact sound attenuation by flooring. (Data from
DELTA, *Målerapport: Boligers lydisolation ved lave frekvenser.* (In Danish). PDØ
870028, DELTA, Aarhus, Denmark, 1999.)

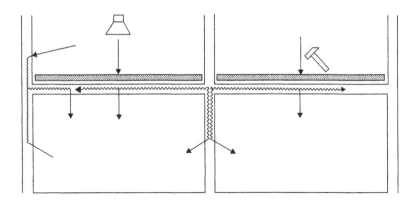

Figure 10.7 Transmission of airborne and impact sound through floors with floating floors. Due to flanking transmission, the attenuation of impact sound is better than the attenuation of airborne sound. (Reproduced from Kristensen, J., Rindel, J. H., *Bygningsakustik – Teori og praksis*, SBI-Anvisning 166 (In Danish), Danish Building Research Institute, Hørsholm, Denmark, 1989. With permission.)

In the following, the attenuation ΔL of the impact sound pressure level is described. The airborne sound insulation for a construction with floating floor may be evaluated by calculation of all possible transmission paths, as described in Chapter 11 on flanking transmission.

10.5.1 Attenuation of impact sound pressure level

The excitation with the standardized tapping machine, as illustrated in Figure 10.8, gives a well-defined pulse. However, the effect that is transferred to the construction will not only depend on the boundary impedance of the construction but also on the impedance of the hammer. Instead of Equation 10.8, the power input to the plate is more generally:

$$P_i = F^2 \operatorname{Re}\left\{\frac{1}{Z_{0,1} + Z_h}\right\} \tag{10.24}$$

where $Z_{0,1}$ is the point impedance of the floating floor slab. The impedance Z_h of the hammer is determined by its mass $m_h = 500\,\mathrm{g}$:

$$Z_h = j\omega m_h \tag{10.25}$$

The transferred power is reduced at high frequencies, where $|Z_h| > |Z_{0,1}|$, i.e. for frequencies:

$$f > f_Z = \frac{2m_1 S_1}{\pi m_h} \cdot \frac{\Delta f}{\Delta N_1} \tag{10.26}$$

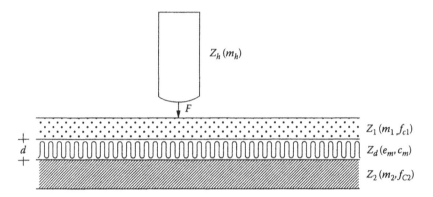

Figure 10.8 Floating floor with excitation from a falling mass. Each element is character-
ized by a mechanical impedance. (Reproduced from Kristensen, J., Rindel, J.
H., *Bygningsakustik – Teori og praksis*, SBI-Anvisning 166 (In Danish), Danish
Building Research Institute, Hørsholm, Denmark, 1989. With permission.)

Here, the point admittance has been inserted from Equation 10.6. For frequencies higher than f_z, an extra contribution ΔL_z for the attenuation of
the impact sound pressure is obtained because of the impedance of the
hammer:

$$\Delta L_Z = 10 \lg\left(1 + \left(\frac{f}{f_Z}\right)^2\right) \qquad (10.27)$$

For $f < f_Z$, the attenuation is $\Delta L_Z \cong 0\,\mathrm{dB}$. However, for lightweight floors this
attenuation has some importance. For example the frequency f_Z is $\cong 700\,\mathrm{Hz}$
for a 22 mm chip board and $\cong 200\,\mathrm{Hz}$ for a 12 mm chip board. Above f_Z,
the attenuation increases by 6 dB per octave in the case of a thin slab performing bending waves, changing to 12 dB per octave in the case of a thick
slab performing shear waves. For measurements in the laboratory of impact
sound attenuation of floating floors, the measurement standard ISO 10140-1
(Annex H.4.1) prescribes an evenly distributed load of 20 kg/m² to 25 kg/m²,
which will diminish the problem of the hammer impedance.

   The attenuation of the impact sound pressure level of a floating floor can
be expressed by the ratio of the velocities in the floor in the two situations
shown in Figure 10.9a and b:

$$\Delta L = 20 \lg \frac{v_{2a}}{v_{2b}} \qquad (10.28)$$

By means of the mechanical impedances Z_1 and Z_2 for the two plates, the
result is:

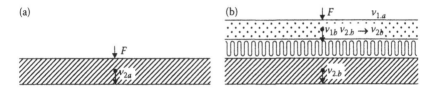

Figure 10.9 Point force excitation of a slab (a) and a floating floor on the same slab (b). (Reproduced from Kristensen, J., Rindel, J. H., *Bygningsakustik – Teori og praksis*, SBI-Anvisning 166 (In Danish), Danish Building Research Institute, Hørsholm, Denmark, 1989. With permission.)

$$v_{2a} = \frac{F}{Z_2} \qquad (10.29)$$

$$v_{1b} = \frac{F}{Z_1} \qquad (10.30)$$

The force F', which is transferred through the elastic layer from plate 1 to plate 2, can be expressed as follows:

$$F' = Z_d(v_{1b} - v_{2b}) \cong Z_d v_{1b} \qquad (10.31)$$

where Z_d is the impedance of the elastic layer. It is presumed that $v_{1b} > v_{2b}$, which means that the supporting slab must be heavy compared to the floating floor. The velocity in the floor due to the force F' is:

$$v_{2b} = \frac{F'}{Z_2} \cong v_{1b}\frac{Z_d}{Z_2} = \frac{FZ_d}{Z_1 Z_2} \qquad (10.32)$$

By inserting the above expression in Equation 10.28, the attenuation of the impact sound pressure level is:

$$\Delta L \cong 20\lg\left|\frac{Z_1}{Z_d}\right| \qquad (10.33)$$

For further analysis, it is necessary to have more information about the impedance Z_d of the elastic layer, which is often made of mineral wool. Furthermore, it is necessary to differ between floors that are more or less locally reacting, as this is of importance for how the sound propagates through the elastic layer. Depending on this, the impedances in Equation 10.33 can be seen either as point impedances or as surface impedances.

10.5.2 Surface-supported floors

The construction shown in Figure 10.9b can be regarded as a double construction, which is characterized by a resonance frequency f_0:

$$f_0 = \frac{1}{2\pi} \sqrt{k''_d \left(\frac{1}{m_1} + \frac{1}{m_2} \right)} \tag{10.34}$$

where k''_d is the dynamic stiffness per unit area of the elastic layer. This equation is the same as Equation 9.4 for a double wall, except that here the dynamic stiffness is not primarily due to the air in the cavity. In many floor constructions, $m_1 \ll m_2$ and, thus, the latter can be removed from Equation 10.34.

The dynamic stiffness per unit area is a combined contribution from the structure of the elastic material and from the air contained in the pores of the elastic material:

$$k''_d = k''_m + k''_a = \frac{\rho_m c_m^2}{d} + \frac{\rho c^2}{qd} \tag{10.35}$$

Here, d is the thickness of the elastic layer and q is the porosity of the material, i.e. the ratio between the pore volume and the total volume. ρ_m and c_m are the density of the material and the speed of longitudinal sound waves, respectively.

An elastic material is characterized by the dynamic modulus of elasticity $\rho_m c_m^2$, which is the same as the stiffness per unit area per thickness unit. Typical values for a selection of elastic materials are stated in Table 10.1. Except from light mineral wool, the contribution from the air is often of minor importance. The dynamic stiffness can be determined experimentally after ISO 9052-1 by measuring the resonance frequency with a load of known mass.

10.5.3 Dynamic stiffness of loaded mineral wool

Figure 10.10 shows the dynamic stiffness of some mineral wool products, and how it increases with the static load. This behaviour is not very surprising since it is a general principle that the speed of propagation of a compression wave in an elastic material increases with increasing static pressure in

Table 10.1 Examples of dynamic modulus of elasticity for materials that can be used for floating floors

Material	Unit: MN/m²
Air	0.14
Mineral wool (50 kg/m³ to 160 kg/m³)	0.02–0.4
Coco mat	0.25–0.5
Polystyrene foam (10 kg/m³ to 20 kg/m³)	0.3–3
Wood fibre concrete	6–17
Cork	10–30

the material. If the results are corrected for the contribution from the stiffness of the air, the following empirical relationship is found:

$$k''_m \cong \text{constant} \cdot \sqrt{m_1} \qquad (10.36)$$

where m_1 is the static load. Because it is mainly the speed of sound c_m that can change with different loads, Equation 10.35 leads to the relationship:

$$c_m \cong \text{constant} \cdot \sqrt[4]{m_1} \qquad (10.37)$$

Table 10.2 gives the data for the same types of mineral wool, as in Figure 10.10. In a special experimental setup, the sound speeds c_m are measured and the calculated dynamic modulus of elasticity correspond to the curves in Figure 10.10.

The speed of sound in mineral wool is approximately one-tenth of the speed of sound in air, which is of significance for the frequency dependency of the impact sound attenuation. The sound propagation in an elastic material, such as mineral wool, will be accompanied by losses. As shown by Cremer and Heckl (1967, p. 169), this can be formulated by a complex sound speed whose imaginary quantity is determined by the loss factor η:

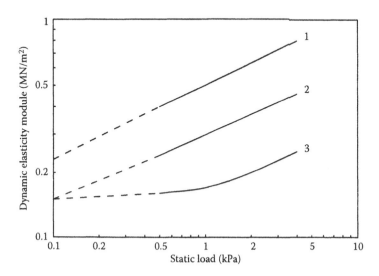

Figure 10.10 Dynamic elasticity as a function of the static load for mineral wool. The curves 1, 2 and 3 are regression lines for the same three types of mineral wool described in Table 10.2. (Adapted from Gudmundsson, S., Sound insulation improvements of floating floors. A study of parameters. Report TVBA-3017, LTH, Lund, Sweden, 1984.)

Table 10.2 Three examples of mineral wool for floating floors (same as in Figure 10.10)

Type (number)	Density, ρ_m (kg/m³)	Speed of sound, c_m (m/s)	Dynamic elasticity (MN/m²)
Rockwool (1)	160	49	0.38
Rockwool (2)	100	35	0.12
Fibre glass (3)	50	22	0.02

Source: Gudmundsson, S., Sound insulation improvements of floating floors. A study of parameters. Report TVBA-3017, LTH, Lund, Sweden, 1984, p. 159.

Note: The speed of sound is measured with a static load of 1 kPa.

$$c_m \rightarrow c_m \left(1 + j\frac{\eta}{2}\right) \tag{10.38}$$

A number of measuring results of the loss factor have been referred by Gudmundsson (1984, p. 162), but there is much uncertainty of this parameter. For Rockwool loaded with 1 kPa, the loss factor has been measured to $\eta = 0.23-0.28$. For heavier loads, loss factors have been measured ($\eta = 0.10-0.15$ for loads of 4 kPa to 6 kPa). In order not to complicate the description further, it is presumed in the following that the loss factor is so small that it will not contribute to the impact sound attenuation.

10.5.4 Resonantly reacting floors

Resonantly reacting floors means floating floors where the whole floor plate contributes efficiently to the transmission of vibration energy through the elastic layer. The impedances in Equation 10.33 should in this case regarded as *surface impedances*. The impedance of the floor is:

$$Z_1 = j\omega m_1 \tag{10.39}$$

where m_1 is the mass of the floor per unit area. For the elastic layer, the impedance is:

$$Z_d = \frac{\rho_m c_m^2}{\sin\left(\frac{\omega}{c_m} d\right)} \tag{10.40}$$

where the denominator goes towards 0 at frequencies where standing waves in the elastic layer may occur. It will be practical to introduce a knee frequency f_d:

$$f_d = \frac{c_m}{2\pi d} \tag{10.41}$$

At low frequencies, the spring effect is not affected by standing waves and $\sin(\omega \, d/c_m) \cong \omega \, d/c_m$. At higher frequencies $f > f_d$ a useful simplification is to set $\sin(\omega \, d/c_m) \cong 1$, leading to:

$$|Z_d| \cong \begin{cases} \dfrac{\rho_m c_m^2}{\omega d} = \dfrac{k''_m}{\omega} & (f \leq f_d) \\[2mm] \rho_m c_m & (f > f_d) \end{cases} \qquad (10.42)$$

At frequencies below the resonance frequency f_0, the construction does not react as a double construction and the attenuation of the impact sound level is insignificant. For floating floors, the expression (Equation 10.34) for the resonance frequency (Equation 10.34) can often be simplified, as $m_1 < m_2$ and $k''_m > k''_a$.

$$f_0 \cong \frac{1}{2\pi} \sqrt{\frac{\rho_m c_m^2}{m_1 d}} \qquad (10.43)$$

The attenuation of the impact sound pressure level in the frequency range $f_0 < f < f_d$ is found by inserting Equations 10.39 and 10.42 in Equation 10.33.

$$\Delta L = 20 \lg \frac{\omega^2 m_1 d}{\rho_m c_m^2} = 40 \lg \frac{f}{f_0} \quad (f_0 < f \leq f_d) \qquad (10.44)$$

This strong frequency dependency of 12 dB per octave continues, however, only to the frequency f_d, which is only a little higher than f_0 because of the low speed of sound in the elastic layer. For $f > f_d$ the result is:

$$\Delta L = 20 \lg \frac{\omega m_1}{\rho_m c_m} = 40 \lg \frac{f_d}{f_0} + 20 \lg \frac{f}{f_0} \quad (f > f_d) \qquad (10.45)$$

Thus, the frequency dependency is reduced to 6 dB per octave in the frequency range above f_d.

In Figure 10.11, there are examples of measured impact sound attenuation for resonantly reacting floating floors of 35 mm concrete with different underlying layers, which give different resonance frequencies and knee frequencies. The knee frequencies for the three materials are indicated in the graph. The curves follow closely the two formulas (Equations 10.44 and 10.45).

Other examples of measured impact sound insulation for resonantly reacting floating floors are shown in Figure 10.12. The only difference between the three results is the different type of mineral wool under the floating floor.

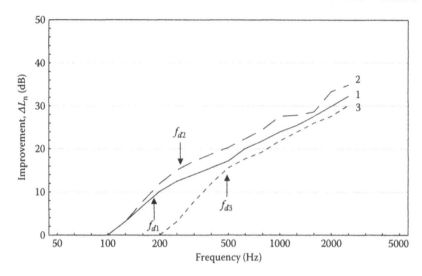

Figure 10.11 Impact sound attenuation of floating floors with three different elastic materials. All curves are averaged results from three floors, measured 3 months and 1 year after inauguration. All floors are built from 150 mm concrete with 35 mm floating concrete floor. Curve 1: 15 mm mineral wool; curve 2: 25 mm mineral wool; curve 3: 15 mm polystyrene foam plate. (Adapted from Kuhl, W., *Acustica*, 51, 103–115, 1982.)

10.5.5 Locally reacting floors

A locally reacting floor means a floor where the force from the hammers of the tapping machine is mainly transmitted to a limited part of the construction close to the excitation point of the force. The difference of the force transmission in resonantly reacting and locally reacting floors is illustrated in Figure 10.13. In this case, the impedances in Equation 10.33 should be regarded as *point impedances*.

The point impedance of the floor is (as in Equations 10.5 and 10.6):

$$Z_{0,1} = 4m_1 S \frac{\Delta f}{\Delta N_1} \cong \frac{4c^2 m_1}{\pi f_{c1}} \tag{10.46}$$

The impedance for a spring element consisting of the active section of the elastic layer can be written as:

$$|Z_{0,d}| = \frac{k_m}{\omega} \tag{10.47}$$

where k_m is the dynamic stiffness of the active section. As shown by Kuhl (1982), it is reasonable to assume that the active section in the floor plank is a cylinder of a diameter, which equals a half bending wave length (see Figure 10.13b).

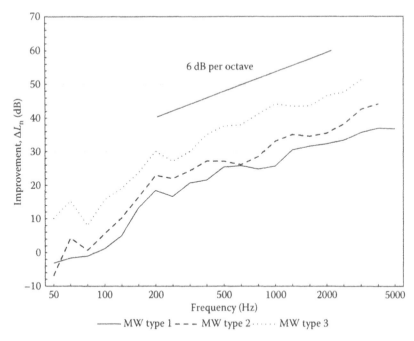

Figure 10.12 Measured impact sound attenuation of resonantly reacting floors made of 50 mm dense concrete laid on different types of mineral wool, 50 mm thick. The curves refer to the material numbers 1, 2, and 3, respectively, in Table 10.2. The structural slab is 160 mm concrete. (Adapted from Gudmundsson, S., Sound insulation improvements of floating floors. A study of parameters. Report TVBA-3017, LTH, Lund, Sweden, 1984.)

If the corresponding active area is called S_a, the result is:

$$k_m = k''_m S_a = k''_m \frac{\pi}{16} \lambda_{B1}^2 \tag{10.48}$$

which by means of Equation 10.35 can be rewritten as:

$$k_m = \frac{\rho_m c_m^2}{d} \cdot \frac{\pi c^2}{16 f f_{c1}} \tag{10.49}$$

The attenuation of impact sound pressure level is found by inserting in Equation 10.33:

$$\Delta L = 20 \lg \left(\frac{128 m_1 d}{\pi \rho_m c_m^2} f^2 \right) \cong 40 \lg \frac{f}{f_0} \quad (f > f_0) \tag{10.50}$$

where f_0 is the resonance frequency (Equation 10.43). The result for locally reacting floors corresponds exactly to the previously found result for

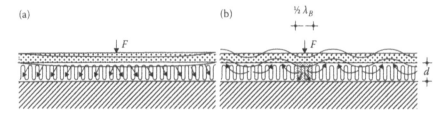

Figure 10.13 Floating floors with a point force excitation. (a) Resonantly reacting floor. (b) Locally reacting floor. (Reproduced from Kristensen, J., Rindel, J. H., *Bygningsakustik – Teori og praksis*, SBI-Anvisning 166 (In Danish), Danish Building Research Institute, Hørsholm, Denmark, 1989. With permission.)

resonantly reacting floors below the knee frequency, but there is no knee frequency for locally reacting floors. The slope of ΔL is 12 dB per octave in the entire frequency range, which is in agreement with the model above for the transmission through the elastic layer. The section of the active cylinder is reduced by increasing frequency and, thus, the stiffness is inversely proportional to the frequency (see Equation 10.49).

Examples of measured attenuation of impact sound pressure level for locally reacting floors are seen in Figure 10.14. The types of applied mineral

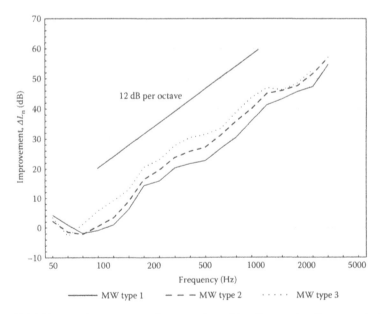

Figure 10.14 Measured impact sound attenuation of locally reacting floors made from 22 mm chipboard laid on different types of mineral wool, 50 mm thick. The curves refer to the material numbers 1, 2 and 3, respectively, in Table 10.2. The structural slab is 160 mm concrete. (Adapted from Gudmundsson, S., Sound insulation improvements of floating floors. A study of parameters. Report TVBA-3017, LTH, Lund, Sweden, 1984. Fig. III.2.3.)

wool are the same as used in Table 10.2 and Figure 10.10. The curves increase by 12 dB per octave in agreement with Equation 10.50.

10.5.6 Point-supported floors

Sometimes, floating floors are constructed with the flooring resting on evenly distributed elastic patches (Figure 10.15). The dynamic stiffness of each patch is called k_{dp} and the number is N_p. As the air contribution to the stiffness is disregarded and the structural slab is assumed to be much heavier than the top floor, the resonance frequency for the floor is:

$$f_0 = \frac{1}{2\pi}\sqrt{\frac{k_{dp} \cdot N_p}{m_1 S}} \tag{10.51}$$

The attenuation of the impact sound by point supported floors that are presumed to be resonantly reacting has been analysed by Ver (1971). With indices 1 and 2 referring to the flooring plate and the structural slab, respectively, his result is:

$$\Delta L = 10\lg\left[\frac{Z_{0,1}}{Z_{0,2}} + \frac{m_1\eta_1}{m_2\eta_2} + 2\pi\eta_1\frac{Z_{0,1}}{k_{dp}} \cdot \frac{f^3}{f_0^2}\right] \tag{10.52}$$

Above a certain frequency, the last term dominates and the impact sound attenuation increases by 9 dB per octave, if the loss factor of the flooring plate η_1 and the stiffness of the patches are frequency independent. Inserting the resonance frequency (Equation 10.51) and the point impedance (Equation 10.46) yields:

$$\Delta L \cong 10\lg\left[\frac{32\pi^3}{k_{dp}^2} \cdot \frac{m_1^2\eta_1}{N_p} \cdot \frac{\Delta f}{\Delta N_1} \cdot f^3\right] \cong 30\lg\frac{f}{f_0} \tag{10.53}$$

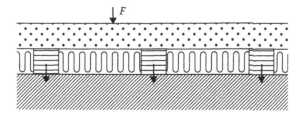

Figure 10.15 Point supported floating floor with a point force excitation. The force is transferred to the slab through a number of elastic patches. (Reproduced from Kristensen, J., Rindel, J. H., *Bygningsakustik – Teori og praksis*, SBI-Anvisning 166 (In Danish), Danish Building Research Institute, Hørsholm, Denmark, 1989. With permission.)

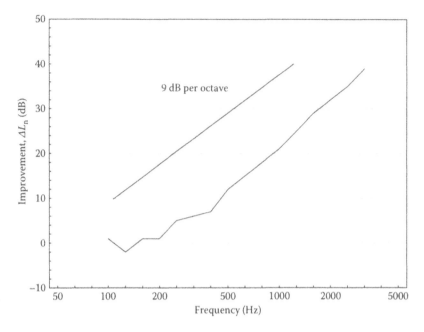

Figure 10.16 Measured impact sound attenuation of a point supported floor made from 22 mm wood on laths on 10 mm fibreboard pads. The structural slab is 200 mm clinker concrete slab with 30 mm finish. Measured data from Figure 5.32.

Especially, the importance of a large loss factor and a low modal density for the flooring plate should be noted. Thus, the flooring should preferably be thick and with high damping.

An example of a flooring that performs in accordance with this model is the traditional wooden floor on laths on resilient pads (Figure 10.16)

10.5.7 The effect of sound bridges

The sound transmission via sound bridges may be of decisive importance for the impact sound pressure of floating floors. In Figure 10.17, examples of measurements in laboratories are seen for a floating concrete floor where mechanical contacts to the floor were made with different number of gypsum cylinders. One small sound bridge is enough to ruin the result, which means that good workmanship is extremely important for a successful result of a floating floor.

10.5.8 Summary on floating floors

The attenuated characteristics of floating floors are so complicated in practice that it is not possible with the available knowledge to draw up a generally acceptable basis of calculation. The survey given in this section is mainly based on analyses by Kuhl (1982) and Gudmundson (1984).

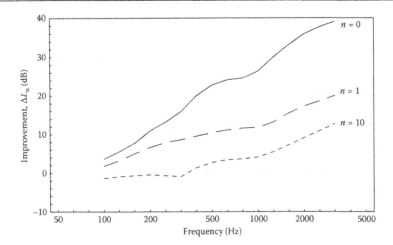

Figure 10.17 Harmful effect of rigid connections in a floating floor. The number of bridges is *n*, and they were made as gypsum cylinders with a diameter of 30 mm. (Adapted from Gösele, K., *Schallbrücken bei schwimmenden Estrichen und anderen schwimmend verlegten Belägen*. Berichte aus der Bauforschung, Heft 35. Wilhelm Ernst und Sohn, Berlin, 1964.)

10.6 SUSPENDED CEILINGS

Compared to floating floors, the suspended ceilings have the advantage that the resonance frequency can be significantly lower (preferably below 50 Hz) because the gap between the slab and the suspended ceiling can be made much wider. Still, the use of suspended ceilings for attenuation of impact sound is not common practice.

The resonance frequency of a suspended ceiling is calculated as for an asymmetric double construction with one heavy mass and one lighter mass:

$$f_0 \cong \frac{1}{2\pi}\sqrt{\frac{\rho c^2}{m_3 d}} \tag{10.54}$$

Here, m_3 is the mass per unit area of the suspended ceiling, d is the suspension, and ρc^2 is the dynamic elasticity of the air in the cavity.

Both perforated and closed gypsum board plates can provide attenuation of impact sound. It is possible, at least to some extent, to combine a room acoustic treatment of the room below with improved impact sound insulation (Seidel and Hengst, 2017).

10.7 MASSIVE WOOD FLOOR CONSTRUCTIONS

Massive wood is an interesting building material with several good properties, especially in relation to ecology. However, due to low density, there are obviously some challenges if good sound insulation shall be obtained. This was

Table 10.3 Overview of measurement results on floor constructions made from massive wood

Construction	Description	R_w (dB)	$R_{w,50}$ (dB)	$L_{n,w}$ (dB)	$L_{n,w,50}$ (dB)
	185 mm wood	44	43	82	75
	115 mm wood 30 mm mineral wool 185 mm wood	56	52	59	60
	185 mm wood 150 mm suspended ceiling 2×13 mm gypsum board	62	59	50	57
	115 mm wood 30 mm mineral wool 185 mm wood 150 mm suspended ceiling 2×13 mm gypsum board	63	59	41	53
	115 mm wood 30 mm mineral wool 50 mm concrete tiles (200×400) mm 185 mm wood	60	57	50	53

(Continued)

Table 10.3 (Continued) Overview of measurement results on floor constructions made from massive wood

Construction	Description	R_w (dB)	$R_{w,50}$ (dB)	$L_{n,w}$ (dB)	$L_{n,w,50}$ (dB)
	115 mm wood 30 mm mineral wool 100 mm sand 185 mm wood	62	60	47	48
	115 mm wood 30 mm mineral wool 100 mm sand 185 mm wood 150 mm suspended ceiling 2×13 mm gypsum board	63	61	33	48

Source: Eriksen, L., Ejlersen, C. V. J., Lydisolering – etageadskillelser af massive træelementer. (In Danish). MSc Thesis, Byg-DTU and Ørsted-DTU, Technical University of Denmark, Lyngby, Denmark, 2003.

studied through a series of laboratory measurements in a project by Eriksen and Ejlersen (2003). A few examples of the results are listed in Table 10.3. The following abbreviations are used: $R_{w,50} = R_w + C_{50-3150}$ and $L_{n,w,50} = L_{n,w} + C_{I,50-2500}$.

The aim of the project was to search for a solution that could fulfil the Danish requirements on airborne and impact sound insulation for floors between dwellings, i.e. minimum $R'_w \geq 55$ dB and $L'_{n,w} \leq 53$ dB, but preferably sound class B: $R'_w + C_{50-3150} \geq 58$ dB and $L'_{n,w} + C_{I,50-2500} \leq 48$ dB. The wood material was glued with the same technique as used for laminated wooden beams. The basic slab was a wooden plate 185 mm thick, and as an upper floor, another wooden plate 115 mm thick was used.

The results show clearly that the 185 mm slab alone is far from meeting the acoustic requirements (the first row in Table 10.3 and Figure 10.18). The critical frequency of the slab is around 110 Hz, and there are very few structural modes below 160 Hz. Above 160 Hz, the transmission is due to resonant shear waves and the impact sound pressure level increases with 3 dB per octave in agreement with Equation 10.16 until 1250 Hz. The attenuation of impact sound at the high frequencies can be due to the relatively soft surface of the wooden slab.

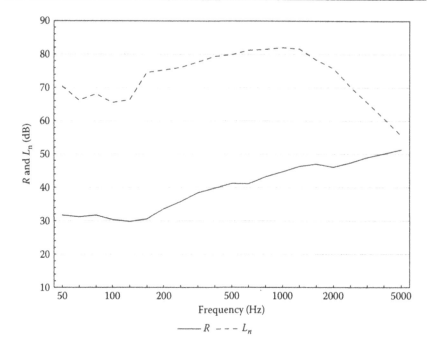

Figure 10.18 Measured transmission loss and impact sound pressure level of a floor of 185 mm massive wood. (Adapted from Eriksen, L., Ejlersen, C. V. J., *Lydisolering – Etageadskillelser af massive træelementer.* (In Danish). MSc Thesis, Byg-DTU and Ørsted-DTU, Technical University of Denmark, Lyngby, Denmark, 2003.)

The combination of a floating floor of 115 mm massive wood on a resilient layer of 30 mm mineral wool gives promising improvements, but the impact sound insulation is still insufficient (see second row in Table 10.3 and Figure 10.19). In order to achieve sufficient impact sound insulation at low frequencies, it is necessary to increase the total mass and this was tried with 50 mm concrete tiles and with 100 mm sand. Figure 10.20 shows the results of the latter. Both solutions are efficient but sand yields results that are a few dB better than results with concrete tiles, probably due to a higher loss factor. Actually, the results with sand meet the goal, at least in the laboratory. When applied in a real building, the sound insulation is always a few dB lower than in the laboratory.

A suspended ceiling was also tried, both alone and in combination with a floating floor. The impact sound attenuation of a 150 mm suspended ceiling is better than that obtained by a floating floor compare (second and third row in Table 10.3). Actually, the suspended ceiling meets the minimum requirements for dwellings with a good margin. The combination of a floating floor and a suspended ceiling is even more efficient as the attenuations are additive (Figure 10.21).

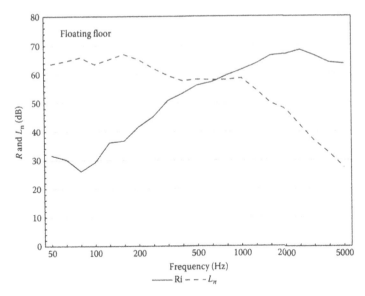

Figure 10.19 Measured transmission loss and impact sound pressure level of a floor of 185 mm massive wood with a floating floor of 115 mm massive wood on 30 mm mineral wool. (Adapted from Eriksen, L., Ejlersen, C. V. J., *Lydisolering – Etageadskillelser af massive træelementer.* (In Danish). MSc Thesis, Byg-DTU and Ørsted-DTU, Technical University of Denmark, Lyngby, Denmark, 2003.)

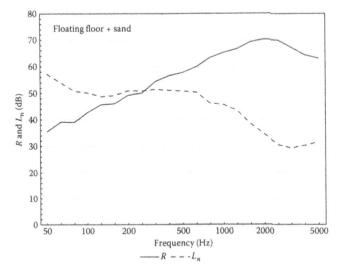

Figure 10.20 Measured transmission loss and impact sound pressure level of a floor of 185 mm massive wood with a topping of 100 mm sand and a floating floor of 115 mm massive wood on 30 mm mineral wool. (Adapted from Eriksen, L., Ejlersen, C. V. J., *Lydisolering – Etageadskillelser af massive træelementer.* (In Danish). MSc Thesis, Byg-DTU and Ørsted-DTU, Technical University of Denmark, Lyngby, Denmark, 2003.)

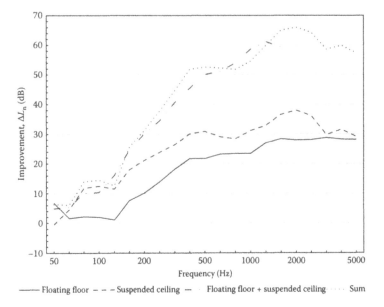

Figure 10.21 Impact sound attenuation of a floating floor of 115 mm massive wood on 30 mm mineral wool or a 15-mm suspended ceiling with 2×13mm gypsum board, or both combined. The supporting floor is 185 mm massive wood. (Data from Eriksen, L., Ejlersen, C. V. J., *Lydisolering – Etageadskillelser af massive træelementer.* (In Danish). MSc Thesis, Byg-DTU and Ørsted-DTU, Technical University of Denmark, Lyngby, Denmark, 2003.)

10.8 SOUND INSULATION IN PRACTICE

10.8.1 A live walker and the ISO tapping machine

The tapping machine was developed to produce impact sound suited for measurements. Hence, the tapping frequency (10 Hz) gives a sufficiently dense line spectrum and the force of the impacts is high enough to allow measurements without too much interference due to background noise. However, the resemblance with real footfall is not obvious. Measurements with the tapping machine are compared with examples of measured footfall noise from a live walker in Figure 10.22. In this example, the construction is a concrete slab with a floating concrete floor. At low frequencies (below 100 Hz), walking *without* shoes creates sound pressure levels surprisingly close to the sound pressure levels from the tapping machine, whereas walking with shoes gives about 10 dB lower levels. The difference with or without shoes can be explained by the difference between a hard and a soft impact; the latter makes an impulse stretched in time, which means more energy at low frequencies. At higher frequencies (above 100 Hz), the sound pressure levels from the tapping machine are 20 dB to 30 dB higher than those from the live walker. Figure 10.22

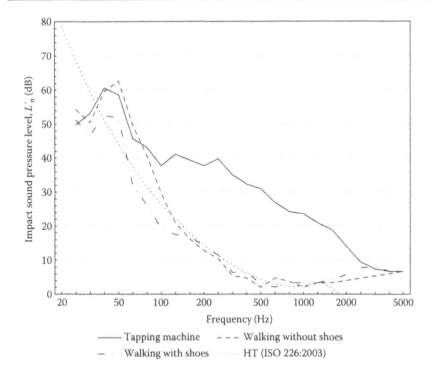

Figure 10.22 Impact sound measured with the tapping machine compared with the average sound pressure level measured with a live walker with or without shoes. The floor is 250 mm concrete slab with 50 mm light concrete topping, 17 mm resilient layer (polyurethane) and 65 mm floating concrete floor. The hearing threshold for pure tones (HT) is shown for comparison. (Adapted from Wolf, M., Burkhart, C., Vermeidungsstrategien und Ansätze einer Vermeidung des Estrichdrönens. *Proceedings of DAGA 2016*, Aachen, Germany, 2016. Figure 3.)

also shows the hearing threshold (HT), although this is for pure tones under special listening conditions. Nevertheless, this indicates that for the actual example, the audible footfall noise is dominated by frequencies below 100 Hz.

The relation between the spectrum from a live walker and from the tapping machine depends to a great deal on the floor construction. Figure 10.23 shows the difference in spectrum for three different floors; a bare concrete (difference increases monotonic with frequency), a floating top surface (the difference is nearly constant around 20 dB above 100 Hz) and a floor with carpet (maximum difference around 50 Hz to 200 Hz). A positive difference means that the tapping machine is louder than the live walker. These results demonstrate that the tapping machine can be used effectively as a testing device up to at least 500 Hz for all types of surface except carpet (Warnock, 1998). With a carpeted floor, the difference curve is fairly close to the others below 200 Hz. Since the impact sound from a carpeted floor

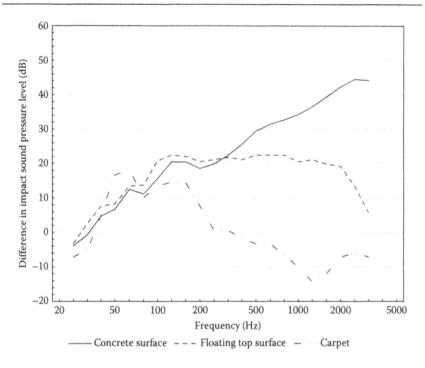

Figure 10.23 Difference between impact sound pressure levels measured with the tapping machine and with a live walker with shoes, shown for three different floorings: Bare concrete, floating floor and carpet. (Adapted from Warnock, A. C. C., Floor research at NRG Canada. *Conference in Building Acoustics "Acoustic Performance of Medium-Rise Timber Buildings"*, Dublin, Ireland, 1998. Figure 3.)

is dominated by low frequencies, it is concluded that a single number rating that includes the low frequencies may correlate well with a rating of the sound from a live walker (Warnock, 1998).

10.8.2 LIGHTWEIGHT AND HEAVYWEIGHT FLOOR CONSTRUCTIONS

Complaints about impact noise in dwellings are very often related to the thudding sound created by footfall. The low-frequency problems of impact sound have been clearly related to lightweight, wood-frame building constructions (Blazier and DuPree, 1994). They found that the peak in the footfall spectrum occurred at the fundamental natural frequency of the floor construction, typically between 15 Hz and 30 Hz. The characteristic difference between impact sound spectra for lightweight and heavyweight floor constructions is illustrated in Figure 10.24. The difference is obvious below 100 Hz.

10.8.3   Typical complaints about impact sound

In multifamily houses, the footfall noise from neighbours is one of the most common causes of complaints. From a technical and acoustical point of view, it is a difficult problem to deal with, one of the reasons being that the measurement and evaluation methods are still insufficient.

In fact, impact noise can be divided into two different types that cover different frequency ranges and may vary independently:

- Low-frequency rumbling, typically below 100 Hz. This is related to the total weight of the floor-ceiling construction.
- High-frequency heel clicks, typically above 400 Hz. This is related to the surface of the floor.

It has been suggested to use two independent evaluation methods to cover these two frequency ranges (LoVerde and Dong, 2017). For the low-frequency range, they suggest using the impact sound pressure level in the octave band including the third octaves 50 Hz, 63 Hz and 80 Hz:

$$L_{50-80} = 10 \lg \left(10^{0.1 \cdot L_{50}} + 10^{0.1 \cdot L_{63}} + 10^{0.1 \cdot L_{80}} \right)$$ (10.55)

For the evaluation of this low-frequency impact sound, the following guide (LoVerde and Dong, 2017) is suggested:

- $L_{50-80} > 70$ dB; the floor is 'thuddy' or 'boomy'; substantial amount of complaints
- $L_{50-80} < 60$ dB; broadly acceptable
- $L_{50-80} < 50$ dB to 55 dB; thudding inaudible; no complaints

In a German investigation on the problems of low-frequency impact sound (Rittig, 2013), a clear difference was found between results from cases of complaints and those from cases of no complaints (Figure 10.25). The question is how to characterize the difference in the best way? In Table 10.4, some objective parameters derived from the two curves in Figure 10.25 are listed. There is no doubt that the frequencies 50 Hz to 80 Hz are very important, which means that we cannot rely on the $L'_{n,w}$ value. In other contexts (noise from Heating, Ventilation and Air Conditioning (HVAC) systems), it is known that rumble causes increased annoyance and this is related to spectral imbalance. In other words, if the energy at mid-frequencies is much lower than the energy at low frequencies, this may cause higher annoyance than noise with a more balanced spectrum (Beranek, 1989).

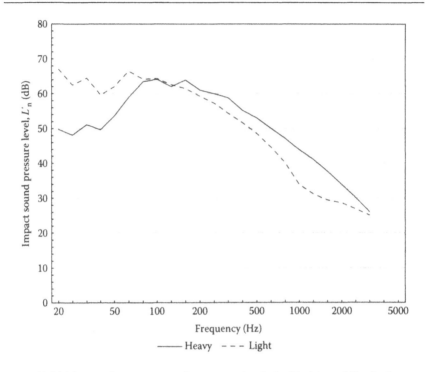

Figure 10.24 Measured impact sound pressure levels in Danish multifamily houses. Average of 10 heavyweight and 8 lightweight floor constructions. Normalization with reverberation time of receiving room is not applied below 50 Hz. (Data from DELTA, *Målerapport: Boligers lydisolation ved lave frekvenser.* (In Danish). PDØ 870028, DELTA, Aarhus, Denmark, 1999.)

The importance of spectrum balance has been found by many researchers, and there seems to be a general agreement about the slope of a potentially annoying sound being around 6 dB to 7 dB per octave or more (Waye, 2011). In the cases of complaints about impact noise, the slope of the spectrum is 7 dB per octave, whereas the other curve with no complaints has a slope of 5 dB per octave. In this connection, we recall that the spectrum from a live walker has more low-frequency energy than that measured with the tapping machine so the spectral imbalance experienced with footfall noise is even more pronounced.

However, it is a myth that low-frequency problems with impact sound are only related to lightweight constructions. Floor constructions that often give rise to complaints about impact noise can be divided into three groups:

- Lightweight timber constructions. The problem is rumbling and the important frequency range may be 20 Hz to 200 Hz. More weight is necessary to solve this problem.

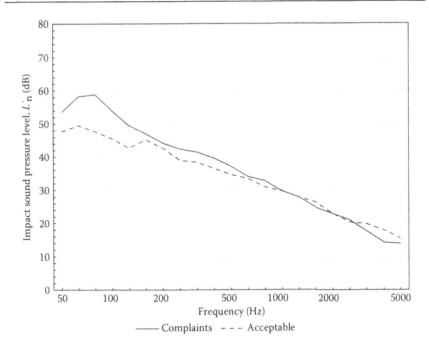

Figure 10.25 Mean values of the impact sound pressure levels measured on site, from 10 cases of quality check and 12 cases of complaints. (Adapted from Rittig, C., Übertragung tieffrequenter trittschallgeräusche auf massivdecken mit schwimmendem estrich. Bachelor thesis, Building Physics, Technical University, Stuttgart, Germany, 2013.)

- Heavyweight concrete constructions with tiles or similar hard surface. The problem is high-frequency clicks due to hard-heeled shoes or similar (chairs drags, dog toenails). The important frequency range is 400 Hz to 3150 Hz (or 5000 Hz). The solution is a soft flooring or a floating floor.
- Heavyweight construction with a heavy floating floor. The problem that can occur is rumbling and the important frequency range may be 50 Hz to 100 Hz. If the floating floor has a resonance frequency above 50 Hz and the resilient layer has a high Q-factor, the impact sound is amplified at frequencies around the resonance frequency. The solution

Table 10.4 Some characteristics of the average impact sound pressure levels in cases of complaints and in cases of no complaints

	$L'_{n,w}$ (dB)	$L'_{n,w,50}$ (dB)	L_{50-80} (dB)	Slope (dB/octave)
Complaints	42	50	62	−7
No complaints	37	41	53	−5

Note: Same data as in Figure 10.25.

can be to bring the resonance frequency below 50 Hz by changing the resilient layer. Alternatively, a resilient material with a low Q-factor may help.

Examples of measured results of constructions from each of the three groups are shown in Figures 10.26 through 10.28. The ISO weighting curve and the unfavourable deviations from the curve are also shown for comparison. While the weighting curve works reasonably well in the example with a hard surface, Figure 10.27, this is not the case in the other examples where low-frequency rumbling is the problem. The parameter with extended frequency range $L'_{n,w,50}$ works fine in the two examples with low-frequency problems, but fails in the example of a hard floor surface (Table 10.5).

The resonance frequency of a floating floor is calculated with Equation 10.43. The level of the resonance peak (negative ΔL) depends on the Q-factor of the resilient layer, see the definition (2.69):

$$\Delta L_0 = -10 \lg\left(1 + Q^2\right) \cong -20 \lg Q \tag{10.56}$$

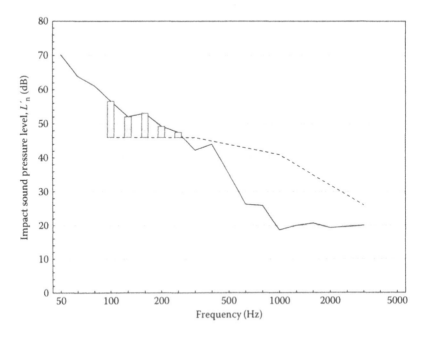

Figure 10.26 Impact sound pressure level of a timber floor. Construction from top to bottom: soft carpet, 22 mm chipboard, 48 mm×198 mm wooden beams with 200 mm mineral wool in the cavity, 48 mm×48 mm lath with resilient mounting of 2×13 mm gypsum board. (Adapted from Hagberg, K., Simmons, C., *Bygg & teknik* 1/97, 17–18, 1997.)

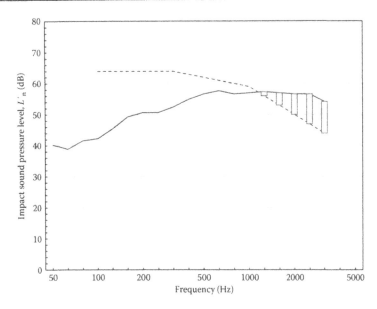

Figure 10.27 Impact sound pressure level of a heavyweight floor with hard surface. Construction from top to bottom: Clinker tiles on 250-mm concrete. (Adapted from Hagberg, K., Simmons, C., *Bygg & teknik* 1/97, 17–18, 1997.)

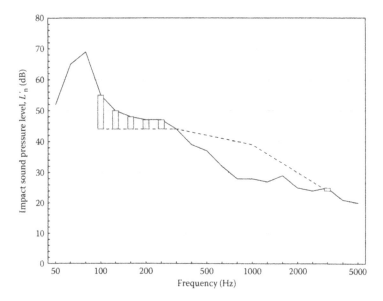

Figure 10.28 Impact sound pressure level of a heavyweight floor with floating floor. Construction from top to bottom: 14 mm parquet, 3 mm ethafoam, 30 mm gypsum, 0.15 mm plastic foil, 15 mm mineral wool, 230 mm concrete. (Data from Multiconsult, Unpublished measurement report. Project Number 103407, 2002.)

Table 10.5 Some characteristics of the impact sound pressure levels in cases of
complaints

Construction	$L'_{n,w}$ (dB)	$L'_{n,w,50}$ (dB)	L_{50-80} (dB)	Slope (dB/octave)
Lightweight timber	44	57	72	−9
Concrete with hard surface	62	52	45	+5
Concrete with floating floor	42	56	71	−7

Note: Same data as in Figures 10.26 through 10.28.

Since ΔL is the positive attenuation of the impact sound, the amplification
at the resonance is a negative dB value. If the resilient layer is some kind of
mineral wool, as described by Schirmer and Schirmer (2016), the Q-factor
can be estimated from the flow resistance:

$$Q = \frac{2\pi m_1 f_0}{r_s d} \tag{10.57}$$

where m_1 is the mass per unit area of the floating floor, r_s is the airflow
resistivity per unit length and d is the thickness of the resilient layer. Typical
values of Q for floating floors are between 2.5 and 10.

Unfortunately, it is difficult to get information about the flow resistance
from the manufacturers of resilient materials for floors; normally, only data for
the dynamic stiffness are available. The relevant measurement standards are
ISO 9052-1 for the dynamic stiffness and ISO 9053 for the airflow resistance.

REFERENCES

L.L. Beranek (1989). Balanced noise-criterion (NCB) curves. *Journal of the Acoustical Society of America* 86(2), 650–664.

L.L. Beranek and I.L. Vér (1992). *Noise and Vibration Control Engineering*. John Wiley & Sons, New York.

W.E. Blazier and R.B. DuPree (1994). Investigation of low-frequency footfall noise in wood-frame, multifamily building construction. *Journal of the Acoustical Society of America* 96 (3), 1521–1532.

L. Cremer and M. Heckl (1967). *Körperschall.* (*Structure-borne sound*, translated by E.E. Ungar 1973). Springer, Berlin, Germany.

DELTA (1999). *Målerapport: Boligers lydisolation ved lave frekvenser.* (Measurement report: Sound insulation of dwellings at low frequencies, in Danish). PDØ 870028, DELTA, Aarhus, Denmark.

L. Eriksen and C.V.J. Ejlersen (2003). *Lydisolering—Etageadskillelser af massive træelementer.* (Sound insulation – floor constructions of massive wood slabs, in Danish). MSc Thesis, Byg-DTU and Ørsted-DTU, Technical University of Denmark, Lyngby, Denmark.

S. Gudmundson (1984). *Sound insulation improvements of floating floors. A study of parameters.* Report TVBA-3017, LTH, Lund, Sweden.

K. Gösele (1964). *Schallbrücken bei schwimmenden Estrichen und anderen schwimmend verlegten Belägen.* (Sound bridges in floating floors and other floating layers, in German). Berichte aus der Bauforschung, Heft 35. Wilhelm Ernst und Sohn, Berlin, Germany.

K. Hagberg and C. Simmons (1997). Nordisk ljudklassningsstandard – skärpta ljudkrav. (Nordic standard for sound classification – increased acoustical requirements, in Swedish). *Bygg & teknik* 1/97, 17–18.

ISO 9052-1 (1989). *Acoustics – Determination of dynamic stiffness – Part 1: Materials used under floating floors in dwellings.* International Organization for Standardization, Geneva, Switzerland.

ISO 9053 (1991). *Acoustics – Materials for acoustical applications – Determination of airflow resistance.* International Organization for Standardization, Geneva, Switzerland.

ISO 10140-5 (2010). *Acoustics – Laboratory Measurement of Sound Insulation of Building Elements – Part 5: Requirements for test facilities and equipment.* International Organization for Standardization, Geneva, Switzerland.

ISO 10140-1 (2010). *Acoustics – Laboratory measurement of sound insulation of building elements – Part 1: Application rules for specific products.* International Organization for Standardization, Geneva, Switzerland.

J. Kristensen and J.H. Rindel (1989). *Bygningsakustik – Teori og praksis.* SBI-Anvisning 166 (Building acoustics – Theory and practice, in SBI-Anvisning 166. Danish). Danish Building Research Institute, Aalborg University Copenhagen.

W. Kuhl (1982). Impedanz von platten endlicher fläche, Körperschallpegeldifferenz punktförmig angeregter paralleler platten und trittschallminderung. (Impedance of plates with finite size, structure-borne sound level difference of parallel plates with point excitation and attenuation of impact sound, in German). *Acustica* 51, 103–115.

J.J. LoVerde and D.W. Dong (2017). A dual-rating method for evaluating impact noise isolation of floor-ceiling assemblies. *Journal of the Acoustical Society of America* 141, 428–440.

Multiconsult (2002). Unpublished measurement report. Project Number 103407.

C. Rittig (2013). *Übertragung tieffrequenter trittschallgeräusche auf massivdecken mit schwimmendem estrich.* (Transfer of low-frequency impact sound of solid slabs with a floating floor, in German). Bachelor's thesis, Building Physics, Technical University, Stuttgart, Germany.

H. Schirmer and W. Schirmer (2016). Schwimmender estrich und unterhangdecke mit schwach gedämpfter eigenfrequenz fast gleicher grösse als ursache für erhöhe tieffrequente trittschallübertragung. (Floating floor and suspended ceiling with weakly attenuated eigenfrequency although the same as reason for increase low-frequency transmission of impact sound, in German). *Proceedings of DAGA 2016*, 4–17 March 2016, Aachen, Germany.

J. Seidel and K. Hengst (2017). Trittschallminderung und einfügungsdämmung von raumakustik-decken. (Impact sound attenuation and insertion loss of room acoustical ceilings, in German). *Proceedings of DAGA 2017*, 6–9 March 2017, Kiel, Germany.

I.L. Ver (1971). Impact noise isolation of composite floors. *Journal of the Acoustical Society of America* 50, 1043–1050.

A.C.C. Warnock (1998). Floor research at NRG Canada. *Conference in Building Acoustics "Acoustic Performance of Medium-Rise Timber Buildings"*, Dublin, Ireland.

K.P. Waye (2011). *Noise and Health – Effects of Low Frequency Noise and Vibrations: Environmental and Occupational Perspectives.* Encyclopedia on Environmental Health (J.O. Nriagu, Editor-in-Chief). Elsevier, Amsterdam.

M. Wolf and C. Burkhart (2016). Vermeidungsstrategien und ansätze einer vermeidung des estrichdrönens. (Strategies for control and attempts to attenuate the thump of floors, in German). *Proceedings of DAGA 2016*, 14–17 March 2016, Aachen, Germany.

Chapter 11

Flanking transmission

The flanking constructions and the junctions between the constructions are of high importance for the sound insulation between rooms. The number of transmission paths to be considered is often quite high, so there is need for a computer program to handle this in a reasonable way. Computation programs for sound insulation normally have the drawback that their results are the maximum values that can be achieved under conditions with perfect workmanship and acoustically correct building details, which can seldom be achieved in practice. Yet, computation programs are useful tools at the design stage for sound insulation.

11.1 FLANKING TRANSMISSION OF AIRBORNE SOUND

11.1.1 Transmission paths

In buildings with traditional heavy constructions of concrete or masonry, about 50 % of the sound transmission between rooms with a common dividing partition is flanking transmission. Of course, the flanking transmission is 100 % between rooms without a common dividing partition.

Figure 11.1 shows the principles of flanking transmission on four different types of building. It is important whether the constructions are heavyweight or lightweight because the sound radiation of free bending waves is different. Heavyweight walls and floors have a low critical frequency and, thus, the radiation efficiency for free bending waves is high (approximately unity). Figure 11.1a shows a building type with heavyweight walls and floors like the precast concrete houses typical from the 1960s and the 1970s. Flanking is both vertical and horizontal, and this may have contributed to the bad reputation of sound insulation in houses from that time. Type B with lightweight walls and heavyweight floors is not unusual today, and flanking is mainly in the horizontal direction. Type C with heavyweight, load-bearing walls and lightweight floors is very common in older buildings with masonry walls and traditional timber floors. These houses

313

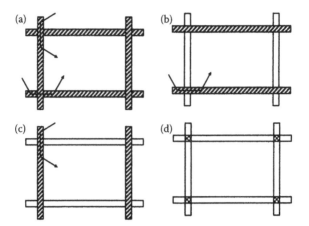

Figure 11.1 Direction of flanking transmission in four building types. (a) Type A – heavy-weight walls and floors; (b) type B – light-weight walls and heavy-weight floors; (c) type C – heavy-weight walls and light-weight floors; and (d) type D – light-weight walls and floors.

are characterized by flanking in the vertical direction. Finally, type D has a load-bearing system of columns and beams, both wall and floor constructions are lightweight, and flanking is not a big issue.

Figure 11.2 shows examples of transmission paths for airborne sound from one room to the adjacent room. For each set of flanking surfaces, there are three flanking transmission paths, i.e. with four flanking surfaces, there are 12 transmission paths in addition to the direct transmission.

11.1.2 Junction attenuation

One flanking transmission path involves two surfaces: one in the source room and another in the receiving room. The pair of plates is connected through a junction (Figure 11.3). The *junction attenuation* is defined as the

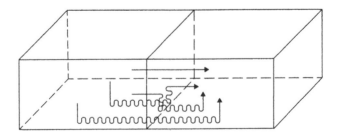

Figure 11.2 Airborne sound transmission horizontally. The transmission paths are shown via one flanking surface (the floor) and directly through the wall. (Reproduced from Kristensen, J. and Rindel, J. H., *Bygningsakustik – Teori og praksis (Building acoustics – Theory and practice, in Danish)*. SBI-Anvisning 166, Danish Building Research Institute, Hørsholm, Denmark, 1989. With permission.)

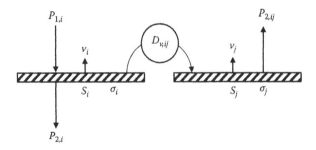

Figure 11.3 Simplified outline for flanking transmission from plate i to plate j at airborne sound exciting. $D_{v,ij}$ is the junction attenuation.

difference in velocity level on the plate i in the source room and that on the plate j radiating into the receiving room:

$$D_{v,ij} = 10 \lg \frac{\langle \tilde{v}_i^2 \rangle}{\langle \tilde{v}_j^2 \rangle} = 10 \lg \frac{1}{d_{ij}} \tag{11.1}$$

The direction is $i \to j$, which is important; the junction attenuation may be different in the opposite direction. The determination of the junction attenuation will be described later in this chapter. Since the plate i is in the source room, both forced and resonant excitation will take place.

11.1.3 Flanking transmission loss

Each transmission path is described by a flanking transmission loss R_{ij} corresponding to the transmission via the surface i of the source room and the surface j of the receiving room. The flanking transmission loss from i to j is:

$$R_{ij} = 10 \lg \frac{P_{1,0}}{P_{2,ij}} \tag{11.2}$$

where $P_{1,0}$ is the incident sound power on the partition wall area S_0 and $P_{2,ij}$ is the power radiated into the receiving room from the transmission path under consideration. The reason for choosing the same reference area S_0 for all flanking transmission losses is that they can easily be added together without further area corrections. This would not be the case if each flanking transmission loss had the incident sound power $P_{1,i}$ on the flanking surface S_i as a reference.

Under the assumption of diffuse sound field in the source room, the flanking transmission loss can be written as (see Equation 8.25):

$$R_{ij} = 10 \lg \frac{\langle \tilde{p}_1^2 \rangle S_0}{4\rho c P_{2,ij}} \tag{11.3}$$

where $\langle \tilde{p}_1^2 \rangle$ is the mean square sound pressure in the source room and S_0 is the area of the common surface between the rooms. In cases where the rooms have no common separation area, S_0 is set to a fixed reference value of $10\,\mathrm{m}^2$.

With a diffuse sound field in the source room, the mean velocity of forced vibrations in the excited plate i can be determined from Equations 8.23 and 5.35:

$$\langle \tilde{v}_{i,\mathrm{for}}^2 \rangle = \frac{2 \langle \tilde{p}_1^2 \rangle}{\left(\omega m_i \right)^2 \left(\left(1 - \left(f / f_{c,i} \right)^2 \right)^2 + \eta_i^2 \right)} \tag{11.4}$$

The resonant vibrations are similarly determined from Equation 7.29:

$$\langle \tilde{v}_{i,\mathrm{res}}^2 \rangle = \frac{\langle \tilde{p}_1^2 \rangle}{\left(\omega m_i \right)^2} \cdot \frac{c^2 \sigma_{i,\mathrm{res}}}{2 \eta_i S_i f} \cdot \frac{\Delta N_i}{\Delta f} \tag{11.5}$$

The total vibrations in plate i are combined from forced and resonant excitation. It can be assumed that both are transferred to the receiving plate j with the same junction attenuation and, thus, the resulting velocity in plate j becomes:

$$\langle \tilde{v}_j^2 \rangle = d_{ij} \cdot \left(\langle \tilde{v}_{i,\mathrm{for}}^2 \rangle + \langle \tilde{v}_{i,\mathrm{res}}^2 \rangle \right) \tag{11.6}$$

The forced vibrations are restricted to the excited plate i. In plate j, only resonant vibrations are possible although the energy of the vibrations is fed from forced and resonant vibrations in plate i. By means of Equations 11.4 through 11.6, the radiated sound power from plate j is:

$$P_{2,ij} = \langle \tilde{v}_j^2 \rangle \rho c S_j \sigma_{j,\mathrm{res}}$$

$$= d_{ij} \cdot \langle \tilde{p}_1^2 \rangle \cdot \frac{\rho c}{(\omega m_i)^2} \cdot S_j \sigma_{j,\mathrm{res}} \cdot \left(\frac{2}{\left(1 - \left(f / f_{c,i} \right)^2 \right)^2 + \eta_i^2} + \frac{c^2 \sigma_{i,\mathrm{res}}}{2 \eta_i S_i f} \cdot \frac{\Delta N_i}{\Delta f} \right) \tag{11.7}$$

The flanking transmission loss R_{ij} can be determined by insertion of this result in Equation 11.3:

$$R_{ij} = R_{0,i} + D_{v,ij} - 10\lg\frac{S_j\sigma_{j,\text{res}}}{S_0} - 10\lg\left(\frac{2}{\left(1-(f/f_{c,i})^2\right)^2 + \eta_i^2} + \frac{c^2\sigma_{i,\text{res}}}{2\eta_i S_i f} \cdot \frac{\Delta N_i}{\Delta f}\right) \quad (11.8)$$

where $R_{0,i} = 20\lg(\omega m_i/2\rho c)$ is the mass law at normal incidence for plate i. From Equation 11.8, it is seen that a high flanking transmission loss can be achieved when the plate being excited has a high transmission loss, i.e. with heavy flanking constructions. Large junction attenuation or small radiation efficiency for the radiating surface j may also contribute to a high flanking transmission loss.

The transmission loss R of the direct transmission path can be written in a form similar to Equation 11.8, using index 1 for this surface:

$$R = R_{0,1} - 10\lg\left(\frac{2\sigma_{1,\text{for}}}{\left(1-(f/f_{c,1})^2\right)^2 + \eta_1^2} + \frac{c^2\sigma_{1,\text{res}}^2}{2\eta_1 S_1 f} \cdot \frac{\Delta N_1}{\Delta f}\right) \quad (11.9)$$

For this direct transmission, the radiation into the receiving room is different for the forced and resonant vibrations, represented by the different radiation efficiencies $\sigma_{1,\text{for}}$ and $\sigma_{1,\text{res}}$. Here, the area is called S_1, but it is normally the same as the reference area S_0.

11.1.4 The total apparent transmission loss

The apparent sound transmission loss R' can be determined by summing up the radiated sound power from all possible transmission paths including the sound power from the direct path $P_{2,0}$:

$$R' = 10\lg\frac{P_{1,0}}{P_{2,0} + \Sigma P_{2,ij}} = -10\lg\left(10^{-R/10} + \Sigma 10^{-R_{ij}/10}\right) \quad (11.10)$$

where each flanking transmission path is calculated according to Equation 11.8 and the direct transmission from Equation 11.9.

11.1.5 The reciprocity method

A simplification of the basis of calculation is possible when it is assumed that sound transmission is independent of the transmission direction, as shown by Gerretsen (1979). This will only be valid for heavy walls with sufficiently low critical frequency so only resonant transmission is considered. Under these assumptions, the flanking transmission loss may be written in terms of the transmission loss R_i of the flanking wall:

$$R_{ij} = R_i + D_{v,ij} + 10\lg\frac{S_0\sigma_{i,\text{res}}}{S_j\sigma_{j,\text{res}}} \tag{11.11}$$

and if it is assumed that this is the same in both directions:

$$R_{ij} = R_{ji} = \tfrac{1}{2}\left(R_{ij} + R_{ji}\right) \tag{11.12}$$

Thus, the radiation efficiencies will disappear from the equation, and

$$R_{ij} = \tfrac{1}{2}R_i + \tfrac{1}{2}R_j + \bar{D}_{v,ij} + 10\lg\frac{S_0}{\sqrt{S_iS_j}} \tag{11.13}$$

where

$$\bar{D}_{v,ij} = \tfrac{1}{2}\left(D_{ij} + D_{ji}\right) \tag{11.14}$$

is the junction attenuation as a mean value for the two directions $i \to j$ and $j \to i$. This reciprocity method has been the basis for the standard ISO 12354-1. However, as explained here, the applications are rather limited since the method is only valid for heavy constructions where the forced transmission can be neglected.

11.2 FLANKING TRANSMISSION OF IMPACT SOUND

Figure 11.4 shows examples of transmission paths for impact sound pressure horizontally and vertically. Horizontally, there will normally only be two significant transmission paths, whereas vertically, there will be one transmission path for each of the four flanking walls, in addition to the direct transmission through the horizontal division.

11.2.1 The impact sound pressure level for one transmission path

The impact sound pressure level with sound transmission from i to j only is:

$$L_{n,ij} = 10\lg\frac{\langle\tilde{p}_{2,ij}^2\rangle A_2}{p_0^2 A_0} = 10\lg\frac{4\rho c P_{2,ij}}{p_0^2 A_0} \tag{11.15}$$

(a)   (b)

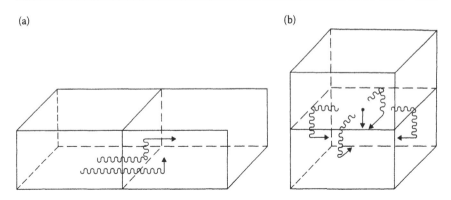

Figure 11.4 Impact sound transmission paths. (a) Horizontal and (b) vertical. (Reproduced from Kristensen, J. and Rindel, J. H., *Bygningsakustik—Teori og praksis (Building acoustics—Theory and practice, in Danish)*. SBI-Anvisning 166, Danish Building Research Institute, Hørsholm, Denmark, 1989. With permission.)

where A_2 is the equivalent absorption area in the receiving room and $A_0 = 10\,\text{m}^2$ is a reference area (see also Figure 5.31). The transmission is illustrated schematically in Figure 11.5. With the excitation force F, the resonant modes in plate i are activated (see Equation 10.10) and the mean velocity square can be found from:

$$\langle \tilde{v}_i^2 \rangle = \frac{\tilde{F}^2 \, Re\{Y_i\}}{2\pi f m_i S_i \eta_i} = \frac{\tilde{F}^2}{8\pi f m_i^2 S_i^2 \eta_i} \cdot \frac{\Delta N_i}{\Delta f} \tag{11.16}$$

where m_i, S_i and η_i are the mass per unit area, area and loss factor of plate i, respectively. Y_i is the mechanical point admittance of the plate, i.e. the inverse of the point impedance $Y_i = 1/Z_i$. The modal density in plate i is $\Delta N_i / \Delta f$, and this is a function of the frequency (different for bending waves and shear waves).

The transmission of the vibration energy from plate i to plate j is described with the junction attenuation $D_{v,ij}$ (Equation 11.1). The radiated sound power $P_{2,ij}$ is a function of the average velocity in plate j and the radiation efficiency:

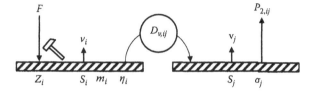

Figure 11.5 Simplified outline for flanking transmission from plate i to plate j by exciting with a tapping machine. $D_{v,ij}$ is the junction attenuation and Z_i is the mechanical point impedance.

$$P_{2,ij} = \langle \tilde{v}_j^2 \rangle \rho c S_j \sigma_{j,\mathrm{res}}$$

$$= \frac{\langle \tilde{v}_j^2 \rangle}{\langle \tilde{v}_i^2 \rangle} \cdot \frac{\tilde{F}^2 \rho c}{8\pi f m_i^2 S_i^2 \eta_i} \cdot \frac{\Delta N_i}{\Delta f} \cdot S_j \sigma_{j,\mathrm{res}} \tag{11.17}$$

The impact sound pressure level for transmission path $i \to j$ can now be determined from:

$$L_{n,ij} = 10\lg \frac{4\rho c P_{2,ij}}{p_0^2 A_0} = 10\lg \left(\frac{\rho^2 c^2}{p_0^2 A_0} \frac{\langle \tilde{v}_j^2 \rangle}{\langle \tilde{v}_i^2 \rangle} \cdot \frac{\tilde{F}^2}{2\pi f m_i^2 S_i^2 \eta_i} \cdot \frac{\Delta N_i}{\Delta f} \cdot S_j \sigma_{j,\mathrm{res}} \right) \tag{11.18}$$

By insertion of the force generated by the hammers of the tapping machine in one-third octave frequency bands $\tilde{F}^2 = 0.91f$ (N²), (Equation 10.4) and other constants and including the possibility of soft flooring with the attenuation $\Delta L_{\mathrm{cover}}$, the result is:

$$L_{n,ij} = 82 - 20\lg m_i - D_{v,ij} + 10\lg \left(\frac{c^2}{\pi S_i^2 \eta_i} \cdot \frac{\Delta N_i}{\Delta f} \cdot S_j \sigma_{j,\mathrm{res}} \right) - \Delta L_{\mathrm{cover}} \tag{11.19}$$

This general result can be simplified if it is assumed that the vibrations in plate i are bending waves, and the appropriate modal density is inserted (see Table 7.1):

$$L_{n,ij} = 82 - 10\lg \left(m_i^2 \frac{\eta_i}{f_{c,i}} \right) - D_{v,ij} + 10\lg \left(\frac{S_j \sigma_{j,\mathrm{res}}}{S_i} \right) - \Delta L_{\mathrm{cover}} \tag{11.20}$$

For more complicated floor constructions, it may be a problem to determine the point impedance. A basis of calculation that considers the effect of the lowest natural frequencies has been described by Breeuweer and Tukker (1976). For a number of floor constructions, including ribbed decks and plate girder systems, some measured admittances are reported in the work of Gerretsen (1986).

11.2.2 Total impact sound pressure level

When the impact sound pressure levels for each transmission path are determined, the total impact sound pressure level is calculated by addition of the sound powers radiated into the receiving room:

$$L_n' = 10\lg \left(\sum 10^{L_{n,ij}/10} \right) \tag{11.21}$$

11.2.3 Combination of airborne and impact transmission

The junction attenuation for the impact sound is assumed to be the same as for the airborne flanking transmission. A double calculation can thus be avoided by combining Equation 11.8 with Equation 11.19. Here, it is simplified by neglecting the forced transmission in Equation 11.8, i.e. for frequencies above $f_{c,i}$ we get:

$$R_{ij} + L_{n,ij} = 82 + R_{0,i} - 20 \lg m_i + 10 \lg \left(\frac{2fS_0}{\pi S_i \sigma_{i,\text{res}}} \right) - \Delta L_{\text{cover}} \tag{11.22}$$

As m_i is also inherent in the mass law $R_{0,i}$ it cancels out and we get:

$$R_{ij} + L_{n,ij} = 38 + 30 \lg f - 10 \lg \left(\frac{S_i \sigma_{i,\text{res}}}{S_0} \right) - \Delta L_{\text{cover}} \tag{11.23}$$

This shows that the simple relation between the airborne sound insulation and the impact sound pressure level for the total transmission (Equation 10.21) also applies to the individual transmission paths. The only difference is that the area correction, which is due to the flanking transmission loss always refers to the area of the partition surface S_0. In case of vertical transmission, $S_0 = S_i$ is the area of the floor in the source room.

11.3 JUNCTION ATTENUATION

11.3.1 Empirical model

One of the major problems when calculating sound insulation in buildings is to provide data for the junction attenuation between the building parts. For homogeneous massive plates, there is an extensive experimental basis, as the data for cross junctions shown in Figure 11.6.

The junction attenuation is approximately 12 dB when the plates have the same mass per unit area. Flanking transmission through a structure intersected by a heavier structure is attenuated more than 12 dB and vice versa.

11.3.2 The structural transmission coefficient

Theoretical information on sound transmission between building parts is available, but normally in another version, expressed as a *structural transmission coefficient* γ_{ij}. This is defined as the ratio between the transmitted energy into plate j and the incident energy to the junction from plate i:

$$\gamma_{ij} = \frac{P_{ij}}{P_{inc}} \qquad (11.24)$$

In Figures 11.7 and 11.8, the theoretical transmission coefficients are rendered for different types of junctions. These are X-junctions or junctions between three or two plates. Applying the numbering of Figure 11.9, the corner joint only consists of plate 1 and plate 2, whereas the T-junction misses either plate 3 or plate 4. The abscissa in Figures 11.7 and 11.8 is the ratio:

$$\psi = \frac{m_2 f_{c1}}{m_1 f_{c2}} \qquad (11.25)$$

where m is the mass per unit area and f_c is the critical frequency. The index refers to the plates with numbers 1 and 2, respectively. In X- and T-junctions, it is presumed that plate 3 equals plate 1, and plate 4 equals plate 2.

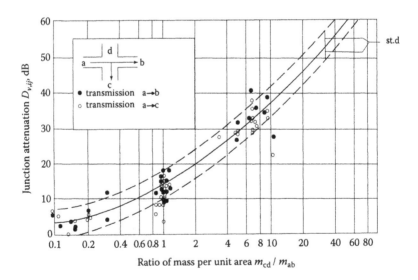

Figure 11.6 Experimental junction attenuation for massive homogeneous junctions shown as a function of the mass ratio (plates c–d/plates a–b). Solid line is a regression line drawn through the measured points. Dashed lines are the standard deviations. (Adapted from Gerretsen, E., Appl. Acoust., 12, 413–433, 1979, Figure 5. With permission from Elsevier.)

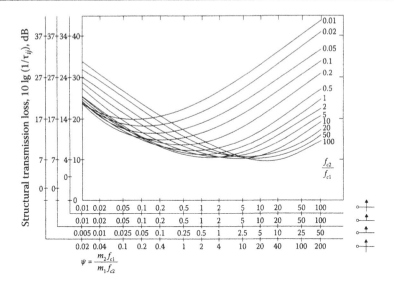

Figure 11.7 Theoretically determined transmission coefficients from plate 1 to plate 2 (γ_{12}) for four different types of junctions. (Adapted from Craik, R. J. M., *Appl. Acoust.*, 14, 347–359, 1981, Figure 2. With permission from Elsevier.)

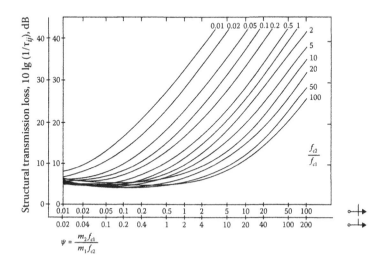

Figure 11.8 Theoretically determined transmission coefficients from plate 1 to plate 3 (γ_{13}) for X- and T-junctions. (Adapted from Craik, R. J. M., *Appl. Acoust.*, 14, 347–359, 1981, Figure 3. With permission from Elsevier.)

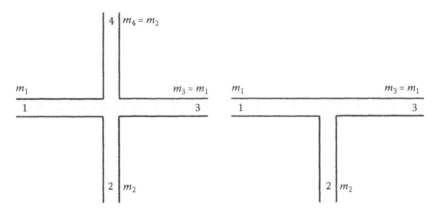

Figure 11.9 Numbering the plates in a junction. Left: Cross-section, right: T-junction.

11.3.3 Relation between junction attenuation and structural transmission loss

The energy in a wave travels with a speed that is called the group speed. In an acoustically thick plate, the phase speed of transverse waves (Mindlin waves) can be expressed by the speed of shear waves c_s and the cross-over frequency for shear waves f_s. From Equation 8.14 and using Equation 1.30, we get:

$$c_{B,\text{eff}} = c_s \frac{f}{f_s} \sqrt{-\frac{1}{2} + \frac{1}{2}\sqrt{1 + \left(\frac{2f_s}{f}\right)^2}} \tag{11.26}$$

However, this is the so-called phase speed. In media with dispersion, the group speed c_g differs from the phase speed (Pedersen, 1995) and we get:

$$c_g = \frac{c_{B,\text{eff}}^3}{c_s^2} \sqrt{1 + \left(\frac{2f_s}{f}\right)^2} \tag{11.27}$$

or approximately:

$$c_g \cong \begin{cases} 2 \cdot c_B = 2 \cdot c\sqrt{f/f_c} & (f < f_s/4) \\ c_s & (f > f_s/4) \end{cases} \tag{11.28}$$

where f_s is the cross-over frequency for shear waves (Equation 1.29).

The group speed is used for the calculation of the incident power of vibrational energy on the connecting junction. Under the assumption that the vibrational energy in plate i travels equally in all directions (Rindel, 1994), the incident sound power on the connecting edge with length l_{ij} is:

$$P_{inc} = \frac{1}{\pi} \ell_{ij} m_i \left\langle \tilde{v}_i^2 \right\rangle c_{g,i} \cong \frac{1}{\pi} \ell_{ij} m_i \left\langle \tilde{v}_i^2 \right\rangle 2 c_{B,i} \qquad (11.29)$$

where $c_{g,i}$ is the *group speed* of the vibrations in plate i, and the approximation is valid for bending waves.

In the receiving plate j, the power transmitted from plate i must equal the absorbed power under steady-state conditions $P_{ij}=P_{j,abs}$ and the steady-state energy in the plate is $E_j=m_j S_j \left\langle \tilde{v}_j^2 \right\rangle$. Thus, we have the energy balance for plate j:

$$P_{ij} = P_{j,abs} = \omega E_j \eta_j = 2\pi f m_j S_j \left\langle \tilde{v}_j^2 \right\rangle \eta_j \qquad (11.30)$$

where S_j is the area and η_j is the total loss factor for plate j. The structural transmission coefficient yields:

$$\gamma_{ij} = \frac{P_{ij}}{P_{inc}} = 2\pi^2 f \cdot \frac{m_j \left\langle \tilde{v}_j^2 \right\rangle S_j \eta_j}{m_i \left\langle \tilde{v}_i^2 \right\rangle \ell_{ij} c_{g,i}} = 2\pi^2 \cdot \frac{m_j \left\langle \tilde{v}_j^2 \right\rangle}{m_i \left\langle \tilde{v}_i^2 \right\rangle} \cdot \frac{2.2 S_j}{\ell_{ij} c_{g,i} T_j} \qquad (11.31)$$

where $c_{g,i}$ is the group speed of transversal waves in plate i and ℓ_{ij} is the length of the edge that adjoins the two plates. In the last expression, the structural reverberation time T_j in plate j is inserted from Equation 2.30. This formula can be used as the basis for measurements of the structural transmission coefficient.

The structural transmission coefficient can be expressed in dB and is called the *structural transmission loss* $R_{s,ij}$:

$$R_{s,ij} = 10 \lg \frac{1}{\gamma_{ij}} = D_{v,ij} + 10 \lg \frac{m_i}{m_j} + 10 \lg \frac{l_{ij}}{a_j} \text{ (dB)} \qquad (11.32)$$

where a_j is the *equivalent absorption length* in plate j:

$$a_j = 2\pi^2 f \cdot \frac{S_j \eta_j}{c_{g,i}} \cong 4.4 \cdot \pi^2 \frac{S_j}{c_{g,i} \cdot T_j} \qquad (11.33)$$

Insertion of the group speed (Equation 11.28) in Equation 11.33 yields the structural transmission loss:

$$R_{s,ij} \cong \begin{cases} D_{v,ij} + 10\lg\dfrac{m_i}{m_j} + 10\lg\left(\dfrac{l_{ij}}{S_j} T_i \sqrt{f/f_{c,i}}\right) + 12\ \text{dB} & (f < f_{s,i}/4) \\[3mm] D_{v,ij} + 10\lg\dfrac{m_i}{m_j} + 10\lg\left(\dfrac{c_{s,i} l_{ij}}{cS_j} T_i\right) + 9\ \text{dB} & (f > f_{s,i}/4) \end{cases}$$

(11.34)

Thus, if we assume that both junction attenuation and structural reverberation time are frequency-independent, the structural transmission loss increases slowly by 1.5 dB per octave for bending waves (at frequencies below $f_{s,i}/4$). At higher frequencies, the structural transmission loss becomes independent of frequency.

11.3.4 Vibration reduction index for rigid connections

The *vibration reduction index* K_{ij} is used to characterize the junction attenuation in the calculation standards ISO 12354-1 and ISO 12354-2. It is defined as the direction-averaged structural transmission loss:

$$K_{ij} = \frac{1}{2}(R_{s,ij} + R_{s,ji}) = \frac{1}{2}(D_{v,ij} + D_{v,ji}) + \frac{1}{2} \cdot 10\lg\left(\frac{l_{ij}}{a_i} \cdot \frac{l_{ij}}{a_j}\right)$$

$$= \overline{D_{v,ij}} + 5\lg\frac{l_{ij}^2}{a_i \cdot a_j}\ (\text{dB})$$

(11.35)

Here, a_i and a_j are the equivalent absorption lengths of plates i and j, respectively, and l_{ij} is the length of the connection.

As examples, we consider a rigid junction of homogeneous plates, either as a cross-junction or a T-junction. See Figure 11.9 for definition of K_{13} and K_{12}. Examples of the vibration reduction index suggested in ISO 12354-1 are shown in Figures 11.10 and 11.11. These are assumed frequency-independent.

11.3.5 Estimated attenuation for elastic connections

The structural transmission loss was measured in the laboratory for a large number of T-junctions by (Pedersen, 1995). Measurements were made with two walls of either 100 mm clinker concrete ($\rho = 1030\,\text{kg/m}^3$) or 100 mm porous concrete ($\rho = 600\,\text{kg/m}^3$) connected to a slab of 180 mm clinker concrete ($\rho = 1700\,\text{kg/m}^3$). For a rigid connection between two clinker concrete walls and the slab, the transmission loss was almost frequency-independent and in reasonable agreement with the prediction model for rigid connections in Figure 11.11.

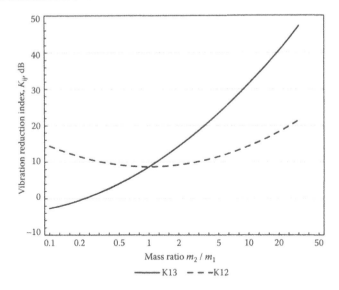

Figure 11.10 Vibration reduction index for a rigid cross-junction. (Adapted from ISO 12354-1, *Building acoustics—Estimation of acoustic performance of buildings from the performance of elements—Part 1: Airborne sound insulation between rooms*, International Organization for Standardization, Geneva, Switzerland, 2017. Figure E1.)

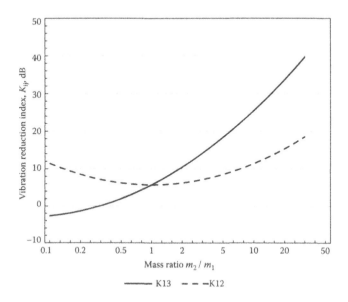

Figure 11.11 Vibration reduction index for a rigid T-junction. (Adapted from ISO 12354-1, *Building acoustics—Estimation of acoustic performance of buildings from the performance of elements—Part 1: Airborne sound insulation between rooms*, International Organization for Standardization, Geneva, Switzerland, 2017. Figure E3.)

With two elastic interlayers, e.g. one below and the other one above the slab, the transmission loss becomes strongly frequency dependent, increasing with frequency above a certain frequency, which according to Pedersen (1995) can be estimated from:

$$f_1 = 2.5 \cdot 10^{-6} \left(\frac{\sqrt{\rho_1 \rho_2}}{G} d \frac{h_1}{w} \right)^{-3/2} \text{(Hz)} \qquad (11.36)$$

where

ρ_1 is the density of plate 1 (kg/m³).
ρ_2 is the density of plate 2 (kg/m³).
G is the shear modulus of the elastic interlayer (Pa).
d is the thickness of the elastic interlayer (m).
h_1 is the thickness of plate 1 (m).
w is the width of the elastic interlayer (m).

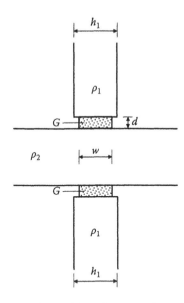

Figure 11.12 Junction with two identical thin elastic interlayers. (Adapted from Pedersen, D. B., Appl. Acoust., 46, 285–305, 1995. With permission from Elsevier.)

The parameters in the above equation are shown in Figure 11.12.

At frequencies below f_1, the transmission loss is similar to that for a rigid connection and frequency-independent. An empirical formula for the structural transmission loss is:

$$R_{s,ij} = R_{s,ij,\text{rigid}} + n \cdot \Delta R_s. \tag{11.37}$$

where $R_{s,ij,\text{rigid}}$ is the structural transmission loss for the equivalent rigid junction, n is the number of resilient layers in the junction (normally 1 or 2) (Pedersen, 1995) and ΔR_s is:

$$\Delta R_s \cong \begin{cases} 0 \text{ dB} & (f < f_1) \\ 10 \lg\left(\dfrac{f}{f_1}\right) \text{ (dB)} & (f > f_1) \end{cases} \tag{11.38}$$

With two elastic interlayers, the transmission loss increases by 6 dB per octave above the frequency f_1. In the case of only one interlayer, the improvement of the transmission loss will be only about half the value for two interlayers. An improvement by 9 dB is obtained with either of these measures:

- halving the shear modulus G
- doubling of the density of the plates (ρ_1 and ρ_2)
- doubling of the thickness of the interlayers d
- doubling of the ratio between thickness of plate 1 and the width of the interlayers h_1/w

An example of a measurement result is shown in Figure 11.13 and compared with the simple prediction method using Equations 11.37 and 11.38. It is a T-junction with two 100 mm porous concrete walls and one 180 mm clinker concrete slab, having two polyurethane interlayers, one below and one above the slab. The measurement was from wall 1 to wall 3, referring to the numbering in Figure 11.9. Speed of shear waves in the porous concrete wall was 750 m/s.

Annex E.3.4 in ISO 12354-1 has a similar formula for the vibration reduction index K_{ij} with flexible interlayers. It is suggested that the attenuation depends on the load on the resilient layer. For one flexible interlayer, the increase with frequency is 3 dB or 6 dB per octave if the load is >750 kN/m² or <80 kN/m², respectively. ISO 12354-2 deals with flanking transmission of impact sound in a similar way.

11.4 EXAMPLE

If the measured sound insulation is less than expected, it can be very difficult to point at the reason. However, the measures needed for an improvement depend very much on which transmission path is most important. An

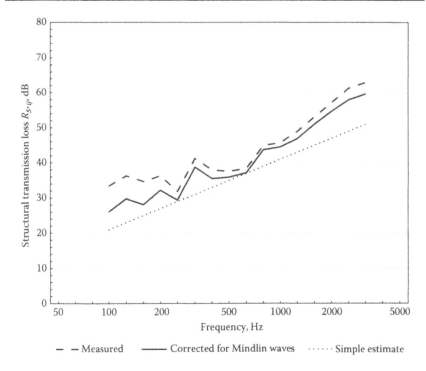

Figure 11.13 Frequency dependency of the structural transmission loss with two elastic interlayers. Measurement results without and with correction for group speed of Mindlin waves (Equation 11.28). (Adapted from Pedersen, D. B., *Appl. Acoust.*, 46, 285–305, 1995, Figure 4. With permission from Elsevier.)

obvious but rather complicated and time-consuming method is the measure the sound intensity radiated from each surface into the receiving room.

In this case, the problem was the airborne sound insulation between dwellings in the vertical direction. The cross-section of floor and internal wall construction is seen in Figure 11.14 and the measurement results for both airborne and impact sound insulation are seen in Figure 11.15. The single-number values are $R'_w = 49$ dB and $L'_{n,w} = 59$ dB. The minimum requirement for airborne sound insulation was $R'_w = 55$ dB, so the result was 6 dB below the requirement and something had to be done about it.

If the floor construction was the cause of the problem, it might be a solution to mount a suspended ceiling or to change the floor to a floating floor. If flanking transmission was the cause, these measures would not help; instead, something should be done with the walls.

In this case, it was helpful to look at the sum $R' + L'_n$ (Figure 11.16). In Section 10.4, we learned that the sum should follow approximately the theoretical straight lines also shown in Figure 11.16, and if that failed, this was an indication of flanking transmission or leakage, i.e. some transmission

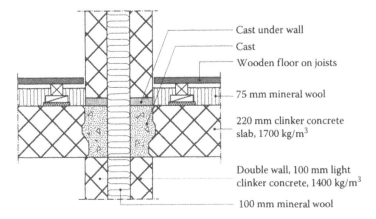

Cast under wall

Cast

Wooden floor on joists

75 mm mineral wool

220 mm clinker concrete
slab, 1700 kg/m³

Double wall, 100 mm light
clinker concrete, 1400 kg/m³

100 mm mineral wool

Figure 11.14 Cross-section of the two-story building.

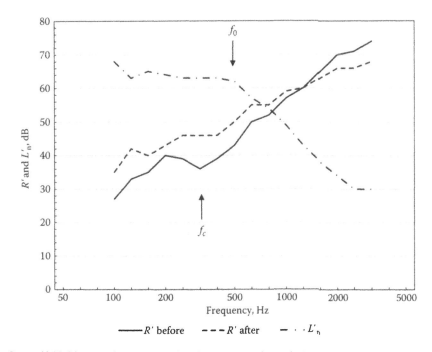

Figure 11.15 Measured impact and airborne sound insulation in vertical direction.
Indicated frequencies are assumed resonance frequency f_0 of the flooring
and critical frequency f_c of the internal wall.

paths that only affect the airborne sound transmission and not the impact
sound transmission. In the current case, we see a clear deviation of the
sum curve at frequencies above 200 Hz and especially around 315 Hz. This

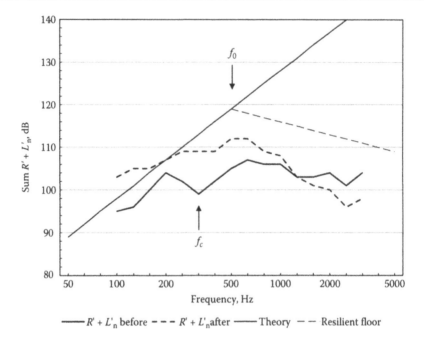

Figure 11.16 Sum of measured $R' + L'_n$ before and after mounting of gypsum board panels on the internal walls. Theoretical curves from Figure 10.6b are shown for comparison.

observation points to the walls, which have a critical frequency around 380 Hz. (From Table 1.1 we have the coincidence number for light clinker concrete $K_c = 38$.) Therefore, the sum-curve gives a strong indication that flanking transmission via the 100 mm light clinker concrete walls is the main cause of the problem.

A plasterboard panel was mounted in front of the light clinker concrete walls and the sound insulation was measured again. The improved airborne sound insulation is seen in Figure 11.15, and the result was $R'_w = 55$ dB, just as required. The new sum-curve also shown in Figure 11.16 indicates that there is still some flanking transmission, but for practical reasons, it was not possible to mount the plasterboard panels on all internal wall surfaces.

REFERENCES

R. Breuwer and J.C. Tucker. 1976. Resilient mounting systems in buildings. *Applied Acoustics* 9, 77–101.

R.J.M. Craik. 1981. Damping of building structures. *Applied Acoustics* 14, 347–359.

E. Gerretsen. 1979. Calculation of the sound transmission between dwellings by partitions and flanking structures. *Applied Acoustics* 12, 413–433.

E. Gerretsen. 1986. Calculation of airborne and impact sound insulation between dwellings. *Applied Acoustics* 19, 245–264.

ISO 12354-1. 2017. *Building acoustics—Estimation of acoustic performance of buildings from the performance of elements—Part 1: Airborne sound insulation between rooms.* International Organization for Standardization, Geneva, Switzerland.

ISO 12354-2. 2017. *Building acoustics—Estimation of acoustic performance of buildings from the performance of elements—Part 2: Impact sound insulation between rooms.* International Organization for Standardization, Geneva, Switzerland.

J. Kristensen and J.H. Rindel. (1989). *Bygningsakustik—Teori og praksis* (Building acoustics—Theory and practice, in Danish). SBI-Anvisning 166. Danish Building Research Institute, Hørsholm, Denmark.

D.B. Pedersen. 1995. Estimation of vibration attenuation through junctions of building structures. *Applied Acoustics* 46, 285–305.

J.H. Rindel. 1994. Dispersion and absorption of structure-borne sound in acoustically thick plates. *Applied Acoustics* 41, 97–111.

Chapter 12

Measurement methods

Measurements related to building acoustics include noise measurements, airborne and impact sound insulation, reverberation time, vibrations, loss factor, radiation efficiency and junction attenuation. Measurements can be made with various degrees of precision, and some measurement standards distinguish between three grades: survey, engineering and precision. The engineering grade measurements are most relevant for field measurements in buildings, but in some cases it may also be useful to take quick measurements at the survey grade as an indication of the acoustical quality. Precision grade measurements are intended for laboratory tests and research, where the measurement uncertainty must be as low as possible, and time is of minor importance. The focus in this chapter is on field measurements in buildings at the engineering grade. Details on the methods are given in the relevant measurement standards and the most important ones are listed in Table 12.1.

12.1 SOUND PRESSURE LEVEL AND LOW-FREQUENCY NOISE

This section describes the principles of sound pressure level measurements in rooms when the sound source is equipment for building services or some outdoor noise source.

12.1.1 Microphone

The sound pressure level in a room is measured with a calibrated sound level meter or a calibrated measurement microphone connected to an analyser. The microphone must be selected to have a flat frequency response in the frequency range of the measurement, and the electrical noise floor must be sufficiently low to allow correct measurements. A microphone with a big diaphragm, e.g. a 1 inch microphone. has a high sensitivity and

Table 12.1 International Organization for Standardization (ISO) and American Society for Testing and Materials (ASTM) standards for measurements of sound insulation and reverberation time

	Building elements measured in a laboratory	Performance measurements in the field
Airborne sound insulation, walls and floors	ISO 10140-2 ASTM E90	ISO 16283-1 ASTM E336
Airborne sound insulation, windows and façades	ISO 10140-2 ASTM E90	ISO 16283-3 ASTM E966
Impact sound insulation	ISO 10140-3 ASTM E492	ISO 16283-2 ASTM E1007
Reverberation time	ISO 3382-2 ASTM E2235	ISO 3382-2 ASTM E2235

an upper frequency limit around 8 kHz, which is sufficient for most cases in building acoustics. However, a smaller ½ inch microphone is often preferred although the noise floor is higher, because it is less expensive and easier to handle.

The frequency response should be flat for diffuse field incidence when measuring in a room, i.e. a pressure microphone should be used, not a free field microphone.

12.1.2 Frequency weighting

Measurements are normally made in one-third octave bands covering the frequency range from 50 Hz to 5000 Hz. In case of measuring noise from service equipment in buildings, octave bands are used as a minimum in the frequency range from 125 Hz to 2000 Hz for survey measurements (ISO 10052) or for engineering grade measurements in an extended frequency range from either 31.5 Hz or 63 Hz to 8000 Hz (ISO 16032). As single-number values, the A-weighted and C-weighted sound pressure levels can be calculated using Equation 1.16.

Low-frequency noise can be particularly annoying for the dwellers in buildings, although the noise source may be far away from the building. Sometimes, it is difficult to identify the origin at low-frequency noise. One evaluation method that works satisfactorily was developed in the Netherlands (Vercammen, 1992) and adapted in Denmark (Jakobsen, 2001; Poulsen and Mortensen, 2002). The one-third octave bands are measured from 10 Hz to 160 Hz and adjusted with the A-weighting filter corrections (Table 1.3) and then the low-frequency noise level $L_{pA,LF}$ is calculated for this frequency range using Equation 1.16.

12.1.3 Averaging time

The energy averaged sound pressure level is determined by:

$$L = 10 \lg \left(\frac{1}{T_m \cdot p_0^2} \int_0^{T_m} p^2 \, dt \right) dB \tag{12.1}$$

where

> p is the sound pressure as a function of time (Pa).
> $p_0 = 20 \, \mu Pa$ is the reference sound pressure.
> T_m is the integration time (s).

Measuring the sound pressure level of service equipment, the averaging time must be long enough to cover typical time variations, but not less than 30 s. For sound insulation measurements, the recommended averaging times depend on the frequency (see Section 12.2.2).

12.1.4 Correction for background noise

For the measurement of service equipment noise, the background noise level should be at least 4 dB, and preferably more than 10 dB, below the measured sound pressure level in each frequency band.

If the background noise level is between 4 dB and 10 dB below the measured sound pressure level, the result is adjusted by:

$$L = 10 \lg \left(10^{L_{sb}/10} - 10^{L_b/10} \right) dB \tag{12.2}$$

where

> L is the adjusted sound pressure level.
> L_{sb} is the measured sound pressure level (signal and background noise combined).
> L_b is the sound pressure level of the background noise.

When the difference is 4 dB, the correction is 2.2 dB. If the difference is less than 4 dB, the correction 2.2 dB is applied and it must be clearly marked that the result is influenced by background noise. For sound insulation measurements, the background noise level should be at least 6 dB below the measured sound pressure level (see Section 12.2.1).

12.1.5 Measurement positions and spatial averaging

The recommended choice of measurement positions for the sound pressure level in a room is very different for frequencies above or below the Schroeder limiting frequency (see Equation 4.21). In Table 12.2, it is shown which volume corresponds to a certain limiting frequency, f_g, assuming the reverberation time to be around 0.5 s as in a typical furnished room in a dwelling, or around 0.8 s as in an unfurnished or sparsely furnished room. It is seen that the limiting frequency is around 200 Hz for a 50 m³ furnished room.

Above the limiting frequency, statistical methods apply and the positions represent samples from a more or less diffuse sound field. The ideal would be to use a large number of microphone positions to find the grand average over the entire volume. However, this is not possible in practice, where a rather limited number of positions must be selected. Alternatively, the squared sound pressure can be averaged using a moving microphone following some path through the middle part or the room.

The single measurements can be considered uncorrelated if the distance from other positions is at least half a wavelength, e.g. 0.86 m at 200 Hz (Table 12.2). Near reflecting boundaries, the sound pressure increases due to interference. This can actually be considered as the microphone being close to the reflected image of the microphone and, thus, the position should be at least a quarter wavelength from reflecting surfaces, e.g. 0.43 m at 200 Hz. Positions should be distributed randomly over the centre part of the room, keeping a minimum distance from the room boundaries. In accordance with statistical principles, the measurement uncertainty is reduced if the number of uncorrelated positions is increased.

Below the Schroeder limiting frequency, the sound field in the room is deterministic, dominated by a few room modes, as described in Section 4.1. This implies that preferred positions are near the corners, where the sound pressure

Table 12.2 Relation between the Schroeder limiting frequency and the room volume when the reverberation time is either 0.5 s or 0.8 s

f_g (Hz)	V (m³)		λ/2 (m)
	T=0.5 s	T=0.8 s	
100	200	320	1.72
125	125	200	1.38
160	80	125	1.08
200	50	80	0.86
250	32	50	0.69
315	20	32	0.55
400	13	20	0.43
500	8	13	0.34

Note: The minimum distance between uncorrelated microphone positions in a diffuse sound field (λ/2) is also given.

level is well defined, whereas the centre region should be avoided because extremely low sound pressure levels occur along the nodal lines. Since statistical methods do not apply at these low frequencies, it is impossible to avoid correlation between the positions, and in contrast to the high-frequency range, increasing the number of positions does not affect the measurement uncertainty.

Measuring the sound pressure level from building service equipment in accordance with ISO 16032, three microphone positions are used; one in a corner and two distributed in the reverberant field. The corner position should be about 0.5 m from the nearest surfaces. Initial measurements are made in all corners using the C-weighting filter, and the corner with the highest C-weighted sound pressure level is chosen as position 1. Positions 2 and 3 shall be at least 1.5 m from position 1 and from any sound source. The distance from room surfaces shall be at least 0.75 m. The result of measured sound pressure levels in the three microphone positions is calculated as the energetic average using Equation 1.34.

12.1.6 Correction for reverberation time

It is known that the sound pressure level in a depends on the absorption area, i.e. on the reverberation time and the volume. Therefore, the measured sound pressure level should be either standardized or normalized.

The *standardized sound pressure level* L_{nT} is adjusted to a reverberation time of $T_0 = 0.5$ s:

$$L_{nT} = L - 10 \lg \frac{T}{T_0} \; (dB) \tag{12.3}$$

It is generally assumed that the reverberation time in furnished dwellings is around 0.5 s, and this has been confirmed by Takala and Kylliäinen (2013).

The *normalized sound pressure level* L_n is adjusted to an equivalent sound absorption area of $A_0 = 10 \, m^2$:

$$L_n = L - 10 \lg \frac{A_0}{A} = L - 10 \lg \frac{c \cdot A_0 \cdot T}{55.3 \cdot V} \; (dB) \tag{12.4}$$

where

  L is the measured sound pressure level (dB).
  T is the reverberation time (s).
  V is the volume (m^3).
  $c = 344 \, m/s$ is the speed of sound in air.

Since the reverberation time is frequency dependent (more or less), the corrections should be made in each frequency band (octave or one-third octave) before the A-weighted or C-weighted levels are calculated.

12.2 AIRBORNE SOUND INSULATION BETWEEN ROOMS

12.2.1 Generation of sound

A loudspeaker is used to generate the sound, which must be at a sufficiently high level in order to get a sound pressure level in the receiving room without the influence of background noise. The directivity of the loudspeaker should be close to omnidirectional, and the frequency response should cover the relevant measurement range, which is at least from 100 Hz to 3150 Hz, but preferably from 50 Hz to 5000 Hz in one-third octave bands.

The traditional measurement technique uses stochastic noise (white or pink), which can either be emitted as a broad-band signal or as a band-limited signal with a bandwidth of at least one-third octave. The latter is, of course, more time consuming, but has the advantage of allowing a higher sound pressure level to be generated without overloading the loudspeaker. Airborne sound insulation can also be measured using sound intensity, see ISO 15186 (Part 1 to 3) and Machimbarrena and Jacobsen (1999).

More advanced measurement techniques use deterministic sound signals, either MLS (maximum-length-sequence) signal or swept-sine signals in combination with either correlation signal processing or deconvolution (see ISO 18233). The resulting impulse responses are squared and integrated in order to get the stationary sound pressure level. Compared to the traditional method, both methods have the advantage of suppressing the influence of background noise. The swept-sine method has a number of advantages over the MLS method:

- Nearly immune against harmonic distortion
- Higher excitation levels may be used, due to a lower crest factor
- More robust against time variance

The *crest factor* of a sound signal is defined as the ratio between the *peak* value and the root-mean-square (RMS) value of the signal. A harmonic signal that is composed of N sine functions has the crest factor $\sqrt{(2N)}$, and a single sine function has the crest factor $\sqrt{2} \approx 1.414$. White noise and pink noise have a much higher crest factor, which means a higher risk that the loudspeaker can be overloaded.

While distortion from the loudspeaker is not really a problem with the traditional measurement technique, it is fatal to the MLS technique. With the swept-sine technique, some (but not all) of the harmonic distortion is separated from the impulse response and does not influence the result. Both methods require time-invariant systems, i.e. nothing must move in the rooms during the measurements; even air movements may cause errors in the measurements and, of course, moving microphones are out of question. While time invariance is mandatory with MLS measurements, the swept-sine is much more robust

Table 12.3 Improved SNR as function of sweep duration using the swept-sine technique

Duration of sweep (s)	60	336	762
Duration per one-third octave (s)	2.9	16	32
Improved SNR (dB)	14	21	24

Note: The reference for SNR is here set to fast (0.125 s).

to this. Some years ago, when the computer capacity was more restricted, the MLS technique had the advantage of requiring less memory, but since this is no longer an issue, the many advantages of the swept-sine technique means that the MLS technique is obsolete for building acoustic applications.

With the swept-sine technique, the advantage of improving the signal-to-noise ratio (SNR) is connected to the length of the sweep signal. Only a single sweep is used in each measurement position, which is in contrast to the MLS technique, which uses the averaging of several repeated sequences. If we compare measurements with the averaging time Fast (0.125 s), the improved SNR with different sweep durations are shown in Table 12.3.

As an example, the sound level difference between two rooms has been measured with different techniques and the results are shown in Figure 12.1. The background noise in one-third octaves was around 43 dB at 125 Hz to 200 Hz, decreasing towards lower and higher frequencies, being around 30 dB at 50 Hz to 80 Hz and around 25 dB at 2500 Hz to 5000 Hz. The influence of the background noise is worst using the broadband excitation and parallel analysis. Some improvement was obtained using band-limited noise excitation because the loudspeaker could be set to a level about 10 dB higher. The best result was obtained with the swept-sine; the excitation signal could be increased another 3 dB, but most importantly, the influence of the background noise in the receiving room was efficiently suppressed. The difference between the methods is clearly seen at medium and high frequencies.

The use of swept sine for measuring airborne sound insulation has been further studied by Satoh et al. (2011). Using very long sweeps (10 min per octave) and a technique for subtraction of background noise, they could measure sound pressure levels 30 dB below the background noise with less than 1 dB error. Løvstad et al. (2014) describe an example of the practical application of the swept-sine technique for a rather extreme case where the sound insulation was 95 dB at 500 Hz.

12.2.2 Measurement positions and spatial averaging

The requirements for measurement of airborne sound insulation in buildings in ISO 16283-1 imply that two source positions shall be used in the source room, the distance from room boundaries being at least 0.5 m and at least 1.0 m from the partition wall separating the source and receiving rooms.

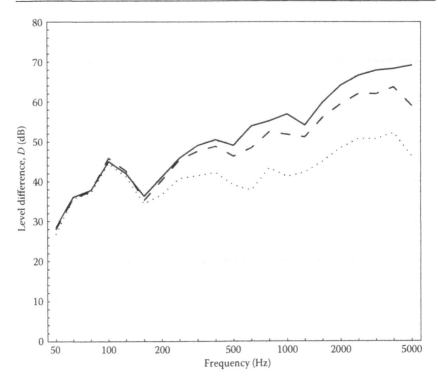

Figure 12.1 Level difference between two rooms measured with different techniques. Solid line: Swept-sine excitation. Dashed line: Serial excitation with noise in one-third octave bands. Dotted line: Parallel excitation with pink noise. (Adapted from Norsonic, *Swept sine measurement technique using the Nor121 analyser.* Application Note, Ed. 3, 10.06, Norsonic, Lierskogen, Norway, 2006. Figure 3.)

If measuring in vertical direction, and if the source is in the upper room, the source position shall be at least 1.0 m above the floor. However, it is advisable that measurements in the vertical direction are made with the lower room as the source room in order to avoid the measurement to be influenced by structure borne transmission from the loudspeaker stand. If the rooms are of different volume, the larger room should normally be chosen as the source room.

Instead of measuring with one loudspeaker in two different positions, it is possible to use two loudspeakers simultaneously, and then calculate the energy-averaged sound pressure levels in the source a nd receiving rooms. If necessary, the receiving room levels are corrected for background noise and, finally, the reverberation time correction is applied to get either the standardized level difference D_{nT} or the apparent sound reduction index R'.

If a single loudspeaker is used, the measurements must be repeated with the source moved to a different position. In this case, the standardized level difference D_{nT} or the apparent sound reduction index R' must be measured and calculated for each source position separately and, then, the result is averaged using one of these formulas:

$$D_{nT} = -10\ \lg\left(\frac{1}{m}\sum_{j=1}^{m}10^{-D_{nT,j}/10}\right)(\text{dB}) \tag{12.5}$$

$$R' = -10\ \lg\left(\frac{1}{m}\sum_{j=1}^{m}10^{-R'_{j}/10}\right)(\text{dB}) \tag{12.6}$$

where

m is the number of loudspeaker positions.
$D_{nT,j}$ is the standardized level difference for loudspeaker position j.
R'_{j} is the apparent sound reduction index for loudspeaker position j.

The microphones shall be calibrated as for normal sound level measurements before the measurements, although strictly speaking, it is sufficient that the same microphone is used in the source room and in the receiving room, since the measurement result is taken from the level difference. In each room, the spatially averaged sound pressure level must be measured, and this can be done in three different ways:

- Fixed microphone positions, at least five in each room with at least 0.7 m between positions (Figure 12.2). No two positions shall lie in the same plane relative to the room boundaries.
- Mechanically moved microphone with radius at least 0.7 m (Figure 12.3). The plane of the traverse shall be inclined at least 10°.
- Manually scanned microphone, following a certain path (Figure 12.4).

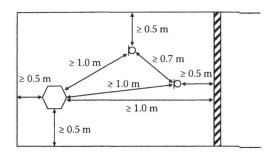

Figure 12.2 Sketch of a source position and microphone positions in the source room with minimum distances indicated.

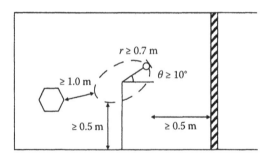

Figure 12.3 Sketch of mechanically moving microphone with minimum radius, inclination and distances from source and room boundaries.

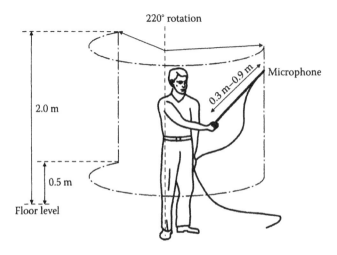

Figure 12.4 Manual scanning with the microphone mounted on a rod and following a cylindrical path. (Modified from ISO 16283-1, *Acoustics – Field measurements of sound insulation in buildings and of building elements – Part 1: Airborne sound insulation,* International Organization for Standardization, Geneva, Switzerland. Figure 1c.)

The intention is to achieve an equivalent of five uncorrelated microphone positions for frequencies above 200 Hz. This is obtained for the mechanically moved microphone with a radius of 0.7 m (see Chapter 13, Table 13.4). The averaging time shall equal the time for one rotation, which is at least 30 s at frequencies 100 Hz to 5000 Hz and at least 60 s at low frequencies 50 Hz to 80 Hz.

For the fixed positions, the minimum distance of 0.7 m corresponds to uncorrelated positions above 250 Hz (the distance must be half a wavelength). This means that with five fixed positions, the equivalent number of uncorrelated positions is less than five at frequencies below 250 Hz. Increasing the minimum distance between microphone positions to 0.85 m would comply with uncorrelated positions above 200 Hz (Table 12.2). The averaging time shall be:

≥ 15 s at frequencies 50 Hz to 80 Hz
≥ 6 s at frequencies 100 Hz to 400 Hz
≥ 4 s at frequencies 500 Hz to 5000 Hz

The manually scanning method was introduced as an option in ISO 16283-1. Four different patterns are suggested, and the most efficient among these is the cylindrical type, as shown in Figure 12.4.

The microphone is mounted at the end of a rod, 0.3 m to 0.9 m long. The path starts about 0.5 m above the floor and approximately 90° to one side; the rod is swept in a circular path parallel to the floor to cover an angle of approximately 220°. The sweep continues vertically upwards along a straight line until the microphone is about 2 m above the floor, but at least 0.5 m below the ceiling. Then, the movement continues with another circular sweep, before descending to the starting point. A speed of the microphone shall be as constant as possible, with a maximum of 0.25 m/s. The minimum averaging times are the same as for the mechanically moved microphone, i.e. 30 s at 100 Hz and above and 60 s below 100 Hz. This procedure is equivalent to five uncorrelated positions at 100 Hz using a 0.3 m rod, and eight positions using a 0.9 m rod (Hopkins, 2015, Figure 2).

12.2.3 Low-frequency procedure

According to ISO 16283-1, a special low-frequency procedure shall be used for the 50 Hz, 63 Hz and 80 Hz one-third octave bands when the room volume is smaller than 25 m³. This applies to both the source room and the receiving room. However, it would make sense to apply this method more generally to the frequencies that are below the Schroeder limiting frequency. The principle is that measurements made in the central zone of the room (the default procedure) are combined with the highest sound pressure level that occurs in a room corner. The microphone position in the corner is at a distance of 0.3 m to 0.4 m from each room boundary (Figure 12.5).

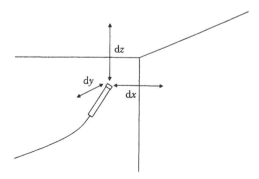

Figure 12.5 Position of a microphone near a corner. The distances dx, dy and dz should all be within 0.3 m to 0.4 m.

Actually, measurements shall be made in four corners, two at floor level and two at ceiling level, but only the highest of these four results is used. Which position gives the highest results may change with frequency. If a single loudspeaker is used, the measurements must be repeated with the source moved to a different position, and the corner sound pressure level in each frequency band is calculated as the energy-averaged sound pressure level using Equation 1.34. The low-frequency energy-averaged sound pressure level L_{LF} is calculated in each frequency band by combining the sound pressure level from the default procedure, L, and the highest sound pressure level from the corners, L_{corner}, using:

$$L_{LF} = 10 \lg\left(\frac{1}{3} \cdot 10^{\frac{L_{corner}}{10}} + \frac{2}{3} \cdot 10^{\frac{L}{10}}\right) (\text{dB}) \tag{12.7}$$

L_{LF} is determined for each loudspeaker position and the resulting D_{nT} or R' values for 50 Hz, 63 Hz and 80 Hz are averaged as for the default procedure. However, the reverberation time is measured in the octave band 63 Hz, and the result is applied in all three one-third octave bands 50 Hz, 63 Hz and 80 Hz. The reason for this pragmatic method is that the modal density may be very low below 100 Hz, especially in small volumes and, thus, measurements in one-third octave bands may be unreliable or even impossible.

12.3 WINDOWS AND FACADES

The purpose of these measurements can be either to evaluate the protection of the façade against outdoor noise sources or to evaluate the transmission loss of a particular façade element, e.g. a window. So, ISO 16283-3 describes two methods, a global method and an element method.

The sound source used for the measurement can be either the existing traffic noise or a loudspeaker. Traffic noise is the preferred source for the global method, while the measurement using a loudspeaker is preferred for the element method.

12.3.1 The global method

The outdoor sound pressure level is measured in a distance of 2.0 m from the façade, and in the receiving room, the time- and space-averaged sound pressure level is measured in five fixed positions or using a moving microphone, exactly as when measuring the sound insulation between two rooms. The measured level difference can be either normalized (adjusted to 10 m² absorption area) or standardized (adjusted to 0.5 s reverberation time). The standardized level difference is:

$$D_{2m,nT} = L_{1,2m} - L_2 + 10 \lg\frac{T}{T_0} (\text{dB}) \tag{12.8}$$

where

$L_{1,2m}$ is the sound pressure level 2.0 m in front of the façade.
L_2 is the spatially averaged sound pressure level in the receiving room.
T is the reverberation time in the receiving room (s).
$T_0 = 0.5$ s is the reference reverberation time.

Equation 12.8 is used both for traffic noise and for loudspeaker measurements. Using a loudspeaker, the position of the loudspeaker is in a distance of $r \geq 7$ m from the centre of the test specimen or at least $D = 5$ m in front of the façade (Figure 12.6). The angle of incidence shall be $45° \pm 5°$ and the loudspeaker can be placed either on the ground (preferred) or as high over the ground as possible.

12.3.2 The element method

The outdoor sound pressure level is measured close to the façade (Figure 12.7). If the sound source is a loudspeaker (the preferred method), it should be placed so that the angle of incidence is close to 45°, as shown in Figure 12.6. The distance shall be $r \geq 5$ m from the centre of the test specimen or at least $D = 3.5$ m in front of the façade. The result of the measurement is the apparent sound reduction index:

$$R'_{45°} = L_{1,s} - L_2 + 10 \lg \frac{S}{A} - 1.5 \, \mathrm{dB} \tag{12.9}$$

where

$L_{1,s}$ is the sound pressure level on the test specimen.
L_2 is the spatially averaged sound pressure level in the receiving room.

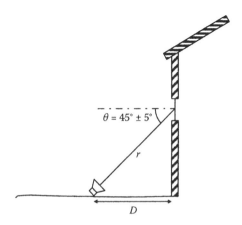

Figure 12.6 Position of the loudspeaker in front of the façade. For the global method, $r \geq 7$ m and $D \geq 5$ m. For the element method, $r \geq 5$ m and $D \geq 3.5$ m.

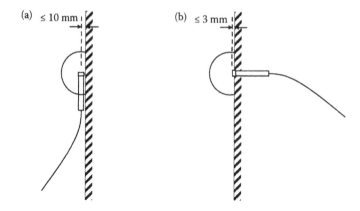

Figure 12.7 Position of the microphone on the test surface for the measurement of $L_{1,s}$. (a) Microphone with diaphragm perpendicular to the surface. (b) Microphone with diaphragm parallel with the surface.

S is the area of the test specimen (m²).
A is the equivalent absorption area of the receiving room (m²).

If, for practical reasons, the element loudspeaker method cannot be used, it is possible to use road traffic as a noise source. Then, it is assumed that the angles of incidence cover a wide range and the formula for the apparent sound reduction index is:

$$R'_{tr,s} = L_{1,s} - L_2 + 10 \lg \frac{S}{A} - 3 \text{ dB} \qquad (12.10)$$

where the symbols are the same as in Equation 12.9.

A major problem with measurements using traffic noise as the sound source is that the background noise in the receiving room can make measurements impossible at high frequencies. This is a lesser problem using a loudspeaker as the sound source, especially if the swept-sine method is used. In that case, fixed microphone positions (at least five) must be used in the receiving room.

12.4 IMPACT SOUND INSULATION

The impact sound insulation is measured in a building using a tapping machine, as described in ISO 16283-2. The tapping machine is placed on the floor under test in at least four positions, randomly distributed, but not closer than 0.5 m to any wall. The tapping machine has five hammers in a line, and in case of floor constructions with beams or ribs, the line through the hammers should be oriented 45° to the direction of the beams or ribs. If the floor under test is larger than $20\,m^2$, the number of tapping machine positions may be increased, and if the area of the floor is larger than $50\,m^2$, the number of tapping machine positions should be eight.

The measurement of spatially averaged sound pressure level and the position of microphones in the receiving room follow the same rules as for airborne sound insulation (see Section 12.2). If fixed microphone positions are used, the number of positions shall equal the number of tapping machine positions (or a multiple of the number of tapping machine positions). Measurements shall be made in at least two different microphone positions for each tapping machine position, if the number of tapping machine positions is less than six. If six or more tapping machine positions are used, it is sufficient to measure in one microphone position for each tapping machine position.

Measurements of impact sound insulation should include the low frequencies down to the $50\,Hz$ one-third octave band. For frequencies below $100\,Hz$, the low-frequency procedure using corner positions apply (see Section 12.2). Averaging times for the measurements and correction for background noise are also the same as for airborne sound insulation.

For each tapping machine position, the result is calculated either as the standardized impact sound pressure level (Equation 12.3) or as the normalized impact sound pressure level (Equation 12.4). Finally, the averaging over all tapping machine positions is made as the energetic average using one of these formulas:

$$L'_{nT} = 10 \lg\left(\frac{1}{m}\sum_{j=1}^{m}10^{L'_{nT,j}/10}\right) \text{(dB)} \tag{12.11}$$

$$L'_{n} = 10 \lg\left(\frac{1}{m}\sum_{j=1}^{m}10^{L'_{n,j}/10}\right) \text{(dB)} \tag{12.12}$$

where

m is the number of tapping machine positions.

$L'_{nT,j}$ is the standardized impact sound pressure level for tapping machine position j.

$L'_{n,j}$ is the normalized impact sound pressure level for tapping machine position j.

The microphone shall be calibrated as for normal sound level measurements before the measurements. This is particularly important for the measurement of impact sound pressure level, as it is the actual sound pressure level that is measured, not the difference between two levels as for airborne sound insulation.

12.5 REVERBERATION TIME

The measurement of reverberation time in the receiving room is a part of most building acoustic measurements. The procedure is described in ISO 3382-2 and two different methods are available, the interrupted noise method and the integrated impulse response method. Independent of the method, the decay curve is evaluated between 5 dB and 25 dB below the initial level, and the reverberation time is denoted as T_{20} in order to indicate the evaluation range 20 dB. Since the reverberation time is defined at the time needed for a 60 dB decay, T_{20} is three times the time interval between the start and stop of the evaluation range.

12.5.1 Measurement positions

For an engineering grade measurement, the number of source-microphone combinations should be at least six, e.g. two source positions combined with three microphone positions. However, when the result is used only as a correction term to other engineering-level measurements, only one source position and three microphone positions are required. Also, under these conditions, a rotating microphone boom may be used instead of multiple microphone positions, but only for the interrupted noise method.

ISO 3382-2 recommends that in small domestic rooms, one source position should be in a corner. Microphone positions should be at least half a wavelength apart and a quarter wavelength from the boundaries. If a rotating microphone boom is used, the radius of the boom shall be at least 0.7 m.

12.5.2 The interrupted noise method

The traditional method uses white or pink noise emitted from a loudspeaker and interrupted after a stationary sound level has been built up in the room. Since the signal is stochastic, the measurements must be repeated, and at least two decays must be measured in each source-receiver combination. Background noise may be a problem if the loudspeaker cannot provide a sufficiently high level, which should be at least 35 dB above the background noise in each frequency band. Although the use of broadband noise and parallel filters for the analyses is very efficient, it may be necessary to use band-limited noise in one-third octave bands in order to get sufficient signal-to-noise level.

12.5.3 The integrated impulse response method

The integrated impulse response method has become a very attractive alternative to the traditional method after the introduction of the swept-sine technique. A single sweep can cover the entire frequency range and the background noise can be suppressed as much as needed, simply by choosing a sufficiently long sweep time. Since the signal is ergodic, there is no need to repeat measurements (they will be identical); one measurement in each position is sufficient. Only fixed source and microphone positions can be used because the technique requires time-invariant systems.

The practical application of the measurement method is discussed by Christensen et al. (2013). They show signal-to-noise level of up to 100 dB with very long sweeps. With 40-s sweep, they got nearly 90 dB signal-to-noise level, which is more than sufficient.

12.5.4 Lower limits for reverberation time

There are several problems with measuring reverberation times at low frequencies, especially the short reverberation times in typical furnished, domestic rooms:

- ringing of the one-third octave band filters
- time constant of the detector
- low modal density, especially in small room volumes
- insufficient signal-to-noise ratio

Normally a filter with bandwidth B has an impulse response with a reverberation time of its own and, of course, this may interfere with the measured decay curve of the room. As shown by Jacobsen and Rindel (1987), the following requirements should be fulfilled for the reverberation time T:

$$BT > 16 \quad (T > 70/f \text{ for one-third octave bands}) \tag{12.13}$$

$$T > 2 \cdot T_{det} \tag{12.14}$$

where f is the centre frequency of the filter and T_{det} is the reverberation time of the detector, assuming exponential averaging of the decaying sound pressure. This implies a lower limit for measurable reverberation times, which is 0.7 s at 100 Hz and 1.4 s at 50 Hz (Table 12.4). However, it was suggested by Jacobsen and Rindel (1987) to overcome these problems by using a time-reversal technique, either by recording the decay in the room and playing it backwards through the filters or by constructing the filters with a time-reversed impulse response meaning a slow built-up and fast decay. With this technique, the minimum requirements for the reverberation time are changed to:

$$BT > 4 \ (T > 17.5/f \text{ for one-third octave bands}) \tag{12.15}$$

$$T > 1/4 \cdot T_{det} \tag{12.16}$$

Thus, the lower limit at 100 Hz one-third octave band is reduced to 0.18 s and 0.35 s at 50 Hz.

Some instrument manufacturers have overcome the problem by the use of reverse filtering, i.e. the one-third octave band filters have been specially crafted to provide short reverberation times and still comply with requirements for the frequency response. Thus, the lower limit for measurement of reverberation time in one-third octave bands can be around 0.1 s in the entire frequency range from 50 Hz to 5000 Hz.

Another problem in small volumes is that the modal density is very low. In the one-third octave bands below 100 Hz, there may be only a few modes, or none. With two or three modes within a band, the decay curve may suffer from strong wavy behaviour due to interference, or double slope due to very different attenuation of the modes. In either case, the evaluation of a reverberation time is difficult and associated with a huge uncertainty. A pragmatic solution to this problem was suggested by Hopkins and Turner (2005), namely, to measure the reverberation time in the 63 Hz octave band and apply this result for the 50 Hz, 63 Hz and 80 Hz one-third octave bands. This idea has been adapted in the ISO 16283 series for sound insulation measurements.

Finally, it is a problem with commonly used sound sources for building acoustic measurements that the sound power in the 63 Hz octave is not sufficient for measuring the reverberation time. The signal-to-noise level should be at least 35 dB and the background noise is often quite high at low frequencies. The best solution is probably to use the swept-sine technique in combination with a loudspeaker that has good performance at low frequencies.

Table 12.4 Minimum measurable reverberation time (in s) using normal one-third octave and octave band filters and either forward or time reversed analysis

f, Hz	Forward		Backward	
	One-third octave	Octave	One-third octave	Octave
50	1.40		0.35	
63	1.11	0.36	0.28	0.09
80	0.88		0.22	
100	0.70		0.18	
125	0.56	0.18	0.14	0.04
160	0.44		0.11	
200	0.35		0.09	

12.5.5 Spatial averaging

The spatial averaging of reverberation time can be made in two different ways. The reverberation times are evaluated in each pair of source-receiver combinations, and the arithmetic average of reverberation times is calculated. Alternatively, the decay curves can be averaged, synchronizing the beginnings and averaging the sound pressure squared in each time slot during the decay. The sound power of the source shall be kept the same for all measurements. The reverberation time is calculated from the ensemble-averaged decay curve. This is the preferred method of spatial averaging in ISO 3382-2.

12.6 VIBRATIONS

Vibrations in buildings are usually measured in the frequency range from 1 Hz to 80 Hz. The transducer can be an accelerometer of a geophone.

The *accelerometer* is an electromechanical transducer that produces a voltage proportional to the acceleration of the vibrations. The most common principle is the compression type piezoelectric accelerometer. A heavy mass resists on two piezoelectric discs; the mass is preloaded by a stiff spring and the whole assembly is mounted in a metal housing with a thick base. When the accelerometer is subjected to vibration, the mass will produce a force on the piezoelectric discs, and the force is proportional to the acceleration. A voltage will be developed across the two piezoelectric discs, and the voltage is proportional to the force, and thus also to the acceleration. The accelerometer has a mass-stiffness resonance, which defines the upper limit of the usable frequency range. The typical range of the resonance frequency is 12 kHz to 70 kHz, i.e. of no practical importance.

The principle of a *geophone* is a magnetic mass mounted on a spring and moving within a wire coil. It generates a voltage proportional to the velocity of vibrations. The sensitivity can be very good and, thus, it can be used for low vibration levels. The geophone has a mass-stiffness resonance, typically in the range from 5 Hz to 50 Hz: Above this frequency, the velocity response is flat, but falls off towards lower frequencies. The resonance is normally heavily damped, so that there is no resonance peak in the frequency response. Still, a correction filter may be necessary to correct for the non-flat response in the frequency range of interest, 1 Hz to 80 Hz.

12.6.1 Calibration

Before any vibration measurement, it is important to calibrate the equipment. Some calibrators give a specified RMS acceleration while other calibrators give a specified *peak* acceleration. An example is the B&K type

4291 calibrator that gives a peak acceleration level of 1 $g=9.81\,\mathrm{m/s^2}$ at 80 Hz. Using the reference acceleration $a_0=10^{-6}\,\mathrm{m/s^2}$, the calibration signal corresponds to the acceleration level:

$$L_a = 20\ \lg\frac{a}{a_0} = 20\ \lg\frac{9.81}{\sqrt{2}\cdot10^{-6}} \cong 136.8\ \ \mathrm{dB\ re}\ \ 1\,\mu\mathrm{m/s^2} \tag{12.17}$$

For a harmonic vibration with the angular frequency ω, the velocity is related to the acceleration by:

$$v = \frac{a}{\omega} = \frac{a}{2\pi f} \tag{12.18}$$

Using the reference velocity $v_0=10^{-9}\,\mathrm{m/s}$ and $f=80\,\mathrm{Hz}$, the corresponding velocity level of the calibration signal yields:

$$L_v = 20\ \lg\frac{v}{v_0} = L_a + 20\ \lg\left(\frac{a_0}{\omega\cdot v_0}\right)$$

$$= L_a + 60 - 20\ \lg\omega \cong 142.8\ \mathrm{dB\ re}\ 1\ \mathrm{nm/s} \tag{12.19}$$

12.6.2 Measurement positions

Vibration measurements can be made both in vertical and horizontal directions. In buildings, the highest vibration levels are usually the vertical vibrations on the floor. The vibration measurements should be made in positions where the highest vibration level is obtained, which is in the middle of the floor, or between the primary beams supporting the floor (Figure 12.8). Additional measurement positions should be chosen more than 1 m from the walls and from other positions.

The mounting of the accelerometer can be with bees' wax on hard surfaces or using a device, as shown in Figure 12.9, in the case of carpets or similar soft floor coverings.

12.6.3 Frequency weighting

Vibration measurements can either be performed in one-third octave bands in the frequency range from 1 Hz to 80 Hz or using the standardized frequency weighting curve (Figure 2.18). It is possible to calculate the weighted acceleration level or the weighted velocity level from the one-third

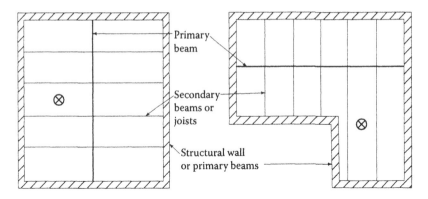

Figure 12.8 Examples of measurement positions where the highest level of vertical vibration is expected. (Reproduced from Jakobsen, J., *Measurement of vibration and shock in buildings for the evaluation of annoyance.* Nordtest project 618–86. Technical Report No. 139, Danish Acoustical Institute, Denmark, 1987. With permission.)

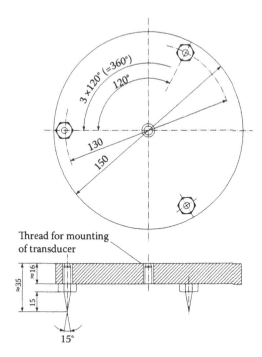

Figure 12.9 Example of mounting device for measurement of vertical vibration on carpet-covered floors. The weight of the device plus attached transducer should be about 2.5 kg. Dimensions are in mm. (Reproduced from Jakobsen, J., *Measurement of vibration and shock in buildings for the evaluation of annoyance.* Nordtest project 618–86. Technical Report No. 139, Danish Acoustical Institute, Denmark, 1987. With permission.)

octave band spectrum using an equation similar to Equation 1.16 for the A-weighted sound pressure level.

12.6.4 Integration time

The instantaneous maximum value of the acceleration level can be measured with an exponential averaging time constant of 1 s ("slow") or with a linear averaging over 2 s.

12.6.5 Noise floor

It is always recommended to ensure that the measured vibration values are not influenced from the background noise of the measurement system. If the accelerometer is placed on a heavy non-vibrating object, the apparent vibration level should be at least 10 dB below the level of the real measurements. If possible, baseline measurements should be obtained to verify the background noise level.

12.6.6 Averaging of results

Whether it is relevant to average the measured results depends on the situation and the purpose of the measurements. In some cases, the spatial average over the measurement positions is calculated as an energetic average (Equation 1.34). In other cases, the result must be reported for each measurement position.

The Norwegian standard NS 8176 deals with vibrations in buildings due to land, based transport, and the statistical maximum velocity must be calculated in each position from at least 15 events of passing vehicles or trains. According to this standard, the average maximum velocity is calculated as the arithmetic average of the frequency-weighted maximum velocities (not the squared velocities). Finally, the statistical maximum value (95 percentile) is calculated as the arithmetic average plus 1.8 times the standard deviation of the frequency-weighted maximum velocities.

12.6.7 The vibration dose value

The Vibration Dose Value (VDV) is used in some countries for assessing human reactions to intermittent vibration (see ISO 2631-1, 1997). Instead of the common RMS acceleration, the root-mean-quad acceleration is applied, which ensures that the VDV is more sensitive to peaks in the acceleration level:

$$\text{VDV} = \sqrt[4]{\int_0^T a_w^4 \, dt} \quad (\text{m/s}^{1.75}) \tag{12.20}$$

where a_w is the frequency-weighted acceleration as a function of time and T is the total measurement period in seconds. Note the unusual unit of VDV. An example of the practical application of VDV is found in the work of Yang et al. (2014).

12.7 LOSS FACTOR

12.7.1 Structural reverberation time

The total loss factor of a building element can be estimated by measuring the structural reverberation time using Equation 2.30:

$$\eta_{\text{total}} = \frac{2.2}{f\,T_s} \qquad (12.21)$$

where f is the centre frequency of a one-third octave band and T_s is the measured structural reverberation time.

The structural reverberation time is measured in accordance with ISO 10848-1 Section 7.3 and ISO 3382-2 using the integrated impulse response method with point excitation and measurement of velocity or acceleration in a number of positions. The excitation can be achieved using either a hammer or a vibration exciter and the MLS technique. See ISO 18233 for details on the latter. At least three excitation points are used, and at least three transducer positions are used for each excitation point.

12.7.2 Upper limits for loss factor

The problems with measuring structural reverberation time are no less than those with measuring reverberation time in a room. The ringing of the filters means that it can be necessary to apply the time-reversed analysis. As loss factor is reciprocal to reverberation time, the ringing of the filters causes an upper limit of measureable loss factors. Equations 12.13 and 12.15 lead to upper limits of the loss factor that can be measured with one-third or octave band filters (Table 12.5). Again, it is an advantage to use time-reversed analysis.

The low modal density is another problem, which is more pronounced in structures than in rooms. A low modal density is a precaution for using an

Table 12.5 Maximum measurable loss factors using normal one-third octave and octave band filters and either forward or time-reversed analysis

Forward		Backward	
One-third octave	Octave	One-third octave	Octave
0.03	0.10	0.13	0.39

alternative method of measuring the bandwidth of single structural modes. If a structural mode is separated in frequency from other modes, it is possible to estimate the loss factor at the natural frequency f_0 by measuring the 3 dB bandwidth B_r and using Equation 2.52:

$$\eta_{\text{total}} = \frac{B_r}{f_0} \tag{12.22}$$

12.8 RADIATION EFFICIENCY

The radiation efficiency of a surface element can be measured if the element can radiate sound to a receiving room with approximately diffuse sound field, preferably in a laboratory. The radiating element should be excited to vibrations, e.g. by a steel framework bolted to the lower part of the element, as described in NT ACOU 033.

The average surface velocity level L_v is calculated as the energetic average from:

$$L_v = 10 \lg \frac{\sum_{i=1}^{n} v_i^2}{n v_0^2} \text{ (dB)} \tag{12.23}$$

where n is the number of measurement positions on the radiating surface, v_i is the RMS velocity in the normal direction in position i and $v_0 = 10^{-9}$ m/s is the reference velocity.

The radiated sound power from the surface to the receiving room is determined by measuring the average sound pressure level and assuming a diffuse sound field. The average sound pressure level L_p is calculated as the energetic average from:

$$L_p = 10 \lg \frac{\sum_{i=1}^{m} p_i^2}{m \, p_0^2} \text{ (dB)} \tag{12.24}$$

where m is the number of measurement positions in the receiving room, p_i is the RMS sound pressure in position i and $p_0 = 20$ µPa is the reference sound pressure.

At least two different positions of the exciter should be used. For each exciter position, the vibration level should be measured in at least ten different positions and the sound pressure level in at least three different positions.

The radiation efficiency σ is calculated from:

$$10 \lg \sigma = L_p - L_v - 10 \lg \frac{S}{A} + 28 \text{ (dB)} \tag{12.25}$$

where S is the area of the radiating surface and A is the equivalent absorption area of the receiving room. This formula can be derived from Equations 6.24 and 4.29.

12.9 JUNCTION ATTENUATION

The measurement of transmission of vibrations through a junction in buildings can be made in accordance with the method in NT ACOU 090, which has been adapted in ISO 10848-1.

The building element i is excited either with an electrodynamic exciter or with a hammer. The vibration level is measured simultaneously on element i and on the element j on the other side of the junction. Measurements are made in both directions, and for each direction at least two exciter positions should be used. For each exciter position, the vibrations are measured in at least eight positions distributed over elements i and j. If using excitation with a hammer, at least ten hammer blows at different positions are used for each of the eight pairs of vibration transducer positions. In addition, the structural reverberation time is measured in each of the two elements under test following ISO 3382-2 and using a 20 dB evaluation range.

The average vibration level difference is determined by:

$$\overline{D_{v,ij}} = \frac{1}{2}\left(D_{v,ij} + D_{v,ji}\right)(dB) \tag{12.26}$$

where

$D_{v,ij}$ is the difference between the average velocity level of element i and j, when element i is excited.

$D_{v,ji}$ is the difference between the average velocity level of element j and i, when element j is excited.

In case of transmission between single-layered structures, the vibration reduction index K_{ij} should be determined. Then, it is necessary to measure the structural reverberation time of both structures. For each of the measured elements (i and j), the equivalent absorption length is estimated by:

$$a_j = \frac{2.2\,\pi^2\,S_j}{T_{s,j}\,c\,\sqrt{\dfrac{f}{f_{ref}}}}\,(m) \tag{12.27}$$

where

S_j is the area of element j (m^2).
$T_{s,j}$ is the structural reverberation time of element j (s).
$c = 344$ m/s is the speed of sound in air.
f is the frequency (Hz).
$f_{ref} = 1000$ Hz is a reference frequency.

The vibration reduction index is calculated by:

$$K_{ij} = \overline{D_{v,ij}} + 10\lg\frac{l_{ij}}{\sqrt{a_i\,a_j}}\ (\text{dB}) \tag{12.28}$$

where

$\overline{D_{v,ij}}$ is the direction-averaged vibration level difference (dB).

$l_{i,j}$ is the length (m) of the junction between the elements i and j.

a_i and a_j are the equivalent absorption lengths (m) of elements i and j, respectively.

REFERENCES

ANSI S1.1 (1994). *American National Standard – Acoustical terminology,* The American National Standards Institute, Inc., New York.

ASTM E90-09 (2009). *Standard test method for laboratory measurement of airborne sound transmission loss of building partitions and elements,* American Society for Testing and Materials, Conshohocken, PA.

ASTM E336-16 (2016). *Standard test method for measurement of airborne sound attenuation between rooms in buildings,* American Society for Testing and Materials, Conshohocken, PA.

ASTM E413-16 (2016). *Classification for rating sound insulation,* American Society for Testing and Materials, Conshohocken, PA.

ASTM E492-09 (2016). *Standard test method for laboratory measurement of impact sound transmission through floor-ceiling assemblies using the tapping machine,* American Society for Testing and Materials, Conshohocken, PA.

ASTM E966-10 (2010). *Standard guide for field measurements of airborne sound insulation of building facades and facade elements,* American Society for Testing and Materials, Conshohocken, PA.

ASTM E989-06 (2012). *Standard classification for determination of impact insulation class (IIC),* American Society for Testing and Materials, Conshohocken, PA.

ASTM E1007-16 (2016). *Standard test method for field measurement of tapping machine impact sound transmission through floor-ceiling assemblies and associated support structures,* American Society for Testing and Materials, Conshohocken, PA.

ASTM E1332-16 (2016). *Standard classification for rating outdoor-indoor sound attenuation,* American Society for Testing and Materials, Conshohocken, PA.

ASTM E2235-04 (2012). *Standard test method for determination of decay rates for use in sound insulation test methods,* American Society for Testing and Materials, Conshohocken, PA.

C.L. Christensen, G. Koutsouris, J.H. Rindel (2013). The ISO 3382 parameters: Can we simulate them? Can we measure them? *Proceedings of ISRA 2013, International Symposium on Room Acoustics,* 9–11 June 2013, Toronto, ON, Canada.

C. Hopkins and P. Turner (2005). Field measurement of airborne sound insulation between rooms with non-diffuse sound fields at low frequencies. *Applied Acoustics* 66, 1339–1382.

C. Hopkins (2015). Revision of international standards on the field measurements of airborne, impact and façade sound insulation to form the ISO 16283 series. *Building and Environment* 92, 703–712.

ISO 717-1 (1996). *Acoustics – Rating of sound insulation in buildings and of building elements – Part 1: Airborne sound insulation*, International Organization for Standardization, Geneva, Switzerland.

ISO 717-2 (1996). *Acoustics – Rating of sound insulation in buildings and of building elements – Part 2: Impact sound insulation*, International Organization for Standardization, Geneva, Switzerland.

ISO 2631-1 (1997). *Mechanical vibration and chock – Evaluation of human exposure to whole-body vibration – Part 1: General requirements*, International Organization for Standardization, Geneva, Switzerland.

ISO 10052 (2004). *Acoustics – Field measurements of airborne and impact sound insulation and of service equipment sound – Survey method*, International Organization for Standardization, Geneva, Switzerland.

ISO 10140-2 (2010). *Acoustics – Laboratory measurement of sound insulation of building elements – Part 2: Measurement of airborne sound insulation*, International Organization for Standardization, Geneva, Switzerland.

ISO 10140-3 (2010). *Acoustics – Laboratory measurement of sound insulation of building elements – Part 3: Measurement of impact sound insulation*, International Organization for Standardization, Geneva, Switzerland.

ISO 10848-1 (2006). *Acoustics – Laboratory measurement of the flanking transmission of airborne and impact sound between adjoining rooms – Part 1: Frame document*, International Organization for Standardization, Geneva, Switzerland.

ISO 15186-1 (2000). *Acoustics – Measurement of sound insulation in buildings and of building elements using sound intensity – Part 1: Laboratory measurements*, International Organization for Standardization, Geneva, Switzerland.

ISO 15186-2 (2003). *Acoustics – Measurement of sound insulation in buildings and of building elements using sound intensity – Part 2: Field measurements*, International Organization for Standardization, Geneva, Switzerland.

ISO 15186-3 (2002). *Acoustics – Measurement of sound insulation in buildings and of building elements using sound intensity – Part 3: Laboratory measurements at low frequencies*, International Organization for Standardization, Geneva, Switzerland.

ISO 16032 (2004). *Acoustics – Measurement of sound pressure level from service equipment in buildings – Engineering method*, International Organization for Standardization, Geneva, Switzerland.

ISO 16283-1 (2014). *Acoustics – Field measurements of sound insulation in buildings and of building elements – Part 1: Airborne sound insulation*, International Organization for Standardization, Geneva, Switzerland.

ISO 16283-2 (2015). *Acoustics – Field measurements of sound insulation in buildings and of building elements – Part 2: Impact sound insulation*, International Organization for Standardization, Geneva, Switzerland.

ISO 16283-3 (2016). *Acoustics – Field measurements of sound insulation in buildings and of building elements – Part 3: Façade sound insulation*, International Organization for Standardization, Geneva, Switzerland.

ISO 18233 (2006). *Acoustics – Application of new measurement methods in building and room acoustics*, International Organization for Standardization, Geneva, Switzerland.

F. Jacobsen and J.H. Rindel (1987). Time reversed decay measurements. *Journal of Sound and Vibration* 117, 187–190.

J. Jakobsen (1987). *Measurement of vibration and shock in buildings for the evaluation of annoyance.* Nordtest project 618–86. Technical Report No. 139, Danish Acoustical Institute, Denmark.

J. Jakobsen (2001). Danish guidelines on environmental low frequency noise, infrasound and vibration. *Journal of Low Frequency Noise, Vibration and Active Control* 20 (3), 141–148.

A. Løvstad, J.H. Rindel, V. Støen, A.T. Windsor, A. Negård (2014). Measurement and simulation of high sound insulation and identification of sound transmission paths in complex building geometries. *Proceedings of Forum Acusticum 2014,* September 2014, Krakow, Poland.

M. Machimbarrena and F. Jacobsen (1999). Is there a systematic disagreement between intensity-based and pressure-based sound transmission loss measurements? *Building Acoustics* 6(2), 101–111.

NT ACOU 033 (1980). *Bulkheads: Sound radiation efficiency – Laboratory test,* Nordtest, Helsinki, Finland.

NT ACOU 090 (1994). *Building structures, junctions: Transmission of vibrations – Field measurements,* Nordtest, Helsinki, Finland.

NT ACOU 102 (1999). *Building elements – Façade elements and façades: Field measurements of airborne sound insulation – Loudspeaker method using MLS noise signals,* Nordtest, Helsinki, Finland.

Norsonic (2006). *Swept sine measurement technique using the Nor121 analyser.* Application Note, Ed. 3, 10.06, Norsonic, Lierskogen, Norway.

NS 8176 (2005). *Vibration and shock. Measurement of vibrations in buildings from landbased transport and guidance to evaluation of effects on human beings,* 2nd edition, Norwegian Standard, Oslo, Norway.

T. Poulsen and F.R. Mortensen (2002). *Laboratory evaluation of annoyance of low frequency noise.* Working Report No. 1, Danish Environmental Protection Agency, Copenhagen, Denmark.

F. Satoh, M. Sano, Y. Hayashi, J. Hirano, S. Sakamoto (2011). Sound insulation measurement using 10 minute swept-sine signal. *Proceedings of InterNoise 2011,* 4–7 September 2011, Osaka, Japan.

C. Simmons (1999). Measurement of sound pressure levels at low frequencies in rooms. Comparison of available methods and standards with respect to microphone positions. *Acustica – Acta Acustica* 85, 88–100.

J. Takala and M. Kylliäinen (2013). Room acoustics and background noise levels in furnished Finnish dwellings. *Proceedings of InterNoise 2013,* 15–18 September 2013, Innsbruck, Austria.

M.L.S. Vercammen (1992). Low-frequency noise limits. *Journal of Low Frequency Noise and Vibration* 11, 7–13.

T. Yang, P. Liu, J. Yin (2014). Research on assessment of fourth power vibration dose value in environmental vibration caused by metro. *Proceedings of ICSV 21,* 13–17 July 2014, Beijing, China.

Chapter 13

Measurement uncertainty

The sound pressure level in a room varies from one point to another, even in an ideal diffuse sound field. These variations are particularly large when the excitation is a pure tone or narrow band noise like a one-third octave band and, thus, some kind of spatial averaging is necessary in order to get a reasonably accurate measurement result. Measurements at low frequencies in small rooms need special consideration. This chapter presents an overview of the theoretical background and methods for estimation of uncertainty in sound pressure level measurements and reverberation time measurements in a room. Chapter 4, Introduction to Room Acoustics, is a basis for this chapter.

13.1 SOME STATISTICAL CONCEPTS

The fundamental principles of measurement uncertainty in building acoustics are found in the ISO Guide to the expression of Uncertainty in Measurement (GUM) and in ISO 12999-1 (2014).

13.1.1 Uncertainty of a measurement result

The following concepts and definitions apply:

Measurand z: The measured quantity, e.g. sound pressure level or reverberation time.
Mean value $\langle z \rangle$: The arithmetic average of all measured z. For sound pressure levels in a room, the arithmetic average of the squared sound pressure in all measurement positions.
Variance σ^2: The mean square deviation from the mean value.
Standard deviation σ: The square root of the variance. For sound pressure levels in a room, the root mean square (RMS) deviation from the mean value of the squared sound pressure in all measurement positions. The standard deviation can either be determined experimentally from the measured data, or it can be derived from assumed probability distributions.

Uncertainty Characterizes the dispersion of the result of a measurement and may be the standard deviation or the half-width of an interval having a stated level of confidence.

Standard uncertainty u Uncertainty of a result expressed as a standard deviation. For a Gaussian distribution $u=\sigma$.

Expanded uncertainty U Quantity defining an interval about the result of a measurement that may be expected to encompass a large fraction of the distribution of values. $U=k\cdot u$.

Coverage factor, k Factor used as a multiplier of the standard uncertainty in order to obtain an expanded uncertainty. It depends on the statistical distribution function, on the confidence level and on whether a one-sided or a two-sided test is considered.

Confidence level, g The statistical probability that the result is within a certain range, either on both sides of the mean value (two sided) or on one side of the mean value (one sided).

A measurand z can be statistically characterized by the *probability density function w(z)*, which in some cases can be a mathematical function of the mean ‹z› and standard deviation σ (or the variance σ^2).

13.1.2 The Gaussian distribution

The *Gaussian distribution* is the most commonly used statistical distribution and is also called the *normal distribution*. If the mean value is μ and the standard deviation is σ, the probability density function is:

$$w(z) = \frac{1}{\sigma\sqrt{2\pi}}\, e^{-\frac{(z-\mu)^2}{2\sigma^2}} \tag{13.1}$$

This is a symmetric distribution and the probability for a random sample to be within the range $\mu\pm\sigma$ is 68 %. The probability to be within an extended range of $\mu\pm2\sigma$ is 95 % (Figure 13.1).

13.1.3 Confidence levels

Assuming a Gaussian distribution, the measurement result is an estimate of the true result, which will be within a range of $<z>\pm k(g)\cdot\sigma(z)$, where $\sigma(z)$ is the standard deviation and $k(g)$ is a coverage factor that depends on the confidence level, g (Table 13.1). This is called a *two-sided test* and is applied when a measurement result is reported without comparison with any target value. For example, with 95 % probability, the true result is within a range of ±1.96 times the standard deviation (the expanded uncertainty U).

When a measured result is compared with a requirement, the *one-sided test* applies. If, for example, a 90 % confidence level is chosen, the coverage factor for a one-sided test is $k=1.28$. This means that the limit z_{max} is not

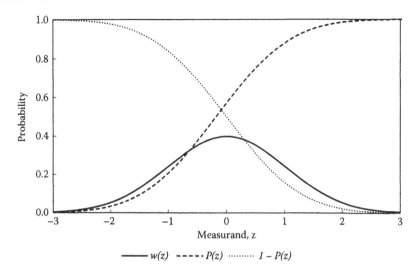

Figure 13.1 Gaussian distribution. Probability density function $w(z)$, cumulative probability function $P(z)$, and the opposite function $1 - P(z)$.

Table 13.1 Confidence levels g and the associated coverage factors k for one- and two-sided tests

Coverage factor, k	1.00	1.28	1.65	1.96	2.58	3.29
Confidence level for two-sided test, g (%)	68	80	90	95	99	99.9
Confidence level for one-sided test, g (%)	84	90	95	97.5	99.5	99.95

exceeded with 90 % probability, if $<z>+k(g)\cdot\sigma(z)=<z>+1.28\cdot\sigma(z) < z_{max}$. It is noted that the coverage factor is smaller for a one-sided test than for a two-sided test with the same confidence level. At the end of this chapter, we shall return to this point with an example.

13.2 VARIANCE OF SOUND PRESSURE IN A ROOM

13.2.1 Pure tone excitation

It is assumed in this section that the frequency is above Schroeder's limiting frequency, so statistical methods apply to the sound field. Thus, the excitation of sound with a single frequency will involve a number of room modes having natural frequencies close to the frequency of excitation. In a measurement point, the sound pressure is the result of contributions from the room modes that have been excited, each mode being represented by its amplitude and phase angle (see Equation 4.6).

If a microphone is moved slowly in a room while a pure tone is emitted, the sound pressure level will vary strongly, and the fluctuations look stochastic (Figure 13.2a).

The range of the fluctuations can be described by the relative spatial variance, defined by:

$$\varepsilon_R^2 = \frac{\sigma^2\left(p^2\right)}{\left\langle p^2\right\rangle^2} = \frac{\left\langle\left(p^2 - \left\langle p^2\right\rangle\right)^2\right\rangle}{\left\langle p^2\right\rangle^2} \tag{13.2}$$

where $\sigma\left(p^2\right)$ is the standard deviation of the squared sound pressure and $<p^2>$ denotes the spatial average. The sound pressure is the RMS value of a time averaging, see Section 13.3.3 about sufficient measurement time.

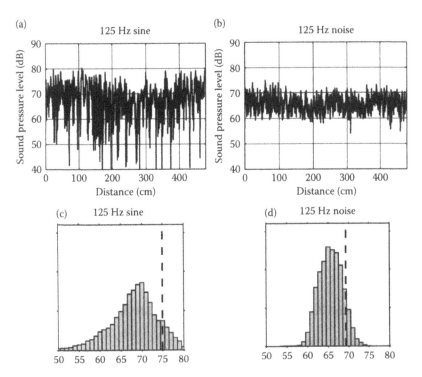

Figure 13.2 Measured distribution of sound pressure level in a 22 m² furnished room. (a) As a function of distance for a 125 Hz sinusoidal signal. (b) As a function of distance for a 125 Hz one-third octave band filtered pink noise. (c) Histogram for a 125 Hz sinusoidal signal. (d) Histogram for a 125 Hz one-third octave band filtered pink noise. The dashed lines in (c) and (d) indicate the 10 percentile. (Adapted from Pedersen, S. et al., J. Low Freq. Noise, Vib. Act. Control 26(4), 249–270, 2007.)

13.2.2 Band-limited noise excitation

A diffuse sound field and frequencies above Schroeder's limiting frequency f_g are assumed (Equation 4.21):

$$f_g = 2000\sqrt{\frac{T}{V}}\,(\text{Hz}) \tag{13.3}$$

where T is the reverberation time in (s) and V is the room volume in (m^3).

If the excitation of a room is with a band-limited noise signal (Figure 13.2b) and the bandwidth is B (Hz), the relative spatial variance (Lubman, 1968) is approximately:

$$\varepsilon_R^2 = \frac{2}{Z}\arctan(Z) - \frac{1}{Z^2}\ln(1+Z^2) \tag{13.4}$$

where $Z=B/B_r$ is the ratio between the bandwidth B of the noise signal and the bandwidth B_r of a typical room mode. With the reverberation time T, the mode bandwidth is $B_r \approx 2.2/T$ (see Equation 4.19). If the bandwidth B is not very small, a good approximation for the relative spatial variance is:

$$\varepsilon_R^2 \cong \left(1+\frac{B}{\pi B_r}\right)^{-1} = \left(1+\frac{BT}{6.9}\right)^{-1} \tag{13.5}$$

where T is the reverberation time. It is seen that the variance decreases with increasing bandwidth of the noise signal and with increasing reverberation time; thus, a wideband noise signal in a highly reverberant room yields very small spatial variations of the squared sound pressure. The approximation (Equation 13.5) is valid within 2.5 % for $BT > 20$. Below that range, Equation 13.4 is used.

In practice, most measurements are performed with a wide band noise signal and a series of band-limiting filters after the microphone. However, it does not matter where in the measurement chain the band-limiting filter is located; the above equation is still a valid approximation. The most common filters are the octave band ($B=0.71\,f$) and the one-third octave band ($B=0.23\,f$), where f is the centre frequency of the band.

In the extreme case of a pure tone (bandwidth $B=0\,\text{Hz}$), we have $\varepsilon_R^2=1$, and this corresponds to a standard deviation on the sound pressure level of 5.6 dB (Schroeder, 1969). Thus, the sound pressure level for a pure tone excitation in a room will vary within a range of approximately 22 dB (with 95 % probability assuming Gaussian distribution). However, these results refer to using the sound pressure level as the measurand, i.e. when the mean value and the standard deviation are determined directly from the dB values. This introduces a bias error in the mean value compared to the correct result calculated from the squared sound pressures. The bias error and the

Table 13.2 Bias error D and standard deviation σ of the sound pressure level in a diffuse sound field as functions of the product BT, B being the measurement bandwidth and T the reverberation time

$1+BT/6.9$	1	3	10	30	100	300	1000
BT	0	14	62	200	683	2063	6893
Bias D (dB)	2.51	0.76	0.22	0.07	0.02	0.00	0.00
Standard deviation σ (dB)	5.57	2.73	1.41	0.80	0.44	0.25	0.14

Source: Lubman, D., *J. Sound Vib.*, 16, 43–58, 1971.

standard deviation in dB can be estimated from the BT product and some values are given in Table 13.2.

The bias means that the correct energetic averaged sound pressure level L_{p^2} (Equation 1.34) is greater than or equal to the simple arithmetic dB-average L_{dB} (Equation 1.31), and the difference is the bias D error:

$$L_{p^2} = L_{dB} + D \ (\text{dB}) \tag{13.6}$$

In Table 13.2 the bias error is 2.5 dB for a pure tone ($B=0$) but decreases with increasing bandwidth. For broadband noise excitation ($BT > 200$), the bias error becomes negligible. The standard deviation of the sound pressure level L from the mean value $<L>$ is 5.6 dB for a pure tone in a diffuse sound field, as first found by Doak (1959).

13.2.3 Variance at low frequencies

Below Schroeder's limiting frequency, the modal overlap index is small (less than 3) and the room modes may be too spread to fill the frequency band that is measured. Thus, the efficient bandwidth of the modes may be less than the measurement bandwidth (see Figure 4.7). For this case, Lubman (1974) suggested to replace the measurement bandwidth B by *the effective noise bandwidth* B_{eff} calculated as the number of modes within B multiplied with the bandwidth of the modes B_r:

$$B_{eff} = B \cdot \frac{\Delta N}{\Delta f} \cdot B_r \cdot \kappa = \Delta N(B) \cdot B_r \cdot \kappa \tag{13.7}$$

where $\Delta N(B)$ is the number of modes within the bandwidth B. The overlap factor κ (≤ 1) is included in order to take account of the fact that the modes have some degree of overlap. Even if the modal overlap index is $M < 3$, there is still more than one mode within the bandwidth of the modes ($M > 1$) in a certain frequency range extending at least one octave below f_g.

So, the relative spatial variance is calculated with $Z=B_{eff}/B_r=\Delta N(B)\cdot\kappa$ in Equation 13.4. Using instead the approximation (Equation 13.5) yields:

$$\varepsilon_R^2 \cong \left(1 + \frac{\Delta N(B) \cdot \kappa}{\pi}\right)^{-1} \text{ for } f < f_g \tag{13.8}$$

The overlap factor κ has been estimated by comparison with empirical data like those in Figure 13.3, and the result is $\kappa = 0.37 = \pi/8.5$. The suggested approximation for the relative spatial variance below the limiting frequency is:

$$\varepsilon_R^2 \cong \left(1 + \frac{\Delta N(B)}{8.5}\right)^{-1} \text{ for } f < f_g \tag{13.9}$$

where $\Delta N(B)$ is the number of room modes within the measurement bandwidth B. This approximation was first suggested by (Rindel, 1981).

Figure 13.3 shows a comparison of measured spatial standard deviations in a room compared with the estimations from Equation 13.9 below the limiting frequency and Equation 13.5 above the limiting frequency. The two theoretical curves cross at the limiting frequency.

In the common case of measurements in one-third octave bands, Equations 13.5 and 13.9 can be modified with insertion of $B = 0.23\ f$ and the modal density (Equation 4.13). Using the approximation (Equation 4.14), the relative spatial standard deviation yields, below and above f_g, respectively:

$$\varepsilon_R \cong \left(1 + \frac{V \cdot f^3}{2.9 \cdot c^3}\right)^{-1/2} \text{ for } f < f_g \tag{13.10}$$

$$\varepsilon_R \cong \left(1 + \frac{T \cdot f}{30}\right)^{-1/2} \text{ for } f \geq f_g \tag{13.11}$$

These are displayed graphically in Figure 13.4. The relative spatial standard deviation is the higher one in the two Equations 13.10 and 13.11. At frequencies below the Schroeder limiting frequency, the room volume is the important parameter, whereas at frequencies above the limiting frequency, the reverberation time is the important parameter. The approximation above the limiting frequency (Equation 13.11) is good, and that under the limiting frequency (Equation 13.10) is also quite good, because the error by using the simple Equation 13.5, instead of Equation 13.4, to some extent, compensates for the error due to the approximate modal density (Equation 4.14) instead of the more correct one (Equation 4.13).

If the volume and reverberation time are known, it is possible to estimate the relative spatial standard deviation for the curves in Figure 13.4. The curve with the highest standard deviation is always the valid one, and the cross-over is at the Schroeder limiting frequency. It should be noted that reducing the reverberation time in a room makes things worse because the standard deviation is increased at frequencies above the limiting frequency.

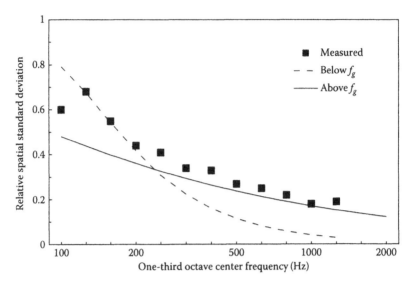

Figure 13.3 Measured and estimated relative spatial standard deviation in one-third octave bands for a 70 m³ room with reverberation time approximately 1.0 s. The Schroeder limiting frequency is 240 Hz. (Measured data from Jacobsen, F., The diffuse sound field. Statistical considerations concerning the reverberant field in the steady state. Report No. 27, The Acoustics Laboratory, Technical University of Denmark, 1979.)

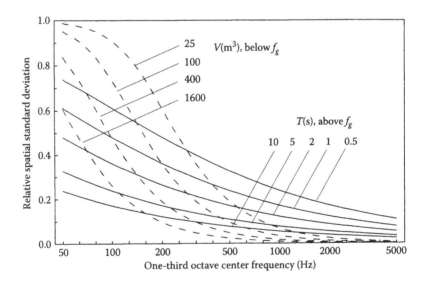

Figure 13.4 Curves for estimation of the relative spatial standard deviation on one-third octave band measurements in a room. At low frequencies, the volume is the determining parameter, at high frequencies it is the reverberation time. For a given volume and reverberation time, the cross-over is at the Schroeder limiting frequency.

It is sometimes claimed that it is preferable to have a low limiting frequency and that the reverberation time should not be too long for that reason; however, it is seen from Figure 13.4 that for a fixed volume, a lower limiting frequency goes hand in hand with a higher spatial standard deviation.

At very low frequencies, the number of room modes is small and the one-dimensional axial modes tend to dominate. There may be none or very few tangential or oblique modes, depending on the ratio between room dimensions.

The calculated frequency response of a room at low frequencies is shown in Figure 13.5. The room is rectangular with dimensions $(l_x, l_y, l_z) = (4.32\,\mathrm{m}, 3.38\,\mathrm{m}, 2.70\,\mathrm{m})$ and volume $39\,\mathrm{m}^3$. The frequency response is shown for a source in one corner and three different receiver positions; in the opposite corner, in the centre of the room, and in a point halfway between the centre and a corner. All modes are at maximum in the corner position while in the centre position, all modes with an uneven number n_x or n_y or n_z are eliminated; so, the first peak in the centre is at the mode $(2,0,0)$ around $80\,\mathrm{Hz}$.

In the very low frequency range $0\,\mathrm{Hz} < f < f_{2,0,0}$, the following observations are made:

- The modal density is very low; in some one-third octave bands, there may be a single mode: in other bands, the sound pressure rides on the skirts of the frequency response of one or two modes with natural frequencies outside the band.
- The range of variation of the sound pressure level, i.e. the difference between the maximum and minimum curves, is high in the region above the first axial mode $(1,0,0)$, but decreases towards zero when the frequency drops towards the cavity mode $(0,0,0)$.

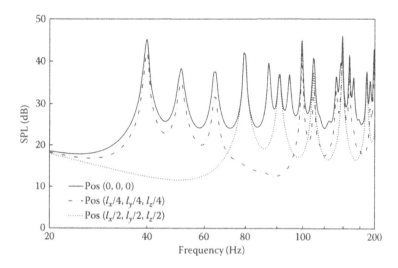

Figure 13.5 Calculated frequency responses in three receiver positions at low frequencies in a $39\,\mathrm{m}^3$ room. The source is in a corner.

- The probability function is asymmetric; in the position halfway between corner and centre, the level is only a few dB lower than in the corner. This reflects the shape of the cosine-shaped axial mode, having maximum at the walls and a node in the centre.
- The spatial variation of the sound pressure is deterministic. Thus, the normal statistical methods applied to a stochastic system are not usable.

These observations lead to the result shown in Figure 13.6. The frequency range is divided into three, and the corresponding limiting frequencies are $f_{2,0,0}$ (the frequency of the second axial mode in the longest room dimension) and f_g (the Schroeder limiting frequency).

The course of the standard deviation in region 1 has been determined on the basis of a number of measured data from some cases: see examples in Figures 13.7 and 13.8. The empirical equation is:

$$\varepsilon_R \cong \frac{l_x \cdot f}{c} \text{ for } f < f_{2,0,0} \tag{13.12}$$

where l_x is the longest dimension of the room (assuming a box-hyphen room) and $f_{2,0,0} = c/l_x$ is the natural frequency of the room mode (2,0,0). Below this frequency, it can be assumed that there are very few room modes, and they are dominated by the three lowest axial modes (1,0,0), (0,1,0) and

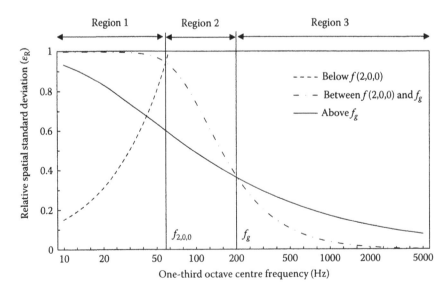

Figure 13.6 Course of the relative spatial standard deviation, and division into three frequency regions.

(0,0,1). The spatial standard deviation in the three regions is determined by Equation 13.12 in region 1, Equation 13.10 in region 2, and Equation 13.11 in region 3, respectively.

The spatial standard deviation in dB is determined as:

$$\sigma(L_p) \cong 10 \lg \frac{\langle p^2 \rangle + \sigma(p^2)}{\langle p^2 \rangle} = 10 \lg(1 + \varepsilon_R)$$

$$\cong 10 \lg(e) \cdot \varepsilon_R \cong 4.34 \cdot \varepsilon_R \text{ for } \varepsilon_R \ll 1 \tag{13.13}$$

The results shown in Figure 13.7 were derived from measurements in a 148 m³ reverberation room with hard surfaces and average reverberation time 2.1 s. Measurement positions were located in a mesh of 7×10 positions in two different heights above the floor, 0.7 m and 1.7 m. The distance between the horizontal positions was 0.7 m. Region 2 is between the frequencies 45 Hz and 238 Hz. The very different course in the three regions is clearly seen in this example, and the peak is noticed at 45 Hz.

The results shown in Figure 13.8 were derived from measurements in a 31 m³ bedroom with furniture and average reverberation time 0.2 s. Measurement positions were located in a mesh of 4×5 positions in two different heights above the floor, 0.7 m and 1.7 m. The distance between the horizontal positions

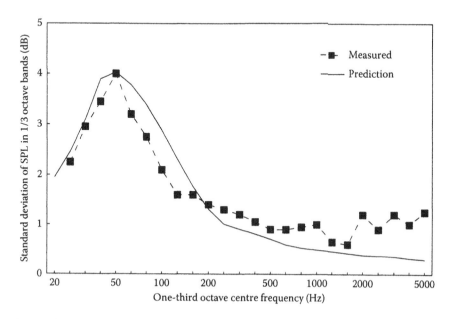

Figure 13.7 Measured and estimated spatial standard deviation in a 148 m³ reverberation room (7.5 m×5.5 m×3.5 m, *T*=2.1 s). (Measured data from Simmons, C., *Noise Control Eng. J.*, 60, 405–420, 2012.)

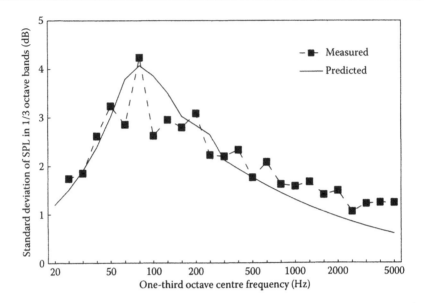

Figure 13.8 Measured and estimated spatial standard deviation in a 31 m³ bedroom (3.8 m×3.3 m×2.5 m, *T*=0.2 s). (Measured data from Simmons, C., *Noise Control Eng. J.*, 60, 405–420, 2012.)

was 0.7 m. In this example, the volume is small and reverberation time is very short. The Schroeder limiting frequency is 160 Hz and $f_{2,0,0}$=90 Hz, so region 2 is quite narrow in this case. The peak is noticed around 90 Hz.

13.3 MEASUREMENT UNCERTAINTY OF SOUND PRESSURE LEVEL MEASUREMENTS

Measurement uncertainty in building acoustics is addressed in ISO 12999-1 (2014), and it is based on the assumption of a Gaussian (normal) distribution of the sound pressure level when measured in a room. Although this is not correct, it may be a sufficiently good approximation at frequencies above the Schroeder limiting frequency. The obvious advantage of this assumption is that the expanded uncertainty can be calculated for a given confidence level.

13.3.1 Probability density functions

In a diffuse sound field, or in practice at frequencies above the Schroeder limiting frequency, it has been shown both theoretically and experimentally that the RMS deviation of the sound pressure level L from the spatial average <L> is 5.6 dB (Doak, 1959; Schroeder, 1969).

However, the probability density function is not the symmetric Gaussian distribution. Instead, the mean squared sound pressure in a room with a pure tone excitation is known to follow a *Poisson distribution*. The probability density function for the sound pressure level relative to the spatial average (Pierce 1989, Equations 6 and 7.16) can be mathematically expressed as:

$$w(z) = e^{z - e^z} \tag{13.14}$$

where

$$z = \frac{\ln(10)}{10}(L - L_0) \cong 4.34\,(L - L_0) \tag{13.15}$$

Here, L is the sound pressure level in an arbitrary point and L_0 is the sound pressure level corresponding to the spatial average of the squared sound pressure, $\overline{p^2}$.

The Poisson distribution is an *asymmetric* function (Figure 13.9). This implies that the spatial averaged sound pressure level (L) is 2.5 dB lower than L_0. The probability for a measured sound pressure level L to be less than L_0 is 63%, whereas the probability to exceed L_0 is only about half of that, 37%. Other characteristic measures that can be extracted from the cumulative probability function of the Poisson distribution are that the 50% (median) level is $L_0 - 1.6$ dB, the 68% confidence interval is (−7.6 dB; 2.7 dB) and the

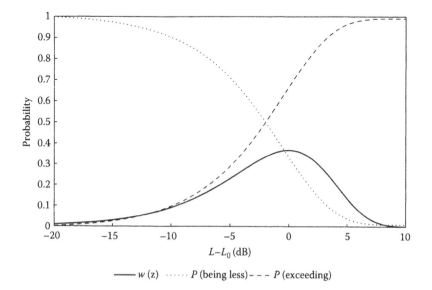

Figure 13.9 Probability functions for the sound pressure level in a diffuse sound field. (Adapted from Pierce 1989, Figures 6.16.)

95% confidence interval is (−16 dB; 5.7 dB). So, if a microphone is moved in a room excited by a pure tone, the sound pressure level fluctuates more than 20 dB, but the maxima will be only about 6 dB above the spatial average L_0. Experimental results for a pure tone excitation in a room are shown in Figure 13.2a and c.

If a diffuse sound field cannot be assumed, which is the case at frequencies below Schroeder's limiting frequency, the sound field can be analysed by assuming a rectangular room with dimensions $l_x \geq l_y \geq l_z$. Then, the squared sound pressure is known as a function of position (x, y, z) at each mode (n_x, n_y, n_z) from Equation 4.4:

$$p^2 = p_0^2 \cdot \cos^2\left(\pi n_x \frac{x}{l_x}\right) \cdot \cos^2\left(\pi n_y \frac{y}{l_y}\right) \cdot \cos^2\left(\pi n_z \frac{z}{l_z}\right) \tag{13.16}$$

It follows that in an oblique mode (all $n \neq 0$), the spatial average covering the entire volume is:

$$\langle p^2 \rangle = p_0^2 \cdot \int_0^{l_x} \cos^2\left(\pi n_x \frac{x}{l_x}\right) dx \cdot \int_0^{l_y} \cos^2\left(\pi n_y \frac{y}{l_y}\right) dy$$

$$\cdot \int_0^{l_z} \cos^2\left(\pi n_z \frac{z}{l_z}\right) dz = \frac{1}{8} \cdot p_0^2 \tag{13.17}$$

This implies that in each of the eight corners, the sound pressure level is 10 lg (8) = 9 dB higher than the sound pressure level L_0 of spatial averaged squared sound pressure. In a similar way, it is found that in the tangential modes (two $n \neq 0$), the highest sound pressure level is 6 dB above the room average and in the axial modes (one $n \neq 0$), the highest sound pressure level is 3 dB above the room average.

The probability density functions of the various types of room modes have been derived by Waterhouse (1970). The cumulative probability curves for the various types of room modes, oblique, tangential, axial and the cavity mode, are shown in Figure 13.10.

The statistics of the *oblique* and *tangential* modes have some similarities; about 40% of all positions in the volume have sound pressure levels below L_0 and about 60% of all positions have a higher sound pressure level. The median level is 1.5 dB to 1.7 dB higher than L_0. These are the most important modes in the frequency range between $f_{2,0,0}$ and f_g (region 2 in Figure 13.6). Although the sound field is not diffuse because of insufficient modal overlap, the sound field is still three-dimensional and several modes will occur together within the one-third octave bands. Thus, it is impossible to predict where the minima will occur.

In the extreme low frequency range – below, the first room mode (1,0,0) – the sound in the room excites the first room mode as well as the

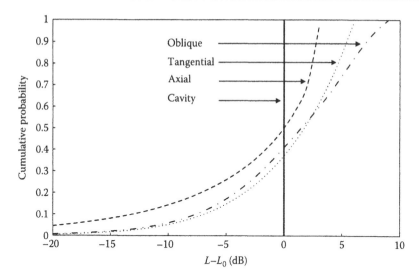

Figure 13.10 Cumulative probability functions for the sound pressure level in various types of room modes.

cavity mode (0,0,0). The latter has a very simple probability density function because all parts of the volume have equal sound pressure, so the variance is zero and the cumulative probability curve is collapsed into a vertical line at L_0 (Figure 13.10). If the frequency decreases from $f_{1,0,0}$ towards 0 Hz, the cumulative probability curve gradually changes from that of the pure axial mode to that of the cavity mode. This means that the variance decreases towards zero. This effect is clearly seen in Figure 13.5.

The statistics of the *axial* modes is different from that of the other mode types; the median level equals L_0, so half of all positions in the room will have sound pressure levels below L_0, and the other half have levels above L_0. However, the probability density function is strongly asymmetric; minimum levels are very low in the centre of the room, whereas maximum levels are only 3 dB higher than L_0. In the low frequency range (region 1 in Figure 13.6), the axial modes (1,0,0), (0,1,0) and (0,0,1) dominate, and in each mode the sound field is one-dimensional and spatially well defined.

In Table 13.3, the median level and the confidence intervals associated with the various spatial distribution functions for pure tone excitation of a room are stated. They are all asymmetric, in contrast to the Gaussian distribution, which is included for comparison.

13.3.2 Zones to be avoided

Below the Schroeder limiting frequency, the sound field in a room is not stochastic and it is possible to choose measurement positions that minimize

Table 13.3 Measurement uncertainty (dB), single frequency excitation, for different confidence levels, calculated for some probability functions

Confidence level (%)		Axial	Tangential	Oblique	Poisson	Gaussian
2-sided	1-sided low / 1-sided high					
Displayed in figure		13.10	13.10	13.10	13.9	13.1
Median (50%)		0.0	1.7	1.5	−1.6	0.0
68	16	−9.2	−4.6	−5.4	−7.6	−4.3
	84	2.7	5.0	6.2	2.7	4.3
80	10	−13.0	−6.9	−7.6	−9.8	−5.5
	90	2.9	5.4	7.1	3.6	5.5
90	5	−18.9	−10.1	−10.8	−12.9	−7.1
	95	3.0	5.7	7.9	4.8	7.1
95	2.5	−24.8	−13.2	−14.0	−16.0	−8.5
	97.5	3.0	5.9	8.3	5.7	8.5
99	0.5	−37.5	−20.5	−21.5	−23.2	−11.1
	99.5	3.0	6.0	8.7	7.2	11.1

the measurement uncertainty. Figure 13.11 shows the sound pressure level along the length of a room; the L_0 level occurs in the distance $l_x/4$ from the walls, and in distances greater than $l_x/3$, the sound pressure level drops more than 3 dB below L_0. As indicated in Figure 13.11, the centre region should

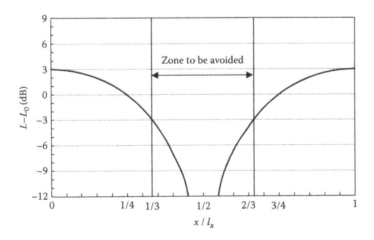

Figure 13.11 Spatial variation of sound pressure level in a room when the first axial mode is excited. Measurements in the central zone more than 1/3 room dimension from the walls should be avoided.

be avoided in order to achieve a low measurement uncertainty. The conclusion is that measurements at low frequencies are preferably made in the corner regions of the room where the spatial variation is limited. The precise distance from the room boundaries is not important.

Above the Schroeder limiting frequency, the picture is different, almost the opposite, in fact. From Chapter 3, Section 3.4, we know that the sound pressure level in a diffuse sound field is influenced by interference effects near the boundaries. With a pure tone excitation, the interference patterns

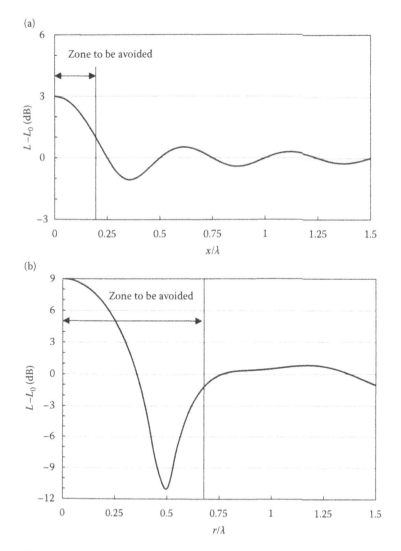

Figure 13.12 Spatial variation of sound pressure level in a near room boundaries in a diffuse sound field excited by a pure tone. (a) Near a wall a zone < 0.2 λ should be avoided. (b) Near a corner a zone < 0.7 λ should be avoided.

near a wall and a corner are shown in Figure 13.12. The size of the interference zones that should be avoided is 0.2 λ from the wall and 0.7 λ from the corner. Measurements are preferably made in the central zone of the room, but in practice it is often impossible to avoid the interference zones unless the room is big.

13.3.3 The influence of the measurement time

The measurement time t may also give a contribution to the measurement uncertainty. For band-limited noise with the bandwidth B, the relative variance of the time-averaged squared sound pressure p^2 is:

$$\varepsilon_t^2 = \frac{2}{Bt} \tag{13.18}$$

where t is the measurement time (or the integration time). The relationship is also shown in Figure 13.13 in terms of the expanded uncertainty in dB. The conversion to dB is used (Equation 13.13).

$$\sigma(L_p) = 4.34 \cdot \varepsilon_t = \frac{U(L_p)}{k(g)} \tag{13.19}$$

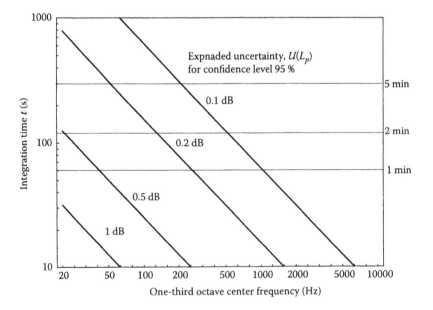

Figure 13.13 Necessary integration time for measurements in one-third octave bands in a room. The parameter on the curves is the expanded uncertainty in dB for 95 % confidence level.

Together with Equation 13.18, we get the measurement time:

$$t = \frac{2}{B}\left(\frac{4.34 \cdot k(g)}{U(L_p)}\right)^2 \tag{13.20}$$

In Figure 13.13, two-sided test and $k(g) = 1.96$ for 95% confidence level is assumed (Table 13.1).

The total relative spatial variance is:

$$\varepsilon^2 = \varepsilon_t^2 + \varepsilon_R^2 \tag{13.21}$$

However, in many cases, it is no problem to avoid the uncertainty contribution due to the measurement time. From Equations 13.5 and 13.18, it follows that $\varepsilon_t^2 \ll \varepsilon_R^2$ if the integration time t is:

$$t \gg 0.29 \cdot T + \frac{2}{B} \tag{13.22}$$

where T is the reverberation time of the room and B is the bandwidth of the measurement. As the last term typically is not significant, a simple rule of thumb is that the integration time in each position should be at least *three times the reverberation time* of the room.

13.3.4 The integrated impulse response method

New and very efficient measuring methods use either the maximum length sequence method or the swept-sine method to generate an impulse response, which is integrated to get the total sound pressure level (see ISO 18233). There are two main advantages of these methods; the signal-to-noise ratio is increased and the stochastic nature of the traditional broad-band noise excitation is avoided. The latter means that the influence of the measurement time on the uncertainty can be neglected (i.e. $\varepsilon_t^2 = 0$). These new measurement methods correspond to measurements with an infinite integration time using noise signals.

13.4 SPATIAL AVERAGING

13.4.1 Correlation coefficient

The correlation coefficient R between the sound pressures p_1 and p_2 taken in two different positions is defined as:

$$R = \frac{\overline{p_1 \cdot p_2}}{\sqrt{\overline{p_1^2} \cdot \overline{p_2^2}}} \tag{13.23}$$

where the horizontal bars mean time averaged values.

In a three-dimensional diffuse sound field, the correlation coefficient (see Kuttruff, 1979) is:

$$R = \frac{\sin(kx)}{kx} \tag{13.24}$$

where $k = 2\pi \, f/c$ is the wave number and x is the distance between the two positions. The shape of the function is shown in Figure 1.4. It is noted that $R = 0$ for $kx = \pi$, and R is relatively small for $kx > \pi$.

For practical applications, it can be concluded that

$$R \cong 0 \text{ for } x \geq \frac{\lambda}{2} \tag{13.25}$$

The distance between the microphone positions in a room must be at least half a wavelength in order to obtain uncorrelated measurement positions. However, this may be difficult or impossible to fulfil at low frequencies, especially in small rooms.

The above consideration may also be applied to the distance from the microphone positions to reflect room surfaces. The reflection from a hard surface can be interpreted as a sound wave that continues through the surface and reaches the image of the microphone that is created by a mirror in the plane of the surface. A measurement position near a reflecting surface should not be correlated with the image position behind the wall. Thus, in order to fulfil Equation 13.25, the distance from the surface must be at least $\lambda/4$.

13.4.2 Discrete spatial averaging

Because of the spatial variations within a room, the sound pressure measurements are usually taken as the average of the squared sound pressure over a number of measurement positions, N. If the distance between the positions is sufficiently large (Equation 13.25), the measurements can be considered uncorrelated and, thus, the relative variance of the averaged result is:

$$\varepsilon^2 = \frac{\varepsilon_R^2}{N} \tag{13.26}$$

If for any reason, the distance between the microphone positions cannot fulfil the minimum requirement (Equation 13.25), it is possible to calculate the equivalent number of uncorrelated positions according to Lubman (1969):

$$N_{eq} = \frac{N}{1 + \dfrac{1}{N}\displaystyle\sum_i \sum_j R^2\left(kx_{ij}\right)} \quad (i \neq j) \tag{13.27}$$

Here, N is the actual number of microphone positions and x_{ij} denotes the distance between positions i and j. So, all possible pair combinations are considered in the denominator and R can be calculated from Equation 13.24 if the sound field can be assumed a diffuse sound field. It is seen that in general $N_{eq} \leq N$. It is also noted that N_{eq} need not be a natural number.

Figure 13.14 shows examples of calculated equivalent number of measurement positions as a function of the frequency and for different distances between the actual positions. At high frequencies, it appears that $N_{eq} \approx N$, but at low frequencies, the positions are not fully independent, and $N_{eq} < N$.

With the positions equally distributed on a circle with radius r as in the example in Figure 13.14, the equivalent number of uncorrelated positions can be estimated by this empirical formula:

$$N_{eq} \cong \min\{N, N_{max}\} = \min\left\{N, \left(1 + \left(\frac{4\pi}{c}rf\right)^3\right)^{-1/3}\right\} \tag{13.28}$$

where f is the centre frequency of the measurement band.

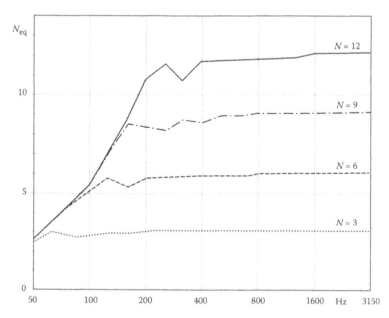

Figure 13.14 Equivalent number of uncorrelated measurement positions per third octave band in a diffuse sound field. The parameter is the actual number of fixed positions, evenly distributed on a circle with radius of 1.5 m. (Adapted from Tuominen, H. T., Proceedings of NAS-80, Åbo, Finland, 67–70, 1980. With permission.)

At very low frequencies in small rooms, it is impossible to obtain sufficient distance between the microphone positions. If the exact positions in the room are not known, it may be sufficient to apply a general rough estimate. A realistic number of equivalent uncorrelated positions is estimated from Equation 13.28 together with an estimated maximum possible radius in the room. For this purpose, it is assumed that the room has a rectangular floor plan with length/width ratio 1.25 and that the circle with measurement positions keeps at least 0.7 m from the walls. This leads to the maximum radius in a room with volume V and height h:

$$r_{max} = \frac{1}{2}\sqrt{\frac{V}{1.25 \cdot h}} - 0.7 \text{ m} \qquad (13.29)$$

As an example, a room with volume 50 m³ and height 2.3 m is considered. This yields the radius $r_{max}=1.39$ m and $N_{eq,max}=5$ at 100 Hz. So, if five fixed microphone positions are used, they actually count at five independent positions at frequencies above 100 Hz. However, another example room has a volume of only 26 m³ and height 2.3 m. In this case, the radius $r_{max}=0.8$ m and $N_{eq,max}=3$ at 100 Hz. Even if the number of microphone positions is more than five, they will only count as 3 at 100 Hz.

13.4.3 Continuous spatial averaging

As an alternative to a number of fixed microphone positions, it is also possible to use continuous spatial averaging. This is usually done by mounting the microphone on a stand that can rotate the microphone slowly with a certain radius, and preferably in a plane that is not parallel to any of the room surfaces.

Assuming a three-dimensional diffuse sound field, i.e. by applying the correlation coefficient (Equation 13.24), the equivalent number of uncorrelated positions has been calculated by Lubman et al. (1973). The results are shown in Figure 13.15 for a straight line segment and for a circular path. The length of the path is L, which in the case of the circle means $L=2\pi r$, where r is the radius. It appears that for the circular path, a good approximation is:

$$N_{eq} \cong \begin{cases} 1 & \text{for} \quad L \leq \frac{\lambda}{2} \\ \dfrac{2L}{\lambda} & \text{for} \quad L > \frac{\lambda}{2} \end{cases} \qquad (13.30)$$

Some formulas to estimate the equivalent number of positions for various kinds of continuous spatial averaging are collected in Table 13.4. The circle path is often used for sound insulation measurements, especially in laboratories, (see Figure 12.3).

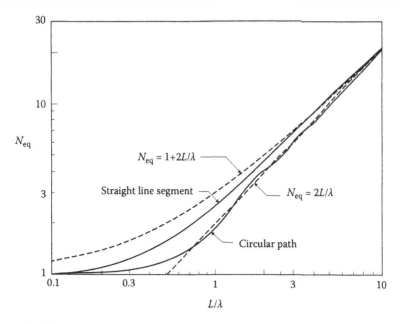

Figure 13.15 Equivalent number of uncorrelated measurement positions in a diffuse sound field using continuous spatial averaging. Results are shown for a straight line segment as well as for a circular path. (Reproduced from Lubman, D. et al., *J. Acoust. Soc. Am.*, 53, 650–659, 1973, Figure 3. With the permission of the Acoustical Society of America).

Table 13.4 Approximate formulae to estimate the equivalent number of measurement positions by continuous averaging over various paths or areas in a diffuse sound field

Region of averaging	Approximate N_{eq}
Line, length L	$1+2L/\lambda$
Circle, circumference L	$2L/\lambda$
Circular disc, area S	$1+S/(\lambda/2)^2$
Rectangular surface, area S	$S/(\lambda/2)^2$
Rectangular box, volume V	$V/(\lambda/2)^3$

Source: Lubman, D., *J. Sound Vib.*, 16, 43–58, 1971.

13.5 MEASUREMENT UNCERTAINTY ON REVERBERATION TIME MEASUREMENTS

13.5.1 General

The uncertainty of decay rate measurements has been studied by Davy et al. (1979) and Davy (1980). The decay rate d in dB per second is related to the reverberation time T in seconds by $d=60/T$.

The measurements can be repeated in the same position, and n denotes the number of decays measured in each position. The ensemble variance in one position is $\varepsilon_e^2(d)$.

The spatial variance is $\varepsilon_s^2(d)$. The number of independent measurement positions is denoted N. Thus, the estimated relative standard deviation of the *average* reverberation time is $(\bar{T})$:

$$\frac{\sigma(\bar{T})}{\bar{T}} = \frac{1}{\sqrt{N}}\sqrt{\frac{\varepsilon_s^2(d)}{d^2} + \frac{\varepsilon_e^2(d)}{n\,d^2}} \tag{13.31}$$

13.5.2 The interrupted noise method

In the work of Davy et al. (1979), the ensemble variance and the spatial variance were derived for the case of measurements using the interrupted noise method. By insertion of the results from this reference in Equation 13.31, we get:

$$\frac{\sigma(\bar{T})}{\bar{T}} = G\sqrt{\frac{1+H/n}{NBT}} \tag{13.32}$$

where

T is the reverberation time, in seconds.
$\sigma\,(T)$ is the standard deviation of T, in seconds.
G and H are constants that depend on the evaluation range.
n is the number of decays measured in each position.
N is the number of independent measurement positions.
B is the bandwidth, in Hz.

The constants G and H depend on the evaluation range D and a parameter $\gamma=T/T_{\text{det}}$, which is the ratio of measured reverberation time and the reverberation time inherent in the measuring apparatus. T_{det} is the reverberation time of the averaging detector. For some typical values of D and γ, the values of the constants G and H can be taken from Table 13.5.

The standard ISO 3382-2 defines three levels of accuracy with a corresponding minimum number of measurement positions and decays in each position (Table 13.6).

Table 13.5 Values of the constants G and H for different evaluation ranges, D

Evaluation range, D (dB)	G	H		
		$\gamma = 3$	$\gamma = 5$	$\gamma = 10$
10	1.75	2.67	3.32	3.87
20	0.88	1.72	1.90	2.04
30	0.55	1.42	1.52	1.59

Table 13.6 Minimum number of positions N and decays *n* for each of the three quality levels of measurements according to ISO 3382-2

	Survey	Engineering	Precision
No. of positions, N	2	6	12
No. of decays, *n* (interrupted noise method)	1	2	3
Equivalent no. of decays, *n* (integrated impulse response method)	10	10	10

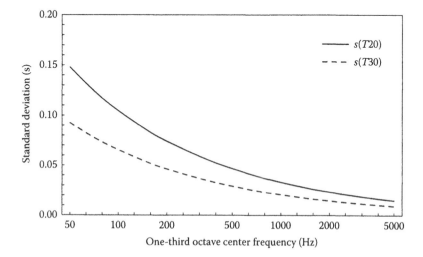

Figure 13.16 Standard deviation of measured reverberation time in one-third octave bands if the reverberation time is 1.0 s, using average of six positions and the interrupted noise method with two decays in each position. Results are shown for T_{20} and T_{30}.

An example of the application of Equation 13.32 together with Table 13.5 is the estimation of the standard deviation of T_{20} with the parameter $\gamma = 5$.

$$\sigma(T_{20}) = 0.88 \cdot T_{20}\sqrt{\frac{1+(1.90/n)}{N\,B\,T_{20}}} \qquad (13.33)$$

This and the analogue result for $\sigma(T_{30})$ are shown as functions of the frequency in Figure 13.16 for the example of $T_{20} = T_{30} = 1.0$ s.

13.5.3 The integrated impulse response method

For the integrated impulse response method, the ensemble variance is theoretically $\varepsilon_e^2(d)=0$. This corresponds to the averaging of an infinite number of excitations in the same position, if the interrupted noise method had been used (see Davy, 1980). For the estimation of the standard deviation of a measurement result, Equation 13.32 may be used with a value of $n=10$.

13.6 MEASUREMENT UNCERTAINTY ON SOUND INSULATION MEASUREMENTS

13.6.1 Combined measurement uncertainty

The sound reduction index is determined from three sets of measurements: the average sound pressure level in the source room, the average sound pressure level in the receiver room and the average reverberation time in the receiver room. The result is calculated from:

$$R = L_{p1} - L_{p2} + 10\lg \frac{S \cdot T}{0.16 \cdot V} \quad \text{(dB)} \tag{13.34}$$

where S is the area of the partition wall or floor (m^2) and V is the volume of the receiver room (m^3). It is noted that the uncertainty calculation will be the same in case of D_{nT}, as the difference is only whether the partition area and the room volume are included or not.

The standard deviation of the last term is estimated from:

$$\sigma\left(10\lg T\right) = 10\lg\left(1 + \frac{\sigma(T)}{T}\right) \cong 4.34 \cdot \frac{\sigma(T)}{T} \quad \text{(dB)} \tag{13.35}$$

where $\sigma(T)$ is from Equation 13.32. The approximation is correct within 10% for $\sigma(T)/T < 0.6$.

The standard deviation for each of the two sound pressure levels is calculated from Equations 13.10, 13.11 and 13.12 and using the equivalent number of positions (Equations 13.28 and 13.29). For the source room, this yields:

$$\sigma(L_{p1}) = 4.34 \cdot \frac{\varepsilon_{R,1}}{\sqrt{N_{eq,1}}} \quad \text{(dB)} \tag{13.36}$$

and similarly for the receiver room. Thus, the standard deviation of the reduction index can be estimated from:

$$\sigma(R) = \sqrt{\sigma^2(L_{p1}) + \sigma^2(L_{p2}) + \left(4.34 \cdot \frac{\sigma(T)}{T}\right)^2} \quad \text{(dB)} \tag{13.37}$$

Similarly, the standard deviation of the impact sound pressure level can be estimated from:

$$\sigma(L_{n,T}) = \sqrt{\sigma^2(L_{p2}) + \left(4.34 \cdot \frac{\sigma(T)}{T}\right)^2} \quad (\text{dB}) \tag{13.38}$$

The measurement uncertainty of impact sound has been studied by Hagberg and Thorsson (2010) with special emphasis on the low frequencies. They found that the reverberation time had nearly normal distribution above 100 Hz, but at lower frequencies the distribution functions were not symmetric.

For a more detailed description, see Machimbarrena et al. (2015). Analysing a large set of measurement data from field measurements of partition walls, it is concluded that the measurement uncertainty of the airborne sound insulation, to some extent, depends on the type of wall construction. Thus, individual uncertainty calculations should be made for each measurement situation instead of applying generalised uncertainty data as given in ISO 12999-1.

13.6.2 Example of airborne sound insulation measurement

Source room, L_{p1}: $V = 500 \, \text{m}^3$, $h = 4 \, \text{m}$, $T = 1 \, \text{s}$, 2 source pos.×5 mic. pos. = 10 positions

Receiver room, L_{p2}: $V = 50 \, \text{m}^3$, $h = 2.3 \, \text{m}$, $T = 0.5 \, \text{s}$, 2 source pos.×5 mic. pos. = 10 positions

Receiver room, T_{20}: $T = 0.5 \, \text{s}$, $2 \times 3 = 6$ positions, interrupted noise, two decays per position

The results are shown in Figure 13.17 and Table 13.7 for the frequency range from 50 Hz to 5 kHz.

The maximum number of independent microphone positions in the receiver room ($50 \, \text{m}^3$) is only about $N = 2.5$ at 50 Hz. Combined with two source positions, this means that in this room $N = 2 \times 2.5 = 5$ at 50 Hz, increasing to $N = 10$ at 125 Hz and higher frequencies. In the source room, $N = 10$ at all frequencies because the volume is sufficiently large.

13.6.3 Measurement uncertainty on single-number quantities

Wittstock (2007, 2009) studied the statistical distribution of measured airborne sound insulation, and he found that the distribution function of the transmission loss R' is very close to the Gaussian distribution, much closer than the distribution function of the transmission coefficient τ. So, it makes sense to apply the statistical methods for uncertainty directly to the transmission loss in dB.

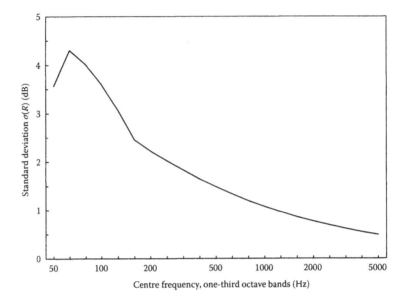

Figure 13.17 Calculated standard deviations in one-third octave bands for the example case of measured airborne sound insulation.

The standard deviation for in situ measurements of airborne sound insulation can be estimated to $\sigma(R'_w)$=0.9 dB to 1.3 dB (ISO 12999-1, Annex B). Similarly, for field measurement of impact sound insulation, the estimated standard deviation is $\sigma(L'_{n,w})$=1.0 dB. However, these estimates may be a little too optimistic. Wittstock and Scholl (2009) have analysed the uncertainty of transmission loss measurements and compared them with calculated uncertainties for the same cases. For 24 heavyweight constructions, the uncertainty was $u(R'_w)$=1.4 dB to 2.1 dB, whereas for 24 lightweight constructions, the uncertainty was $u(R'_w)$=1.5 dB to 2.7 dB.

13.6.4 Application of uncertainties in relation to requirements for sound insulation

The application of measurement uncertainty is highly relevant in connection with field measurements of sound insulation when the purpose, is to prove fulfilment of sound insulation requirements. For this purpose, we will use an example, adapted from Wittstock (2009).

In this case, the airborne sound insulation shall fulfil the limit $R'_{w,req}$=53 dB. Then, the measured transmission loss R'_w must fulfil the equation:

$$R'_w - k \cdot u \geq R'_{w,req} \tag{13.39}$$

Table 13.7 Calculated standard deviations in one-third octave bands for each of the terms in Equation 13.37

Frequency (Hz)	$\sigma(L_{p1})$ (dB)	$\sigma(L_{p2})$ (dB)	$\sigma(10 \log T)$ (dB)	$\sigma(R)$ (dB)
50	1.0	3.3	0.9	3.6
63	0.9	4.1	0.8	4.3
80	0.7	3.9	0.7	4.0
100	0.7	3.5	0.6	3.6
125	0.6	3.0	0.6	3.1
160	0.5	2.3	0.5	2.5
200	0.5	2.1	0.5	2.2
250	0.5	1.9	0.4	2.0
315	0.4	1.8	0.4	1.8
400	0.4	1.6	0.3	1.6
500	0.3	1.4	0.3	1.5
630	0.3	1.3	0.3	1.3
800	0.3	1.1	0.2	1.2
1000	0.2	1.0	0.2	1.1
1250	0.2	0.9	0.2	1.0
1600	0.2	0.8	0.2	0.9
2000	0.2	0.7	0.1	0.8
2500	0.1	0.7	0.1	0.7
3150	0.1	0.6	0.1	0.6
4000	0.1	0.5	0.1	0.5
5000	0.1	0.5	0.1	0.5

Note: Input data according to the example.

According to ISO 12999-1, Table 3, the standard uncertainty for in situ measurements can be estimated to $u = \sigma_{situ} = 0.9$ dB. The coverage factor k is found in Table 13.1 for a one-sided test, but we have to agree on which confidence level is relevant for the case. If the confidence level is 90%, the coverage factor is $k = 1.28$. Therefore, the extended uncertainty is $U = k \times u = 1.28 \times 0.9 \approx 1.2$ dB.

A measurement is made and the result is $R'_w = 53.7$ dB. However, it cannot be decided whether the requirement is fulfilled or not because $R'_w - 1.2$ dB $= 52.5$ dB < 53 dB.

A second *independent* measurement is made, i.e. with different equipment used by different staff. The results is $R'_w = 54.1$ dB. Still, this is not sufficient by itself (54.1 dB − 1.2 dB = 52.9 dB < 53 dB). However, with independent measurements, it is possible to use the mean value of the transmission loss, and the combined uncertainty is reduced by $\sqrt{n}$, where $n = 2$ is the number of independent measurements. The result is:

$$R'_{w,mean} - \frac{k \cdot u}{\sqrt{n}} = \frac{53.7 + 54.1}{2} - \frac{1.2}{\sqrt{2}} = 53.9 - 0.8 = 53.1 \, dB > 53 \, dB$$

The requirement is found to be fulfilled with 90 % confidence level.

REFERENCES

J.L. Davy, I.P. Dunn, P. Dubout (1979). The Variance of Decay rates in Reverberation Rooms. *Acustica* 43, 12–25.

J.L. Davy (1980). The variance of impulse decays. *Acustica* 44, 51–56.

P.E. Doak (1959). Fluctuations of the sound pressure level in rooms when the receiver position is varied. *Acustica* 9, 1–9.

GUM (1995). *Guide to the Expression of Uncertainties in Measurement*, 2nd Edition, International Organization for Standardization, Geneva, Switzerland.

K. Hagberg and P. Thorsson (2010). Uncertainties in standard impact sound measurement and evaluation procedure applied to light weight structures. *Proceedings of ICA 2010*, August 23–27, 2010, Sydney, Australia.

ISO 3382-2 (2008). *Acoustics – Measurement of room acoustic parameters – Part 2: Reverberation time in ordinary rooms,* International Organization for Standardization, Geneva, Switzerland.

ISO 12999-1 (2014). *Acoustics – Determination and application of measurement uncertainties in building acoustics – Part 1: Sound insulation,* International Organization for Standardization, Geneva, Switzerland.

ISO 18233 (2006). *Acoustics – Application of new measurement methods in building and room acoustics,* International Organization for Standardization, Geneva, Switzerland.

F. Jacobsen (1979). *The diffuse sound field. Statistical considerations concerning the reverberant field in the steady state.* Report No. 27. The Acoustics Laboratory. Technical University of Denmark.

H. Kuttruff (1979). *Room Acoustics,* 2nd Edition, Applied Science Publishers, London.

D. Lubman (1968). Fluctuations of sound with position in a reverberant room. *Journal Acoustical Society America* 44, 1491–1502.

D. Lubman (1969). Spatial averaging in a diffuse sound field. *Journal Acoustical Society America* 46, 532–534.

D. Lubman (1971). Spatial averaging in sound power measurements. *Journal of Sound Vibrations* 16, 43–58.

D. Lubman (1974). Precision of reverberant sound power measurements. *Journal of Acoustical Society America* 56, 523–533.

D. Lubman, R.V. Waterhouse, C.S. Chien (1973). Effectiveness of continuous spatial averaging in a diffuse sound field. *Journal Acoustical Society America* 53, 650–659.

M. Machimbarrena, C.R.A. Monteiro, S. Pedersoli, R. Johansson, S. Smith (2015). Uncertainty determination of in situ airborne sound insulation measurements. *Applied Acoustics* 89, 199–210.

S. Pedersen, H. Møller, K. Persson Waye (2007). Indoor measurements of noise at low frequencies – Problems and solutions. *Journal of Low Frequency Noise, Vibration and Active Control* 26(4), 249–270.

A.D. Pierce (1989). *Acoustics, An Introduction to Its Physical Principles and Applications*. 2nd Edition. Acoustical Society of America, New York.

J.H. Rindel (1981). *Stationære lydfelter i rum*. (Stationary sound fields in rooms, in Danish). Lecture Note S. The Acoustics Laboratory. Technical University of Denmark.

M.R. Schroeder (1969). Effect of frequency and space averaging on the transmission responses of multimode media. *Journal Acoustical Society America* 46, 277–283.

C. Simmons (2012). Uncertainties of room average sound pressure levels measured in the field according to the draft standard ISO 16283-1. *Noise Control Engineering Journal* 60, 405–420.

H.T. Tuominen (1980). Practical methods for predicting and checking the confidence intervals of sound insulation measurements. *Proceedings of NAS-80*, June 10–12, 1980, Åbo, Finland, 67–70.

R. Waterhouse (1970), Sampling statistics for an acoustic mode. *Journal Acoustical Society America* 47, 961–967.

V. Wittstock (2007). On the uncertainty of single-number quantities for rating airborne sound insulation. *Acta Acustica – Acustica* 93, 375–386.

V. Wittstock (2009). Uncertainties in applied acoustics – Determination and handling. *Proceedings of NAG/DAGA 2009*, March 23–26, Rotterdam, The Netherlands, 22–29.

V. Wittstock and W. Scholl (2009). Determination of the uncertainty of predicted values in building acoustics. *Proceedings of NAG/DAGA 2009*, March 23–26, Rotterdam, The Netherlands, 920–923.

Chapter 14

Noise effects and subjective evaluation of sound insulation

In this chapter, the sound insulation in buildings is analysed from the point of view of the users. The discussion is restricted to dwellings and, in particular, multi-unit houses. The most serious problem may be the exposure to traffic noise and the need for sound insulation of windows and façades. Noise from neighbours and the need for sound insulation of internal walls and floors are also very important and lead to the question: Which level of sound insulation should be required to obtain satisfactory acoustical conditions?

14.1 NOISE AND HEALTH

14.1.1 Noise descriptors

International standard ISO 1996-1 defines noise descriptors applied to environmental noise, for example, for noise regulations and noise mapping. They are all some kind of time-averaged, A-weighted sound pressure level in the free field. The latter means that even if the sound is measured or calculated close to a building, the result is corrected for the possible influence of reflections from that building.

If we do not take into account the distribution over day and night, the sound energy is integrated over 24 h and we get $L_{p,A,24h}$. However, preferred descriptors do take into account this distribution. The day-evening-night level L_{den} is defined as:

$$L_{den} = 10 \lg \left[\frac{1}{24 \text{ h}} \left(t_{day} \cdot 10^{0.1 \, L_{day,12}} + t_{evening} \right. \right.$$
$$\left. \left. \cdot 10^{0.1 \, (L_{evening,4}+5 \text{ dB})} + t_{night} \cdot 10^{0.1 \, (L_{night,8}+10 \text{ dB})} \right) \right], \text{ dB} \quad (14.1)$$

where t_{day}, $t_{evening}$, t_{night} are the number of hours of the periods (24 h in total), and L_{day}, $L_{evening}$ and L_{night} are the A-weighted, time-averaged sound pressure levels in the same periods. Note the penalty of 5 dB in the evening and

Table 14.1 Relation between some noise descriptors valid for road traffic noise with typical example of traffic distribution 78 % (day, 12 h), 11 % (evening, 3 h) and 11 % (night, 9 h)

	Level (dB)	Difference (dB)
$L_{p,A,24h}$	55	0
$L_{day,12h}$	57	2
$L_{day,15h}$	57	2
$L_{evening,3h}$	54	−1
$L_{night,9h}$	50	−5
L_{den}	58	3
L_{dn}	58	3

10 dB in the night. The definition of the three periods may differ from one country to another; for example, the EU member states use 3 h evening and 9 h night periods. Sometimes, the evening period is not used at all, and we get the day-night level L_{dn} defined as:

$$L_{dn} = 10 \lg\left[\frac{1}{24\,h}\left(t_{day} \cdot 10^{0.1\,L_{day,15}} + t_{night} \cdot 10^{0.1\,(L_{night,9}+10\,dB)}\right)\right]\ dB \quad (14.2)$$

For road traffic noise having typical traffic distribution, the relation between the noise descriptors is shown in Table 14.1.

14.1.2 Sleep disturbance

The World Health Organization has identified neighbour noise as a health problem based on the findings from a large survey in eight European cities from 2002 to 2003, the Large Analysis and Review of European housing and health Status (LARES) project (WHO, 2007). Information was collected in 3382 households from 8539 people by interviews and health questionnaire. Of all responding residents, 24 % reported that noise exposure at night was a main reason for sleep disturbance. Traffic noise was the dominant cause of sleep disturbance, closely followed by noise from neighbouring flats (Figure 14.1).

While traffic noise has since long been known to be a serious source of annoyance, the LARES investigation showed that also chronic annoyance due to neighbour noise is associated with hypertension, depression, and migraine (Maschke and Niemann, 2007). They concluded that neighbour noise annoyance is a highly underestimated risk factor for healthy housing.

Noise from neighbours may affect people in different ways. Figure 14.2 illustrates how some people may react with curiosity, changing to annoyance and anger and, in severe cases, ending with hatred and other similar

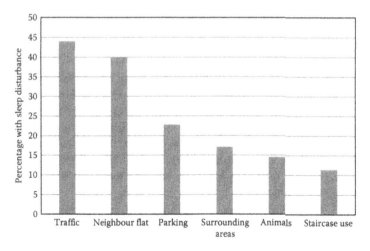

Figure 14.1 Sleep disturbance in adult residents and main sources of disturbing noise. Total for eight European cities (*n*=8539). (Adapted from WHO, *Large Analysis and Review of European Housing and Health Status (LARES)*. Preliminary overview, WHO Regional Office for Europe, Copenhagen, Denmark, 2007, Figure 37. With permission.)

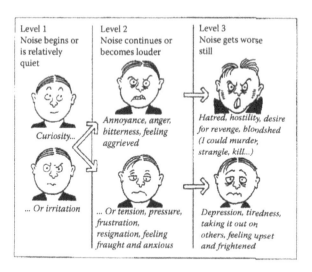

Figure 14.2 Noise reaction process. (Reproduced with permission from Grimwood, C.J., Effects of environmental noise on people at home. BRE information paper IP 22/93, 1993. Public sector information licensed under the Open Government Licence v3.0.)

reactions. Other people may react with irritation, growing to tension and depression. Noise annoyance is related to health (Niemann et al., 2006): *annoyance → negative emotional reaction → neuro-vegetative-hormonal regulatory disturbances → illness.*

14.1.3 Dose-effect relationships

Risk ratio (RR) and odds ratio (OR) are used in statistics, for instance, to calculate dose-effect relationships. With reference to Table 14.2, the investigated people are divided into groups (e.g. based on different level of noise exposure), and the data are collected to show how many exhibit a certain effect or not (e.g. stroke). Group 0 is a reference group, in which the dose-effect relationship is assumed to be negligible. In each group, the risk of being affected is a number between 0 and 1, and the RR is the ratio between the risk in that group and the risk in the reference group. For example, RR=1.05 means that there is an increased risk of 5 % compared to the reference group.

The odds for being affected in a group are simply the ratio between the number being affected and the number not affected. The OR in a group is then the odds in that group divided by the odds in the reference group. If the effect is sufficiently rare (*Yes* << *No*), the risk in all groups may be much less than unity and then RR≈OR.

Annoyance data for 8.6 million people living in London between 2003 and 2010 were analysed in relation to road traffic noise exposure in noisy areas, L_{day} (07–19) > 60 dB relative to quiet areas L_{day} < 50 dB (Halonen et al., 2015). Deaths were 4 % more common among adults and elderly in noisy areas compared with quiet areas (Table 14.3). Adults living in noisy areas were 5 % more likely to be admitted to hospital for stroke compared to those living in quiet areas and went up to 9 % in the elderly population.

In addition to annoyance, sleep disturbance and health problems, traffic noise may have negative effects on children's behaviour. In a large study, in Denmark, 46 904 children with behavioural problems at the age of 7 years were identified, and address from birth to 7 years was used to model the traffic noise exposure. It was found that road traffic noise in early childhood

Table 14.2 Calculation of RR and OR for being affected (symptom, disease etc.) when belonging to a certain group (e.g. based on level of noise exposure)

| | Effect | | | | | |
	Yes	No	Risk	Odds	RR	OR
Group 0 (reference)	Y_0	N_0	$\dfrac{Y_0}{(Y_0 + N_0)}$	$\dfrac{Y_0}{N_0}$	1	1
Group 1	Y_1	N_1	$\dfrac{Y_1}{(Y_1 + N_1)}$	$\dfrac{Y_1}{N_1}$	$\dfrac{Y_1(Y_0 + N_0)}{Y_0(Y_1 + N_1)}$	$\dfrac{Y_1 N_0}{Y_0 N_1}$
Group 2	Y_2	N_2	$\dfrac{Y_2}{(Y_2 + N_2)}$	$\dfrac{Y_2}{N_2}$	$\dfrac{Y_2(Y_0 + N_0)}{Y_0(Y_2 + N_2)}$	$\dfrac{Y_2 N_0}{Y_0 N_2}$

Table 14.3 Risk ratio of death or stroke due to road traffic noise $L_{day} > 60\,dB$ relative to $L_{day} < 50\,dB$

		Risk ratio (RR)	Confidence interval 95 %
Death	Adults (>25)	1.04	(1.00–1.07)
	Elderly (>75)	1.04	(1.00–1.08)
Stroke	Adults (>25)	1.05	(1.02–1.09)
	Elderly (>75)	1.09	(1.04–1.14)

Source: Halonen, J.I. et al., *Eur. Heart J.*, 36(39), 2653–2661, 2015.

(0–7 years) may be associated with hyperactivity and inattention symptoms (Hjortebjerg et al., 2016: Figure 14.3). The results were adjusted for several possible confounders.

The results shown in Figure 14.3 are divided into groups of noise exposure, L_{den} (<50, 50–55, 55–60, 60–65, ≥65) dB, the first one being the reference group. There is a monotonic increase of the odds ratio until 60 dB to 65 dB, but the curve levels out and decreases for the highest exposure group. For noise levels due to road traffic between 50 dB and 65 dB, the relation with the children's behavioural problems is strong and significant.

A possible explanation for the weaker relationship above 65 dB may be that the noise exposure in this investigation was calculated without screens or noise

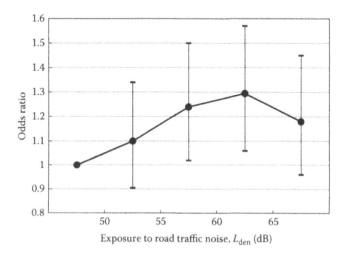

Figure 14.3 Relation between exposure to road traffic noise at childhood and abnormal scores on total difficulties shown as odds ratio with 95 % confidence intervals in four noise exposure categories. The reference category is L_{den} < 50 dB. (Reproduced from Hjortebjerg, D. et al., *Environ. Health Perspect.*, 124(2), 228–234, 2016. With permission.)

barriers; but, in fact, such measures for noise protection of housing areas with high noise exposure are common in Denmark since the 1970s. In the act of environmental protection, it was stated that the number of dwellings exposed to noise levels more than 65 dB from road traffic should be reduced to a minimum of 50 000 before 2010. The typical attenuation of road traffic noise due to a noise barrier is 8 dB to 15 dB. If the result in the highest noise group in Figure 14.3 is shifted downwards by 12 dB, it fits precisely on the curve.

14.2 NOISE FROM NEIGHBOURS

14.2.1 Acoustical comfort

The concept of acoustical comfort in relation to buildings was first used by Cummins (1978). Acoustical comfort is characterised as follows (Rindel and Rasmussen, 1995):

- Absence of unwanted sound
- Presence of wanted sound of desired level and quality
- Opportunities for acoustic activities without annoying or disturbing other people and without being heard by unauthorised persons

In order to achieve acoustical comfort in a building, certain requirements have to be fulfilled concerning the airborne and impact sound insulation and the noise level from traffic and building services.

It is important to observe that acoustic comfort for a person is related to the person not only as a receiver of sound but also as a source of sound. It can be annoying to be exposed to noise from the neighbours, but it can be equally annoying to know that your activities can be heard by other people and may cause annoyance. Poor sound insulation between dwellings can be a cause of conflicts and a cause of restraints of activities.

Playing a musical instrument is an interesting example. Many countries have music schools for children, and it is important that children have the possibility to learn to play a musical instrument. However, if the family lives in a multi-unit house with poor sound insulation, it can be a serious problem to let the children practice their instruments at home.

Another example is a family with small children living in an apartment house with mediocre sound insulation. Normal activity of children includes running, jumping, shouting and crying. However, a noise-sensitive neighbour may bang the wall or the ceiling to signalize disturbance by the noise. So, the parents will try to keep the children quiet. If this becomes a concern all the time, the family can eventually feel forced to move and find another place to live. While sound conditions usually play a minor role when choosing a new home, they can often be the main reason for moving away.

A study was undertaken in England and Wales 1992–1994 in order to investigate complaints about poor sound insulation between dwellings

(Grimwood, 1997). 40 cases of complaints were examined by sound insulation measurements and interviews with complainants and adjacent residents. Most common types of noise sources heard, and the percentage of cases where reported were: Music, television, radio (98 %), footfalls on floors (95 %), voices (78 %), doors and cupboard doors (68 %).

In a social survey of noise conditions in dwellings in Norway (Løvstad et al., 2016), some questions were related to whether people were concerned about disturbing their neighbours when having visitors or children playing, when listening to music or playing a musical instrument, and when walking on the floor. Other questions were dealing with noise annoyance. The results are shown in Figure 14.4.

The main sources of noise annoyance are traffic noise and impact sound (footfall), both with 20 % annoyed (moderately + very + extremely). Airborne noise from neighbours (speech, TV, music, etc.) caused around 10 % annoyance. However, comparing the degree of annoyance with the degree of concern about disturbing, it is remarkable that more than 30 % were concerned (to some extend or very much) about disturbing their neighbours with speech, children, music, and other airborne sounds

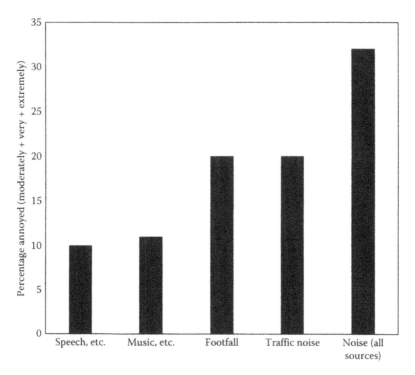

Figure 14.4 Percentage of people being annoyed by various noise sources in their home (*n* = 702). (Adapted from Løvstad et al., Sound quality in dwellings in Norway – A socio-acoustic investigation. *Proceedings of BNAM 2016,* Stockholm, Sweden, 2016.)

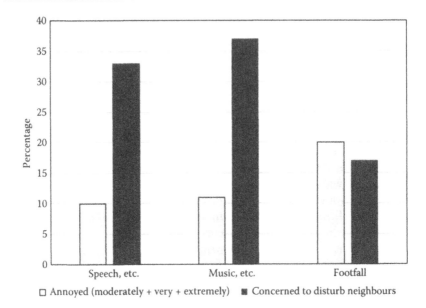

□ Annoyed (moderately + very + extremely)   ■ Concerned to disturb neighbours

Figure 14.5 Percentage of people being annoyed by various noise sources in their home, and percentage of people being concerned that they may disturb their neighbours by various activities (*n*=702). (Adapted from Løvstad et al., Sound quality in dwellings in Norway – A socio-acoustic investigation. *Proceedings of BNAM 2016*, Stockholm, Sweden, 2016.)

(Figure 14.5). More than 15% were concerned (to some extent or very much) about disturbing their neighbours due to footfall noise. These findings clearly demonstrate that the constraints dwellers put on their own activities due to insufficient sound insulation are at least as important as the annoyance from noise. Noise from neighbours is like a troll with two heads; the fear of disturbing the neighbour is half the problem.

14.2.2   Sound insulation and social disruption

An interesting survey was carried out in Canada (Bradley, 1983). The investigation comprised 49 party walls and interviews of the persons living on either side of the walls, in total 98 persons distributed over 11 localities, mainly in two-story terraced houses, 16% multi-storey dwellings. Measured sound insulation of the party walls varied from 39 dB to 60 dB. The measured values were sound transmission class that is approximately equal to R'_w (see Section 5.8.1). The measurements included the recording of A-weighted noise levels for one 24 h period in each subject's living room. The survey illuminates many interesting aspects of the identified noise problems. For example, it appears from the investigation that the residents' perception of whether a neighbour shows consideration or not has a significant correlation with the measured sound insulation, but not with the actual noise level at the neighbours. Bradley (1983) states:

> 'This is strong evidence that in this study at least the inadequacy of the party wall was a source of social disruption in that neighbors were thought to be inconsiderate when it was really the party wall that was at fault.'

14.2.3   Influence of age and lifestyle

Noise protection in multi-family houses from the point of view of the resident is the title of a survey in Germany (Weeber et al., 1986). The investigation comprised 16 housing estates of varying ages, where measurements were carried out and 471 residents were interviewed. The airborne sound insulation R'_w varied between 42 dB and 61 dB and the impact sound insulation $L'_{n,w}$ of the separating floors varied between 70 dB and 26 dB. The houses were of different age, built between 1921 and 1982. The acoustical conditions in new and older dwellings were assessed using very different criteria. It was found that the age and condition of the building had influences on the expectations and thereby also on the residents, assessment.

The main conclusion from this investigation is that to achieve conditions with no disturbance in normal situations, the airborne sound insulation of the floors should be $R'_w \geq 60$ dB, whereas $R'_w = 50$ dB would cause about

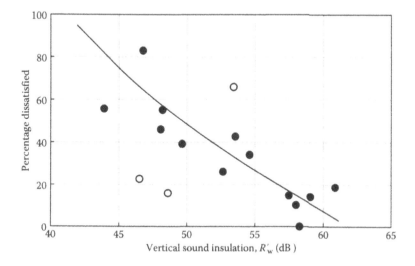

Figure 14.6 Proportion of residents in multi-storey flats who rate the sound insulation as poor, presented as a function of the measured airborne sound insulation through the separating floors. Average for 16 housing estates and regression line ($n=471$). (Weeber, R. et al., *Schallschutz in Mehrfamilienhäusern aus der Sicht der Bewohner* (Sound insulation in multi-family houses from the point of view of the resident, in German). F 2049, IRB Verlag, Stuttgart, Germany, 1986.)

30 % of the residents to be frequently disturbed by the neighbour's TV. Impact sound insulation $L'_{n,w}$ of the separating floors at $L'_{n,w} = 50$ dB would cause about 15 % to be frequently disturbed by the neighbour walking on the floor. Figure 14.6 shows the relation between the percentage of people finding the sound insulation insufficient and the airborne sound insulation of the separating floors. Each dot represents one of the 16 houses. Three outliers were identified (the open dots) and could be explained as follows: One site with higher dissatisfaction than the main trend was an old, refurbished house with new, young tenants with children; little outdoor noise. The other two outliers with less dissatisfaction were houses with elderly tenants, no children, and in addition relatively high levels of outdoor noise.

14.2.4   Willingness to pay for good sound insulation

The Canadian investigation mentioned above (Bradley, 1983) also included a question about the money, and the result showed a high degree of correlation between the measured sound insulation and the amount of money which residents were prepared to pay extra per month to eliminate disturbing noise. The regression line found reaches zero only at a sound insulation of 60 dB.

In 1995, an investigation was made in Sweden in order to find what level of sound insulation new dwellings should have (Wibe, 1997). The minimum requirements for sound insulation in Sweden at that time were $R'_w \geq 55$ dB (row houses), 53 dB (flats, vertical), 52 dB (flats, horizontal) and $L'_{n,w} \leq 58$ dB (between dwellings), 64 dB (from stairways and corridors). A total of 2 322 questionnaires were used for the analysis: 65 % of the participating people lived in multi-family houses, 20 % in single houses, 10 % in row houses and 5 % in other kinds of houses. One of the main questions was about the willingness to pay a higher rent if the sound insulation of the apartment could be significantly improved. The average answer was about 2 500 (Sweden Kroner) per year. In summary, it was found that if the sound insulation of the dwelling could be improved, around 60 % of the population were willing to pay on average a 10 % higher rent.

A survey in Norway 2015 also included questions about willingness to pay for better sound insulation (Løvstad et al., 2016). Among 702 replies, around 25 % would not pay extra, but around 30 % were willing to pay from 1 200 to 12 000 (Norway Kroner) extra per year for better sound insulation. The corresponding minimum requirements for sound insulation in Norway were $R'_w \geq 55$ dB and $L'_{n,w} \leq 53$ dB. To the opposite question, less sound insulation for a reduced monthly payment, 66 % replied that this would be out of question; only 10 % would accept lower sound insulation against a lower payment around 12 000 NOK per year.

14.3   DOSE-RESPONSE FUNCTIONS

14.3.1   Statistical methods

Two different statistical models are usually considered, *probit* or *logit*. The probit model is based on the normal distribution (the Gaussian distribution: see also Chapter 13). However, when cumulative probability functions are evaluated and adapted to observed data, it is a disadvantage that these functions are in integral forms. Instead, the logit model is based on slightly different distributions often assumed in social surveys. The principle behind the logit model is to consider the data to be analysed in proportions, preferably with equal uncertainty attached to each proportion. The logit transformation achieves this. It is defined by Altman (1991, p. 145) as follows:

$$t = \text{logit}(p) = \ln\left(\frac{p}{1-p}\right) \tag{14.3}$$

$$\Rightarrow p = \frac{e^t}{1+e^t} \tag{14.4}$$

where p is a proportion and t is the logit. The logit transformation stretches out the lower and upper parts of the distribution in the same way as in the Gaussian distribution (Table 14.4). They both belong to the group of *sigmoid functions* (S-shaped curves), which are used to model dose-response relationships.

The *logistic function* has the probability density function:

$$w(z) = \frac{e^{-\frac{z-\mu}{s}}}{s\left(1+e^{-\frac{z-\mu}{s}}\right)^2} \tag{14.5}$$

where the mean value is μ and s is a parameter proportional to the standard deviation σ. The cumulative distribution function is:

Table 14.4 Effect of logit transformation of a proportion p

p	$\text{logit}(p)$
0.01	−4.60
0.05	−2.94
0.10	−2.20
0.25	−1.10
0.50	0.00
0.75	1.10
0.90	2.20
0.95	2.94
0.99	4.60

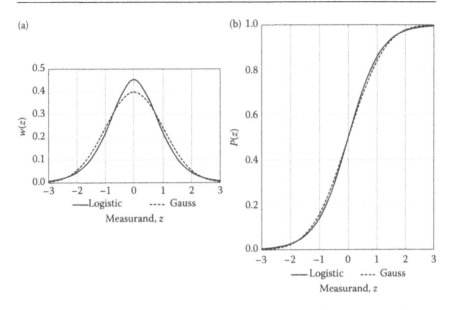

Figure 14.7 Logistic function and comparison with the Gaussian distribution. (a) Probability density functions; (b) cumulative distribution functions.

$$P(z) = \frac{1}{1 + e^{-\frac{z-\mu}{s}}} = \frac{1}{2} + \frac{1}{2}\tanh\left(\frac{z-\mu}{2s}\right) \qquad (14.6)$$

This is a symmetric distribution with the variance $\sigma^2 = \frac{1}{3}\,s^2\,\pi^2$. The functions are shown in Figure 14.7, and it is seen that the cumulative distribution is very close to that of a Gaussian distribution. An advantage of the logistic function is that the hyperbolic tangent can be used for calculations, whereas the Gaussian cumulative distribution can only be expressed in integral form.

With the logistic function, the probability for a random sample to be within the range $\mu \pm \sigma$ is 72 %, i.e. somewhat higher than with the Gaussian distribution (68 %). The probability to be within an extended range of $\mu \pm 2\sigma$ is 95 % with both distributions (Table 14.5).

The proportion of the distribution in the extreme ranges is called the *tails* of the distribution. The proportion outside ±3 standard deviations is 0.9 % in the logistic distribution, but only 0.3 % in the Gaussian

Table 14.5 Comparison of the logistic and the Gaussian distributions

Range	$\pm 1\ \sigma$	$\pm 2\ \sigma$	$\pm 3\ \sigma$
Logit (%)	72.0	94.8	99.1
Gaussian (%)	68.3	95.4	99.7

Note: Probability of being within a range of ±1, ±2 or ±3 standard deviations.

distribution (Table 14.5). Thus, the logistic distribution has heavier tails than the Gaussian distribution, and this makes it more robust to inaccuracies in the model or to errors in the data. This is another advantage of the logit model compared to the probit model.

14.3.2 Traffic noise annoyance

For road traffic noise, it is well established that an outdoor noise level $L_{p,A,24h}$ below 55 dB in a housing area means that ~ 15 % to ~ 20 % of the people living in the houses are annoyed by the noise. An early investigation gave the results shown in Figure 14.8. The investigation is described below as a good example of the challenges we are facing in this kind of social surveys. The fact that it was made as early as 1972 has the unique advantage that the sound insulation of windows and façades can be assumed the same in all the investigated houses. The use of special sound insulating windows in houses with high noise exposure started shortly after 1972. The same kind of investigation is not possible today because houses exposed to high levels of external noise have windows with better sound insulation than houses in quiet areas.

Twenty-eight housing areas in Copenhagen were selected in pairs, so half of them were clearly exposed to road traffic noise and the other half were in more quiet areas. The pairing also considered the age, quality, and price

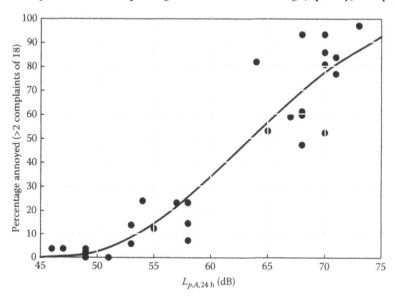

Figure 14.8 Percentage of people annoyed by noise in their dwellings as function A-weighted sound pressure level of outdoor noise from road traffic (free field values). Results from 28 locations in Copenhagen 1972. Statistical analysis ($n=960$), fraction of explained variance 0.91. (Kragh, J., *Analysis of the Correlation between Reactions to Road Traffic Noise and Physical Indices of the Noise*. Report No. 5 (In Danish). The Acoustical Laboratory, The Danish Academy of Technical Sciences, Lyngby, Denmark, 1977.)

level of the houses. It was thought to be very important for the later statistical analysis that there was a sufficient spread in the noise exposure, and that there were none in the middle range. Noise measurements were made with 24 h integration time and the microphone on a building façade. Free field $L_{p,A,24h}$ results were derived by subtraction of 5 dB due to the microphone position. This is the noise parameter shown on the abscissa in Figure 14.8.

In total, 960 personal interviews were made among the dwellers in August and September 1972. The purpose was to find out how traffic noise might influence the psychical and physical health of the inhabitants, but this purpose was hidden. The investigation contained in total 75 questions, out of which 18 were related to annoyance due to road traffic noise, e.g. if it was difficult to fall asleep in the evening, or if they did not open the windows because of the noise. In a preliminary analysis of the results, it was found that a good indicator of noise annoyance could be the number of complaints among the 18 relevant questions and the median of all questionnaires was two complaints out of 18. Thus, the annoyance parameter on the ordinate shown in Figure 14.8 is the percentage in each housing area that gave more than two complaints.

The annoyance data were averaged for each housing area (on average 34 interviewed persons in each area). A regression analysis was made using the results from the 28 housing areas, as plotted in Figure 14.8. A sigmoid function was assumed, and the arcsine function was used as an approximation. The curve shows, for example, that if the free field $L_{p,A,24h}$ increases from 55 dB to 65 dB, the percentage annoyed increases statistically from 14 % to 55 %. In the middle range between 20 % and 80 % the slope of the dose-response curve is ~4 % points per dB.

The dose-response relationships can also be derived from laboratory experiments. The advantages are that the parameters can be chosen to cover a wide range of exposure, which is important for a good statistical analysis. The connection between exposure and respondent can be controlled, and there is little risk of confounders that could lead to errors. The drawbacks of laboratory experiments include lack of realism, only short-time exposure is possible, and the number of test persons is normally more limited than in a field investigation. Figure 14.9 is an example of a dose-response relationship found in a laboratory experiment, where a recording of road traffic noise was used at seven different levels in 5 dB steps from 40 dB to 75 dB. The test persons were (among many other things) asked whether the noise would be acceptable if they worked in an office with that noise. Probit analysis was used to estimate the dose-response curve. The results show that in an office, 10 % would be dissatisfied if the noise level exceeds 45 dB. The average slope of the curve in the middle range is 4 % points per dB.

14.3.3 Socio-acoustic surveys

In a socio-acoustic survey, the questions asked about noise problems are very important for the quality of the results. The way in which such questions are

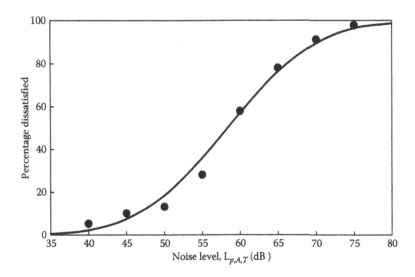

Figure 14.9 Percentage dissatisfied as a function of function A-weighted sound pressure level of road traffic noise. Laboratory experiment on indoor climate in an office (n = 112). The curve is the result of probit analysis. (Adapted from Clausen, G. et al. Indoor Air, 3, 255–262, 1993.)

asked and the wording of the possible answers should follow the recommendations in Technical Specification ISO/TS 15666. Questions and possible answers about how much the person is bothered/disturbed/annoyed can be like this:

"*Thinking about the last 12 months, when you are at home, how much does noise from traffic annoy you? Not at all – Slightly – Moderately – Very – Extremely?*" A numerical rating scale may also be used, e.g. with numbers from 0 to 10, where 0 means not at all and 10 means extremely. In addition, there must be a possibility for answering "do not know". Researchers in this field do not agree on using a numerical scale. However, ISO/TS 15666 is considered the best practice agreed on. The standard also contains an Annex B with recommended wordings of the questions in several languages. The words used for the answers have been balanced in such a way that the rating between the words is about 25 %.

An example is shown in Figure 14.10, which is one of the results from a socio-acoustic survey (Amundsen et al., 2011). From the replies of 738 people in combination with known traffic noise exposure, the four curves in Figure 14.10 were derived by logit analysis. Five regions are indicated corresponding to the distribution of the replies from the interviews. For example at a noise level of 75 dB, 10 % were not annoyed, 20 % slightly annoyed, 30 % moderately annoyed, 31 % very annoyed, and 9 % extremely annoyed (Table 14.6). The often used outdoor noise limit $L_{p,A,24\,h}=55$ dB corresponds approximately to 60 % not annoyed and 15 % annoyed (moderately + very + extremely). Each of the four curves have the same shape, since a logistic distribution was

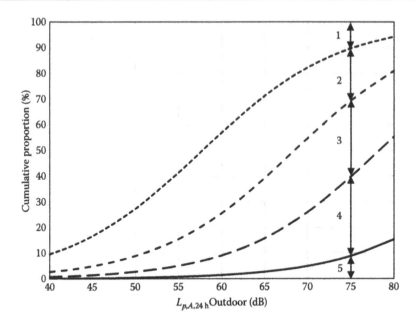

Figure 14.10 Cumulative annoyance curves for road traffic noise in a dwelling as a function of the outdoor A-weighted sound pressure level. The curves are the result of logit analysis (*n* = 738). Region 1: Not annoyed. Region 2: Slightly annoyed. Region 3: Moderately annoyed. Region 4: Very annoyed. Region 5: Extremely annoyed.

Table 14.6 Annoyance due to road traffic noise heard in the dwelling at selected levels of the outdoor noise

		Outdoor sound pressure level $L_{p,A, 24h}$			
	Degree of annoyance	45 dB (%)	55 dB (%)	65 dB (%)	75 dB (%)
1	Not annoyed	83	59	29	10
2	Slightly annoyed	12	56	32	20
3	Moderately annoyed	3	10	23	30
4	Very annoyed	1	4	13	31
5	Extremely annoyed	0	1	3	9

Source: Amundsen, A.H. et al., *J. Acoust. Soc. Am.*, 129, 1381–1389, 2011.

Note: Data from logit analysis in Figure 14.10.

assumed, and the curves are shifted along the dB-axis by approximately 10%, 10% and 15%, respectively. The slope of the curves in the middle range is ~3% points per dB.

In the same investigation, the indoor noise level due to road traffic noise was estimated and the result of the analysis is seen in Figure 14.11. The often used indoor limit of $L_{p,A,24\,h} = 30\,dB$ corresponds approximately to 40% not annoyed and 30% annoyed (moderately+very+extremely).

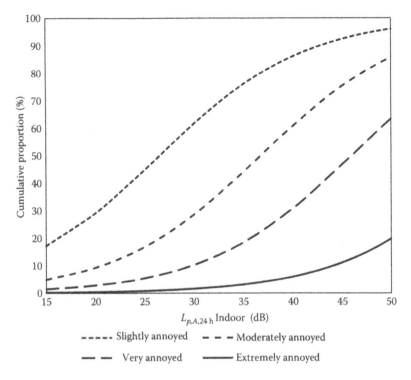

Figure 14.11 Cumulative annoyance curves for road traffic noise in a dwelling as a function of the indoor A-weighted sound pressure level. The curves are the result of logit analysis (*n* =738). (Adapted from Amundsen, A.H. et al., *J. Acoust. Soc. Am.*, 129, 1381–1389, 2011.)

14.3.4 Different traffic noise sources

On the basis of a large amount of data collected from various socio-acoustic surveys, an analysis based on data collected by Miedema (2004) is presented in ISO 1996-1 (2016), Annex H. In contrast to the method used in logit and probit analyses, these data have been analysed by using a model where annoyance is assumed to be proportional to loudness times duration. Annoyance is known to be proportional to sound pressure squared raised to the power of 0.3. Using the day-night level L_{dn} as input, the general equation for percentage highly annoyed becomes:

$$p_{HA} = 100 \cdot \exp\left[-\left(10^{0.1\,(L_{dn}-L_{ct}+5.3\text{ dB})}\right)^{-0,3}\right], \% \qquad (14.7)$$

where the *community tolerance level* L_{ct} is defined as the day-night sound level at which 50% of the people are predicted to be highly annoyed by noise exposure (Schomer et al., 2012; ISO 1996-1, 2016). The mean value of the community tolerance level for different traffic noise sources is given in Table 14.7, together with the standard deviations. The 95% prediction intervals can be estimated as ±2 standard deviations. The dose-response curves for aircraft, road traffic, and railroad are shown in Figure 14.12. Because of the different character of the noise sources (especially different variation with time), the annoyance score is quite different for the same day-night level. Aircraft is more annoying than road traffic (corresponding to +5 dB), while railroad is less annoying (corresponding to –9.5 dB). However, in cases where noise from railroad is accompanied by strong

Table 14.7 Community tolerance level L_{ct} (mean and standard deviation) for different traffic noise sources

	Aircraft	Road traffic	Railroad (low vibration)	Railroad (high vibration)
Mean L_{ct}, dB	73.3	78.3	87.8	75.8
Standard deviation, dB	7.1	5.1	3.5	4.2
Difference from road traffic, dB	5.0	0.0	–9.5	2.5

Source: ISO 1996-1, *Acoustics – Description, measurement and assessment of environmental noise – Part 1: Basic quantities and assessment procedures.* International Organization for Standardization, Geneva, Switzerland, 2016.

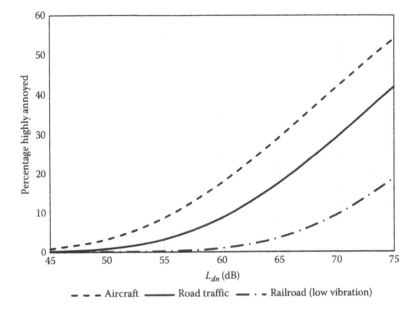

– – – Aircraft   ——— Road traffic   — · — Railroad (low vibration)

Figure 14.12 Dose-response curves for aircraft, road traffic and railroad (low vibration). (Adapted from ISO 1996-1, *Acoustics – Description, measurement and assessment of environmental noise – Part 1: Basic quantities and assessment procedures.* International Organization for Standardization, Geneva, Switzerland, 2016.)

vibrations, the annoyance score is much higher, between that for aircraft and road traffic (not shown in Figure 14.12).

It is interesting to compare the curve for road traffic in Figure 14.12 with the previous Figures 14.8 and 14.10. First, we recall that for a typical day-night distribution of traffic, the outdoor $L_{p,A,24h} = 55$ dB corresponds to $L_{dn} = 58$ dB (Table 14.1). Figure 14.12 then shows 6 % highly annoyed from road traffic. In comparison, Figure 14.8 shows 15 % annoyed. Figure 14.10 also shows 15 % annoyed (moderately + very + extremely) but only 5 % highly annoyed (very + extremely); the latter being close to the 6 % highly annoyed in Figure 14.12. This demonstrates clearly the importance of how we define 'annoyed' or 'highly annoyed'. In future socio-acoustic investigations, this problem may be avoided if the method in ISO/TS 15666 for such surveys is applied.

14.3.5   Airborne sound insulation

In the previously mentioned Norwegian socio-acoustic investigation on noise annoyance in dwellings (Løvstad et al., 2016) measured sound insulation data were collected from the actual addresses where the interviewed persons live. Airborne sound insulation was measured between dwellings in horizontal and vertical directions. The questionnaire included two questions related to airborne sound insulation: one about annoyance due to speech, TV, etc., and another about annoyance due to music with bass and drums. Table 14.8 gives an overview of the results. Both normalized (R'_w) and standardized ($D_{nT,w}$) results were calculated without and with the 50 Hz spectrum adaptation term (indicated by index 50).

Table 14.8 Data from measured airborne sound insulation and degree of annoyance due to speech and music

	Horizontal				Vertical			
	R'_w	$R'_{w,50}$	$D_{nT,w}$	$D_{nT,w,50}$	R'_w	$R'_{w,50}$	$D_{nT,w}$	$D_{nT,w,50}$
Measurements								
Number, N	355	346	296	296	394	354	366	349
Min, dB	46	45	45	44	50	50	52	51
Max, dB	64	63	65	64	69	68	70	69
Mean, dB	56.8	53.8	58.6	55.4	61.6	58.7	61.0	57.8
Standard dev., dB	3.0	2.8	4.0	3.0	4.3	2.6	4.2	2.4
Annoyed due to speech etc., (%)	19.9	20.1	20.0	19.9	22.7	22.0	21.9	21.8
Annoyed due to music with bass (%)	27.8	26.4	25.4	24.8	27.3	25.4	25.4	26.5

Source: Rindel, J.H. et al., Dose-response curves for satisfactory sound insulation between dwellings, *Proceedings of ICBEN 2017*, Zürich, Switzerland, 2017.
Note: Percentage annoyed includes all levels (extremely, very, moderately, slightly).

The upper part of Table 14.8 shows the number of measurements, the range of variation, the mean value, and the standard deviation. It is noted that airborne sound insulation is generally better in the vertical direction than in the horizontal direction; mean values of R'_w and $R'_{w,50}$ are about 5 dB better for floors than for walls. This suggests that annoyance from speech or music should be related mostly to the horizontal sound insulation. If this is correct, it means that the dose-response results for airborne sound insulation in the vertical direction cannot be reliable. In the subjective response from the questionnaire, there is no distinction between horizontal or vertical direction of the annoying sound.

The degree of annoyance from speech is less than that from music. It is also noted that the annoyance from speech seems to be higher in the vertical direction than in the horizontal direction. However, this is simply a consequence of the fact, that the average sound insulation of the floors is about 5 dB better than that of the walls. The degree of annoyance (all levels) corresponding to the mean value of sound insulation is on average 26 % for music and 21 % for speech. The cumulative annoyance curves were quite similar for speech and music, and only the latter are shown here in Figure 14.13.

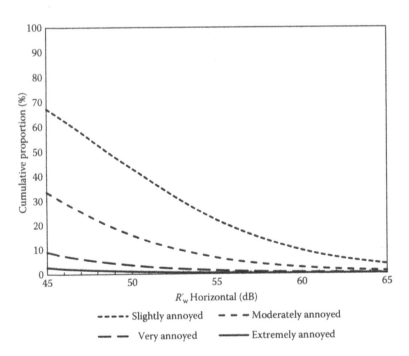

Figure 14.13 Cumulative annoyance curves due to music with bass as a function of the normalised level difference between dwellings in the horizontal direction. The curves are the result of logit analysis ($n = 702$). (Adapted from Rindel, J.H. et al., Aiming at satisfactory sound conditions in dwellings – The use of dose-response curves. *Proceedings of BNAM 2016*, Stockholm, Sweden, 2016.)

14.3.6   Impact sound insulation

The above-mentioned investigation also included impact sound. One question was about annoyance from the neighbour above due to walking, running, jumping, moving furniture, etc. Correlation was found with measured impact sound pressure level, either normalised or standardized, but only when the 50 Hz adaptation term was included. Impact sound omitting the frequencies below 100 Hz had no correlation with the subjective response. The result for normalised impact sound is shown in Figure 14.14. The range of variation of $L'_{n,w}+C_{I,50-2500}$ was 44 dB to 61 dB, and mean±standard deviation was 53.7 dB±3.0 dB. This implies that 95 % of the data can be assumed within a range of 12 dB. The results for the standardised impact sound $L'_{nT,w}+C_{I,50-2500}$ is not shown here; the correlation with subjective data was equally good, but the mean ± standard deviation of the data was 50.1 dB ±2.0 dB, i.e. a very narrow range of variation; the 95 % range being only 8 dB.

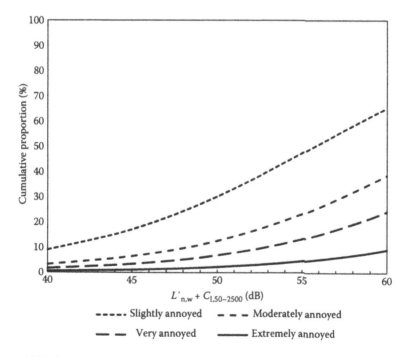

Figure 14.14 Cumulative annoyance curves due to footfall noise from above as function of the normalized impact sound pressure level including the 50 Hz adaptation term. The curves are the result of logit analysis ($n = 702$). (Adapted from Rindel, J.H. et al., Aiming at satisfactory sound conditions in dwellings – The use of dose-response curves. *Proceedings of BNAM 2016*, Stockholm, Sweden, 2016.)

14.3.7   Vibrations

A survey on vibrations in dwellings due to traffic (road and rail) was carried out in Norway (Klæboe et al., 2003; Turunen-Rise et al., 2003). About 1500 people were interviewed and one question was: *"Can you in your dwelling notice shaking or vibration caused by <source>? If yes, is the shaking/these vibrations highly annoying, somewhat annoying, a little annoying or not annoying for you?"*

Measurements of vibrations were performed in the dwellings and outside on the ground. A semi-empirical model was used to determine the statistical maximum RMS vibration velocity using the W_m weighting filter (ISO 2631-2) (see Figure 2.18). The statistical maximum is defined as the 95 percentile, i.e. the value not exceeded in 95 % of the time. The results of the logit analysis using the velocity level 20 lg (v/v_{ref}) dB are shown in Figure 14.15.

The curves show that about 7 % will be highly annoyed at a weighted vibration velocity of 0.3 mm/s. At a velocity of 0.5 mm/s, 10 % will be highly annoyed and about 40 % will annoyed slightly + moderately + highly.

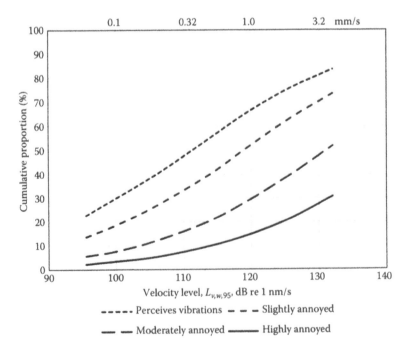

Figure 14.15 Cumulative annoyance curves for vibrations in the dwelling due to road and rail traffic. The curves are the result of logit analysis ($n = 1427$). (Adapted from Turunen-Rise, I.H. et al., *Appl. Acoust.*, 64, 71–87, 2003.)

14.4 LOW-FREQUENCY NOISE FROM NEIGHBORS

Problems with insufficient low-frequency sound insulation and music played on powerful hi-fi equipment were described by Craik and Stirling (1986). The background was a very strong increase in the number of complaints about music from neighbours and, in particular, from discotheques. About 25 % of all complaints about noise were concerning amplified music. Measurements and investigations in 40 selected dwellings showed that the degree of annoyance was related to the signal-to-noise level, and even when the music was below the background noise, the dwellers could hear the music clearly. Still, at a level 5 dB below the background noise, 20 % were bothered by the music.

An investigation on sound in dwellings was performed by Adnadevic et al. (2011). Measurements showed that the highest sound levels were produced by music, singing, playing a musical instrument, or music reproduction using modern audio equipment. The $L_{p,A,T}$ of these sources was between 80 dB and 90 dB. The measured maximum values were around 88 dB to 92 dB for listening to music, and around 91 dB to 98 dB for watching opera on TV. Loud music listening was found to be the only source with significant energy below 100 Hz and the measured spectra in the source rooms were close to spectrum 2 of ISO 717-1 (the traffic noise spectrum).

The importance of music as a sound source in residential buildings was studied by Lang and Muellner (2013). They made a questionnaire investigation in Austria, Germany and Switzerland, which showed that music, especially pop and rock, is an important sound source in residential buildings. It was found that high volume of the music is highly preferred, and 85 % uses bass loudspeaker or subwoofer to enforce low frequencies. The preferred type of music has a nearly flat frequency response down to 50 Hz and, thus, the authors find it reasonable that the frequency range to be observed for sound insulation has to include the 50 Hz to 80 Hz one-third octave bands.

A laboratory experiment has been conducted with the purpose of systematically investigating the influence of the low-frequency content in noise from neighbours (Mortensen, 1999). The experiment was carried out in the listening room at the Technical University of Denmark. Three sound signals were used: music from a neighbouring room, footfall noise from a male walker in the room above and footfall noise from two children running and playing catch in the room above.

The frequency spectrum of each of the three sound signals was modified in order to simulate five different types of building constructions. For airborne sound, the slope of the spectrum between 50 Hz and 160 Hz was varied in order to simulate the sound transmission through different constructions ranging from light to heavy. Above 160 Hz, the spectrum was kept constant. For impact noise, a similar procedure was used; only the slope of the spectrum between 50 Hz and 125 Hz was varied. The sound examples were presented to 25 test persons through loudspeakers. Some of the main results are shown in Tables 14.9 and 14.10.

Table 14.9 Results of laboratory experiment with airborne sound and simulated constructions with different sound insulation below 160 Hz

Type of construction	R'_w (dB)	$R'_w + C_{50-3150}$ (dB)	$L_{p,A,T}$ (dB)	Music, percent annoyed
Light	56	49	48	98
Light-medium	56	53	43	90
Medium	57	55	37	80
Medium-heavy	57	56	36	83
Heavy	57	56	35	83

Source: Mortensen, F.R., *Subjective Evaluation of Noise from Neighbours with Focus on Low Frequencies*. Main report. Publication no. 53, Department of Acoustic Technology, Technical University of Denmark, 1999.

Table 14.10 Results of laboratory experiment with impact sound and simulated constructions with different sound insulation below 125 Hz

Type of construction	$L'_{n,w}$ (dB)	$L'_{n,w} + C_{I,50-2500}$ (dB)	Walker, $L_{p,A,T}$ (dB)	Children, $L_{p,A,T}$ (dB)	Walker, percent annoyed	Children, percent annoyed
Light	55	62	39	41	71	81
Light-medium	55	58	32	35	51	78
Medium	55	56	27	30	36	47
Medium-heavy	55	55	25	28	28	51
Heavy	54	54	24	25	20	47

Source: Mortensen, F.R., *Subjective Evaluation of Noise from Neighbours with Focus on Low Frequencies*. Main report. Publication no. 53, Department of Acoustic Technology, Technical University of Denmark, 1999.

From the results of this investigation, it was concluded that the spectrum adaptation terms in ISO 717-1 and ISO 717-2 with extended frequency range down to 50 Hz give a better correlation between subjective and objective evaluation of sound insulation.

While the importance of the extended frequency range down to 50 Hz is clear in relation to impact sound, and other researchers have come to a similar conclusion (Warnock, 1998; Späh et al., 2014), the case of airborne sound insulation is not so clear. Only when the sound is music with strong contents of low frequencies, the 50 Hz adaptation term gives an improved correlation with subjective impressions; but for other kinds of noise from neighbours like speech, TV, etc., there is no need to include the frequencies below 100 Hz for airborne sound insulation (Kylliäinen et al., 2016).

The spectrum and the frequency range are very different in music and speech. A comprehensive investigation based on listening tests using speech and music in combination with 20 partition walls with quite different acoustic characteristic was performed (Park et al., 2008; Park and

Bradley, 2009a,b). Three music samples were selected for the tests, all of them being potentially annoying but of different musical styles. It was found that the optimum range of included frequencies in the single-number quantity is different for music sounds and speech sounds. Some measures were strongly related to ratings of speech sounds and others to ratings of music sounds, but none were highly successful for both types of sounds. The best correlation with annoyance from music sounds was a parameter defined as the A-weighted sound pressure level difference with pink noise, 63 Hz to 6300 Hz ($R^2=0.978$). For speech sounds, the best correlation with annoyance was a parameter defined as the A-weighted sound pressure level difference with pink noise, 200 Hz to 6300 Hz ($R^2=0.977$). It was also found that the signal-to-noise ratio was well related to annoyance ratings and audibility ratings of transmitted music sounds. Almost all subjects found that sounds were just audible at signal-to-noise levels as low as $-10\,dB$. The thump of the rhythmic beating, which is typical in loud music, may explain this finding.

From the point of view that sound insulation requirements are meant to ensure a reasonable degree of protection against all typical sounds from neighbours, the one-third octave bands 50 Hz to 80 Hz are important and should be included in the evaluation of airborne sound insulation in dwellings (Rindel, 2017).

14.5 DESIGN CRITERIA FOR SATISFACTORY SOUND INSULATION

In many countries, the building regulations for sound insulation were established a long time ago (1940–1960) and mainly from a consideration of the traditional building practice at that time, rather than from an analysis of the needs for sound insulation. Although the purpose of the building regulations was to avoid new buildings with clearly unacceptable acoustical conditions, the result was very often that these minimum requirements were considered a satisfactory goal for the acoustic design. After decades with a very stable situation in building acoustics, changes in this field in many countries started in the 1990s, and the development seems to go in a direction where acoustic conditions are taken more seriously as a relevant and important design criterion for modern buildings. So, the hope for the future is that the building technology will change in such a way that houses can offer acoustical comfort to the people they are built for.

In the national building codes, it is often stated as a basic requirement for new dwellings that the acoustical conditions shall be satisfactory. So, the big question is how to define 'satisfactory'. In recent years, the method of socio-acoustic investigations has matured, the methodology has been standardised (ISO/TS 15666, 2003; ISO 1996-1, 2016), and results from several new investigations have been published. We know that the degree of annoyance or satisfaction is described by a sigmoid function, which means that we can never expect 100 % satisfaction. So, it is merely a political decision as to what is satisfactory. Is it enough if 50 % are not annoyed, or should we aim at 70 % or 80 % being satisfied (not annoyed)? Or should we rather look at the percentage of those who are highly annoyed/extremely annoyed? Since the survey paper by Schultz (1978), it has been common practice to use the fraction "highly annoyed" for predicting the community annoyance to noise, and this is also used in ISO 1996-1 (2016). However, the dose-response curves, of which we have seen several examples in this chapter, have a very flat slope with values close to 0 % if we look at the lowest curve that gives the percentage indicating extremely annoyed persons. The next curve for very+extremely annoyed also has a flat slope. This means that it is very uncertain which dB-value actually corresponds to a certain percentage highly annoyed, e.g. 3 %, 5 % or 7 %. The result will be much more precise if we look at the upper curve, which normally has the steepest slope in the range of interest; this is the curve representing all degrees of annoyance (slightly, moderately, very, and extremely). In addition, the proportion above this curve represents the percentage who is satisfied (not annoyed). This is suggested as a better and more precise way to read the dose-response curves.

Table 14.11 presents relations between the expected percentage satisfied/annoyed and acoustic criteria for sound insulation and noise. In order to create this table, the dose-response results from various investigations have been slightly modified in such a way that the slope of the cumulative logistic distribution is 4 % points per dB in the middle range, as suggested by Rindel (1999) and Rindel et al. (2016).

Table 14.11 Acoustic design criteria for dwellings and the corresponding expected percentage of people finding conditions satisfactory (not annoyed) or not satisfactory (annoyed to some degree)

Satisfied (not annoyed), %		90	80	70	60	50	40	30	20
Annoyed (moderately+ very+ extremely), %		3	5–8	9–12	15–18	20–25	27–35	37–45	50–55
Airborne sound insulation	R'_w, dB	64	59	56	53.5	51	48.5	46	43
	$R'_{w,50}$, dB	62	57	54	51.5	49	46.5	44	41
	$D_{nT,w}$, dB	66	61	58	55.5	53	50.5	48	45
	$D_{nT,w,50}$, dB	64	59	56	53.5	51	48.5	46	43
Impact sound level	$L'_{n,w,50}$, dB	42.5	47.5	50.5	53	55.5	58	60.5	63.5
	$L'_{nT,w,50}$, dB	39	44	47	49.5	52	54.5	57	60
Noise indoor, road traffic	$L_{p,A,24h}$, dB	16	21	24	26.5	29	31.5	34	37
Noise outdoor, road traffic	$L_{p,A,24h}$, dB	47	52	55	57.5	60	62.5	65	68
	L_{den}, dB	50	55	58	60.5	63	65.5	68	71

The following abbreviations are used in Table 14.11: $R'_{w,50} = R'_w + C_{50-3150}$, $D_{nT,w,50} = D_{nT,w} + C_{50-3150}$, $L'_{n,w,50} = L'_{n,w} + C_{I,50-2500}$.

The current minimum requirement in the Nordic countries (Table 5.2) is close to the column with 60 % satisfied. This level of acoustic quality may be labelled 'acceptable', while we should aim somewhat higher for 'satisfactory' conditions. The values corresponding to 80 % satisfied (and 5 % to 8 % annoyed) may be candidates for 'satisfactory' acoustical conditions and thus suggested as a target for future development of better houses.

REFERENCES

M. Adnadevic, M. Mijic, D. Sumarac-Pavlovic, D. Masovic (2011). Noise in dwellings generated in normal home activities – general approach. *Proceedings of Forum Acusticum 2011*, 27 June – 1 July 2011, Aalborg, Denmark, 1335–1340.

D.G. Altman (1991). *Practical Statistics for Medical Research*. Chapman and Hall, London.

A.H. Amundsen, R. Klæboe, G.M. Aasvang (2011). The Norwegian Façade Insulation Study: The efficacy of façade insulation in reducing noise annoyance due to road traffic. *Journal of the Acoustical Society of America* 129, 1381–1389.

J.S. Bradley (1983). Subjective rating of party walls. *Canadian Acoustics* 11(4), 37–45.

G. Clausen, L. Carrick, P.O. Fanger, S.W. Kim, T. Poulsen, J.H. Rindel (1993). A comparative study of discomfort caused by indoor air pollution, thermal load and noise. *Indoor Air*, 3, 255–262.

R.J.M. Craik and J.R. Stirling (1986). Amplified music as a noise nuisance. *Applied Acoustics* 19, 335–356.

D.E. Cummins (1978). Classes of acoustical comfort in housing. *Proceedings of Inter-Noise 78*, 8–10 May 1978, San Francisco, USA, 631–636.

C.J. Grimwood (1993). *Effects of environmental noise on people at home*. BRE information paper IP 22/93.

C.J. Grimwood (1997). Complaints about poor sound insulation between dwellings in England and Wales. *Applied Acoustics* 52, 211–223.

J.I. Halonen, A.L. Hansell, J. Gulliver, D. Morley, M. Blandiardo, D. Fecht, M.B. Toledano, S.D. Beevers, H.R. Anderson, F.J. Kelly, C. Tonne (2015). Road traffic noise is associated with increased cardiovascular morbidity and mortality and all-cause mortality in London. *European Heart Journal* 36(39), 2653–2661.

D. Hjortebjerg, A.M. Nybo Andersen, J. Schultz Christensen, M. Ketzel, O. Raaschou-Nielsen, J. Sunyer, J. Julvez, J. Forns, M. Sørensen (2016). Exposure to road traffic noise and behavioral problems in 7-year-old children: A cohort study. *Environmental Health Perspectives* 124(2), 228–234.

ISO/TS 15666 (2003). *Acoustics – Assessment of noise annoyance by means of social and socio-acoustic surveys*. Technical specification. Organization for Standardization, Geneva, Switzerland.

ISO 1996-1 (2016). *Acoustics – Description, measurement and assessment of environmental noise – Part 1: Basic quantities and assessment procedures*. International Organization for Standardization, Geneva, Switzerland.

ISO 717 (1996). *Acoustics – Rating of sound insulation in buildings and of building elements. Part 1: Airborne sound insulation. Part 2: Impact sound insulation.* International Organization for Standardization, Geneva, Switzerland.

R. Klæboe, I.H. Turunen-Rise, L. Hårvik, C. Madshus (2003). Vibration in dwellings from road and rail traffic – Part II: exposure-effect relationships based on ordinal logit and logistic regression models. *Applied Acoustics* 64, 89–109.

J. Kragh (1977). *Analysis of the Correlation between Reactions to Road Traffic Noise and Physical Indices of the Noise.* Report No. 5 (In Danish). The Acoustical Laboratory, The Danish Academy of Technical Sciences, Lyngby, Denmark.

M. Kylliäinen, J. Takala, D. Oliva, V. Hongisto (2016). Justification of standardized level differences in rating of airborne sound insulation between dwellings. *Applied Acoustics* 102, 12–18.

J. Lang and H. Muellner (2013). The importance of music as sound source in residential buildings. *Proceedings of Inter Noise 2013*, 15–18 September 2013, Innsbruck, Austria.

A. Løvstad, J.H. Rindel, C.O. Høsøien, I. Milford, R. Klæboe (2016). Sound quality in dwellings in Norway – A socio-acoustic investigation. *Proceedings of BNAM 2016*, 20–22 June 2016, Stockholm, Sweden.

C. Maschke and H. Niemann (2007). Health effects of annoyance induced by neighbour noise. *Noise Control Engineering Journal* 55(3), 348–356.

H.M.E. Miedema (2004). Relationship between exposure to multiple noise sources and noise annoyance. *Journal of the Acoustical Society of America* 116, 949–957.

F.R. Mortensen (1999). *Subjective Evaluation of Noise from Neighbours with Focus on Low Frequencies.* Main report. Publication no. 53, Department of Acoustic Technology, Technical University of Denmark.

H. Niemann, X. Bonnefoy, M. Braubach, K. Hecht, C. Maschke, C. Rodrigues, N. Röbbel (2006). Noise-induced annoyance and morbidity results from the pan-European LARES study. *Noise & Health* 8(31), 63–79.

H.K. Park, J.S. Bradley, B.N. Gover (2008). Evaluating airborne sound insulation in terms of speech intelligibility. *Journal of the Acoustical Society of America* 123, 1458–1471.

H.K. Park and J.S. Bradley (2009a). Evaluating standard airborne sound insulation measures in terms of annoyance, loudness, and audibility ratings. *Journal of the Acoustical Society of America* 126, 208–219.

H.K. Park and J.S. Bradley (2009b). Evaluating signal-to-noise ratios, loudness, and related measures as indicators of airborne sound insulation. *Journal of the Acoustical Society of America* 126, 1219–1230.

J.H. Rindel (1999). Acoustic quality and sound insulation between dwellings. *Journal of Building Acoustics* 5, 291–301.

J.H. Rindel (2017). A comment on the importance of low frequency airborne sound insulation between dwellings. *Acta Acustica/Acustica* 103, 164–168.

J.H. Rindel, A. Løvstad, R. Klæboe (2016). Aiming at satisfactory sound conditions in dwellings – The use of dose-response curves. *Proceedings of BNAM 2016*, Stockholm, Sweden.

J.H. Rindel, A. Løvstad, R. Klæboe (2017). Dose-response curves for satisfactory sound insulation between dwellings. *Proceedings of ICBEN 2017*, 18–22 June 2017, Zürich, Switzerland.

J.H. Rindel, B. Rasmussen (1995). *Buildings for the future: The concept of acoustical comfort and how to achieve satisfactory acoustical conditions with new buildings*. (18 pages). COMETT-SAVOIR Course, CSTB, Grenoble.

P. Schomer, V. Mestre, S. Fidell, B. Berry, T. Gjestland, M. Vallet, T. Reid (2012). Pole of community tolerance level (CTL) in predicting the prevalence of the annoyance of road and rail noise. *Journal of the Acoustical Society of America* 131, 2772–2786.

T.J. Schultz (1978). Synthesis of social surveys on noise annoyance. *Journal of the Acoustical Society of America* 64, 377–405.

M. Späh, K. Hagberg, O. Bartlomé, L. Weber, P. Leistner, A. Liebl (2014). Subjective and objective evaluation of impact noise sources in wooden buildings. *Noise Notes* 13, 25–42.

I.H. Turunen-Rise, A. Brekke, L. Hårvik, C. Madshus, R. Klæboe (2003). Vibration in dwellings from road and rail traffic – Part I: A new Norwegian measurement standard and classification system. *Applied Acoustics* 64, 71–87.

A.C.C. Warnock (1998). Floor research at NRC Canada. *Proceedings of COST Action E5 Workshop, Acoustic performance of medium-rise timber buildings*, 3–4 December 1998, Dublin, (10 pages).

R. Weeber, H. Merkel, H. Rossbach-Lochmann, K. Gösele, E. Buchta (1986). *Schallschutz in Mehrfamilienhäusern aus der Sicht der Bewohner* (Sound insulation in multi-family houses from the point of view of the resident, in German). F 2049, IRB Verlag, Stuttgart, Germany.

WHO (2007). *Large Analysis and Review of European Housing and Health Status (LARES)*. Preliminary overview. WHO Regional Office for Europe, Copenhagen, Denmark.

S. Wibe (1997). *Efterfrågan på tyst boende* (The Demand for Silent Dwellings, in Swedish). Anslagsrapport A4:1997. Byggforskningsrådet, Stockholm, Sweden.

Chapter 15

Experimental buildings with high sound insulation

In the previous chapter, the question was which level of sound insulation should be required to obtain satisfactory acoustical conditions for dwellings in multi-unit houses. In this final chapter, we shall look at some examples of experimental buildings with particularly good sound insulation. The examples are restricted to multi-storey buildings with dwellings. Of course, the idea of making experimental buildings with very good sound insulation comes from the ambition that dwellings for the future should have satisfactory conditions, also for acoustics.

In some cases, the purpose has been to demonstrate in full scale that it is possible to give priority to sound insulation in the building design and to achieve practical experience with the suggested technical solution. It is remarkable that the very high acoustic quality achieved in these projects does not necessarily lead to higher construction costs, and one of the solutions was actually developed with the purpose of saving construction time.

15.1 'THE QUIET HOUSE' LUND, SWEDEN

15.1.1 General

The building was erected 1988–1989 in Kv. Bollen, Lund, Sweden. It has three stories and 29 apartments. Builder: JM Byggnads och Fastighets AB. Consultants: Sterner Akustik AB, Bentech Ingenörs AB. The project was supported by The Swedish Council for Building Research and The Development Fund of the Swedish Construction Industry.

Figure 15.1 shows a part of the façade of the experimental building.

15.1.2 Purpose of experiment

The project was initiated by Lars Holmgren, Cementa AB. The purpose was to find out what could be achieved acoustically with a traditional heavy-weight building construction. Well-known solutions and construction details were applied, but with an unusual focus on the acoustical needs.

Figure 15.1 External view of a part of façade of the experimental building. (Adapted from Byggforskningsrådet., *Det tysta huset. Bätre ljudkomfort i nyproducerade bostäder* (The quiet house. Improved acoustical comfort in new-built dwellings, in Swedish), Statens råd för byggforskning, T9:1991, Stockholm, Sweden, 1991.)

The project report from Byggforskningsrådet (1991) contains a catalog of technical details and acoustical solutions.

In the initial stage of the project, quite ambitious requirements for airborne and impact sound insulation were established. This was done by definition of special reference curves in the extended frequency range from 50 Hz to 5000 Hz at a level about 10 dB better than the usual minimum requirement (see Figure 15.2 and 15.3). The evaluation method of a measurement result is the same as for the ISO reference curve; i.e. the sum of unfavourable deviations must not exceed 32 dB.

At the end of the project, the extra costs of future 'quiet houses' built as this experimental building was estimated to be 2 % to 3 % compared to a traditional building.

15.1.3 Building concept

The main constructions are 240 mm concrete walls and 290 mm concrete slabs cast on site. The floors were either carpet or wood on cork. This is type A in Figure 11.1. Flanking transmission is possible in all directions, but of little importance due to the high weight of the constructions.

The project report mentions, as a drawback to the building concept, that the time needed for the concrete to harden is longer than usual because of the thickness of walls and slabs.

15.1.4 Acoustical results

An overview of the measurement results is presented in Table 15.1.

The following abbreviations are used in Table 15.1: $R'_{w,50}=R'_w+C_{50-5000}$, $L'_{n,w,50}=L'_{n,w}+C_{I,50-2500}$.

Examples of measured airborne sound insulation in the horizontal and vertical directions are shown in Figure 15.2. In both directions, the transmission loss is very high; at frequencies above 250 Hz, the transmission loss is better in the vertical direction than in the horizontal direction.

The measured impact sound insulation in the vertical direction is shown in Figure 15.3. The results are very good and the special project

Table 15.1 Single-number results of measured sound insulation in the 'Quiet House'

Frequency range (Hz)	Airborne, horizontal (dB)	Airborne, vertical (dB)	Impact, vertical (dB)
100–3150	$R'_w=62$	$R'_w=65$	$L'_{n,w}=43$
50–5000	$R'_{w,50}=61$	$R'_{w,50}=61$	$L'_{n,w,50}=46$

Note: Same as in Figures 15.2 and 15.3.

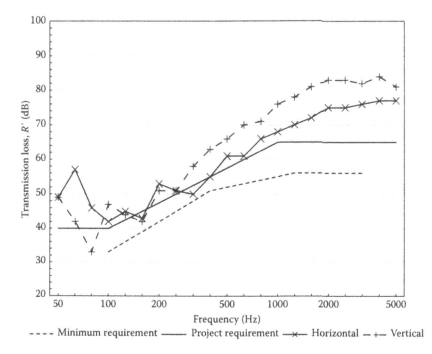

Figure 15.2 Measured airborne sound insulation in the horizontal and vertical directions. The full line is the project requirements in the extended frequency range. The dashed line is the normal minimum requirement at the time of construction.

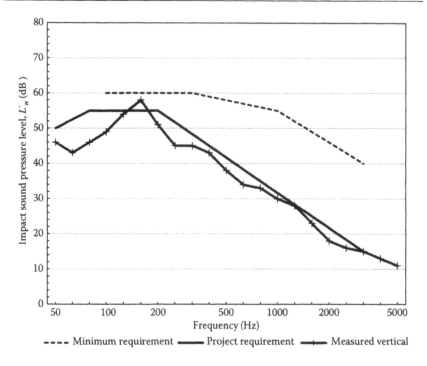

Figure 15.3 Measured impact sound insulation in the vertical direction. The full line is the project requirements in the extended frequency range. The dashed line is the normal minimum requirement at the time of construction.

requirements are fulfilled (recalling that the reference curve may be passed with up to 32 dB of unfavourable deviations).

15.2 'OPEN HOUSE URBAN BUILDING SYSTEM', DENMARK

15.2.1 General

The 'OPEN HOUSE Urban Building System' has been developed by Dominia in cooperation with the architect firm Landskronagruppen, Landskrona, Sweden.

The building concept has been used in several places in Denmark. The experimental buildings described here were erected 1993 and are located at Færgeparken III, Frederikssund, Denmark. There are three 3-storey buildings; in total, 36 dwellings. Architect: Mangor & Nagel, Frederikssund. Builder: Arbejdernes Kooperative Byggeforening (a cooperative building association). Consultant: Dominia, Copenhagen.

15.2.2 Purpose of experiment

The purpose was to gather experience with the newly developed building system, which is a light-weight system with short building time. Improved sound insulation was not a purpose of the experiment.

15.2.3 Building concept

The 'OPEN HOUSE Urban Building System' is a building system based on factory-produced room modules and a site-produced concrete skeleton. Each room module consists of a steel framework enclosing one or more rooms. Walls, floors, and ceilings consist of gypsum boards mounted on steel ribs. Compared to conventional concrete housing, the weight is approximately 35 %, the building time at site is 25 % and the building time, including production of modules, is 70 %. Different stages in the building process are shown in Figure 15.4.

The construction principle is a column-beam skeleton with light-weight walls and floors, type D in Figure 11.1. The concrete columns are cast on site, while the concrete beams are prefabricated (Figures 15.5 and 15.6).

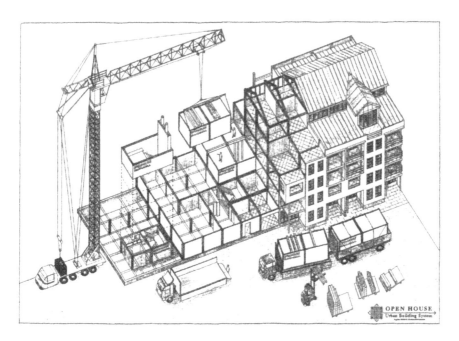

Figure 15.4 Schematic view of different stages in the building process. (From Dominia. With permission from Dominia, Copenhagen, Denmark.)

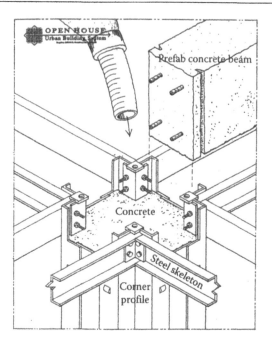

Figure 15.5 Detail of a junction. The steel profiles at the edges of the prefabricated room modules create a form for in situ casting of the concrete columns. The prefabricated concrete beams are fixed in the steel profiles. (With permission from Dominia, Copenhagen, Denmark.)

15.2.4 Acoustical results

An overview of the measurement results is presented in Table 15.2. The measured sound insulation results are 8 dB to 10 dB better than the minimum requirements in Denmark at the time of building. Measurements did not include frequencies below 100 Hz.

The measured airborne sound insulation is shown in Figure 15.7. Above 200 Hz, the transmission loss in the horizontal direction is higher than that in the vertical direction, and the frequency course is typical for a double wall, decreasing below 200 Hz towards a mass-spring-mass resonance frequency that may be just below 100 Hz. Although no measurement results are available below 100 Hz, it seems that the airborne sound insulation at low frequencies may be an issue, both vertically and horizontally.

Table 15.2 Single-number results of measured sound insulation in the 'OPEN HOUSE Urban Building System'

Airborne, horizontal (dB)	Airborne, vertical (dB)	Impact, vertical (dB)
$R'_w = 62$	$R'_w = 61$	$L'_{n,w} = 49$

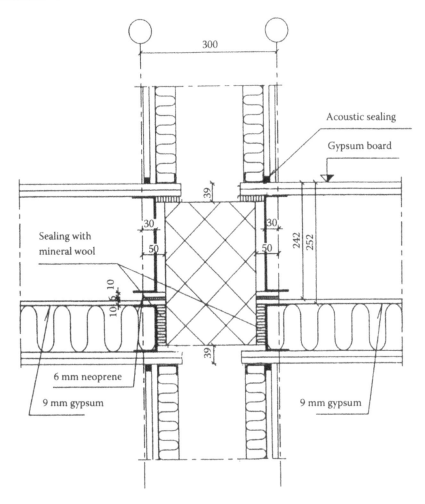

Figure 15.6 Detail of a junction. (With permission from Dominia, Copenhagen, Denmark.)

The measured impact sound insulation is shown in Figure 15.8. The frequency course suggests that there may be an issue with the impact sound below 100 Hz. Otherwise, the results are quite satisfactory.

15.3 'ECO-HOUSE', DENMARK

15.3.1 General

The ECO-House building system was developed in connection with a competition from 1987 on the question: 'How can Danish building industry make the next great jump forward?' Due to the falling price on steel, it was considered advantageous to develop an alternative to the

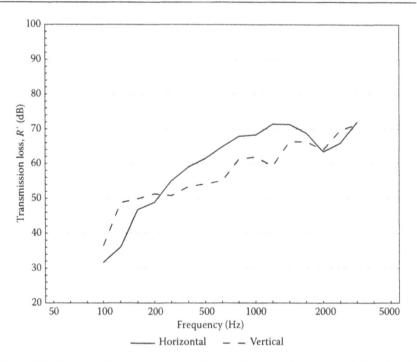

Figure 15.7 Measured airborne sound insulation in horizontal and vertical directions in the 'OPEN HOUSE Urban Building System'.

concrete-based prefabricated building methods that had been dominating since the 1960s. The companies behind the new building system were architect: Skaarup & Jespersen A/S, builder: KKS Enterprise A/S, and consultant: Birch & Krogboe A/S (KKS Enterprise, 1992; Gath and Karstoft, 1992). The first versions of the building system were called 'Scanhouse'.

The building concept has been used in several places in Denmark. The experimental buildings described here were erected in 1992 and are located in Egeskoven, Glostrup, Denmark. These are three-storey buildings, 49 apartments in total.

15.3.2 Purpose of experiment

The basic idea was to maximize prefabrication in order to minimize the building time without compromising economy or quality. Acoustically, there was no ambition to exceed the Danish minimum requirements for sound insulation at the time of construction.

15.3.3 Building concept

The supporting structure is a framework of steel profiles. This is type D in Figure 11.1. Floor constructions and walls are prefabricated elements of gypsum board and mineral wool in thin-plate steel profiles. On the building site,

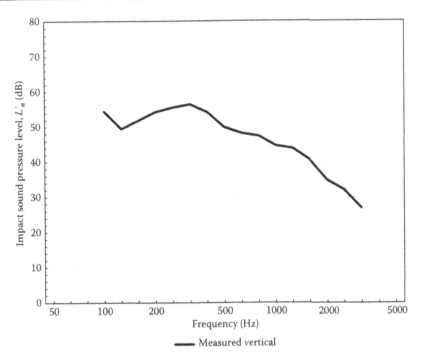

── Measured vertical

Figure 15.8 Measured impact sound insulation in the vertical direction in the 'ECO-HOUSE'.

the steel framework is erected first. Then, the floor and façade elements and the roof are mounted. Building materials needed for the internal parts are brought into the building before the raw house is closed. The rest of the building process can take place in a dry environment protected from rain and wind. The total building time is 70% of that for more traditional building systems.

A construction detail is seen in Figure 15.9. The steel hat profile is a structural beam that also provides support for the floor elements.

15.3.4 Acoustical results

An overview of the measurement results is presented in Table 15.3. The average of measured sound insulation results is 5 dB to 10 dB better than the minimum requirements in Denmark at the time of building. The measured

Table 15.3 Single-number results of measured sound insulation in the 'ECO-HOUSE'

Airborne, horizontal (dB)	Airborne, vertical (dB)	Impact, vertical (dB)
$R'_w = 58–63$	$R'_w = 56–61$	$L'_{n,w} = 46–50$

Source: Gath J., Karstoft, N.O., *ECO-HOUSE. Et nyt dansk byggesystem.* (ECO-HOUSE. A new Danish building system, in Danish). Dansk ståldag 1992, Birch & Krogboe A/S. (9 pages), 1992.

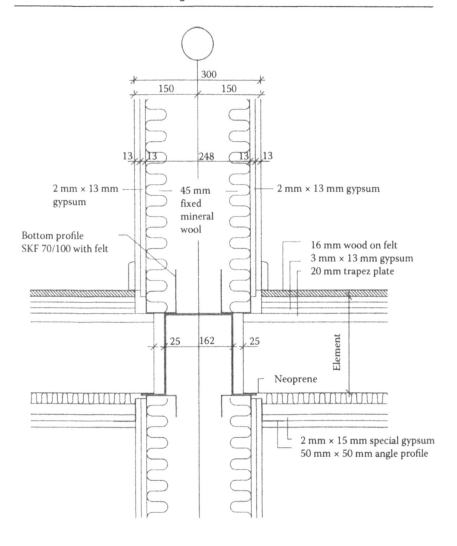

Figure 15.9 Detail of the junction between floor and partition wall in the 'ECO-HOUSE'.

impact sound insulation in the vertical direction is shown in Figure 15.10. There is no available information about the airborne sound insulation in the low frequency range from 50 Hz to 80 Hz.

15.4 'BRF KAJPLATSEN', STOCKHOLM, SWEDEN

15.4.1 General

The building was erected in 1994 in BRF Kajplatsen, Kv Trålen, Stockholm, Sweden. It has seven stories and 26 apartments. The experimental building

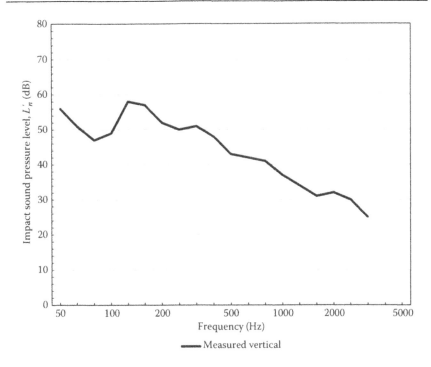

Figure 15.10 Measured impact sound insulation in vertical direction in the 'ECO-HOUSE'. The single-number values for this measurement are $L'_{n,w} = 46$ dB and $L'_{n,w} + C_{I,50-2500} = 49$ dB.

was part of a larger building project with a total of 176 apartments. Architect: Åsberg & Buchmann Arkitekter AB. Builder: JM Byggnads och Fastighets AB. Consultants: IIC, NCAB, Scandiakonsult Inst. AB, Plan & Mark AB, J A Jonsson., Bröderna Skoogs Vent AB, Björn Lindström Elektriska. Figure 15.11 shows the façade of the experimental building and one of the other buildings.

15.4.2 Purpose of experiment

Ljunggren (1993, 1995) has shown theoretically that it should be possible to increase the sound insulation in the vertical direction significantly by using a column-plate structure with a large slab supported by load-bearing columns instead of load-bearing walls (type B in Figure 11.1). The idea was that the flanking transmission through the slab could be turned into an advantage by distributing the energy of structural vibrations in a plate with a large area. This implies that the walls should be light-weight double constructions so that the junction attenuation would be negligible. The principle is explained by Statistical Energy Analysis (SEA) in Section 7.6.

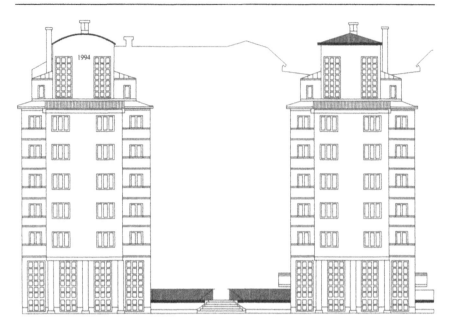

Figure 15.11 External view of two buildings, one of them being the experimental building on 'BRF Kajplatsen'.

The purpose of the experimental building was to prove that the *area effect* could work in practice. The building costs are estimated not to exceed those of a conventional building by more than 1 %.

The theory of this project was developed by Professor Sten Ljunggren in research projects supported by The Swedish Council for Building Research and The Development Fund of the Swedish Construction Industry.

15.4.3 Building concept

Three similar buildings were erected, two of them with traditional load-bearing walls of 180 mm concrete and 240 mm thick concrete slabs. The experimental building has 260 mm thick concrete slabs supported by 100 mm×100 mm steel columns in the façade and 400 mm×400 mm concrete columns hidden in the walls inside the apartments (Figure 15.12). The double walls used as partition walls between the apartments are shown in Figure 15.13. All buildings have concrete walls around the stairwell and elevator in the middle of the building. The experimental building is type B in Figure 11.1, while the other buildings are type A.

The acoustical principle is illustrated in Figure 15.14. The total area of the concrete slab is 300 m². The area of the floor in one on the bedrooms

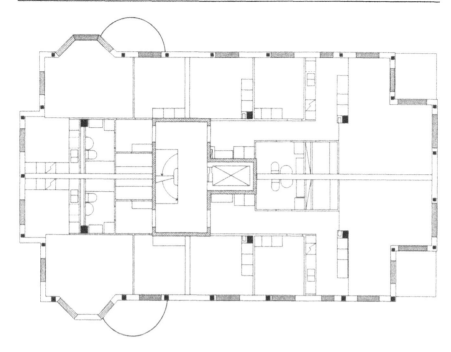

Figure 15.12 Plan of the experimental building on 'BRF Kajplatsen'. The small black squares are the columns, six from concrete inside the building and 25 from steel in the façade. (Reproduced from Ljunggren, S., *wksb, Zeitschrift für Wärmeschutz, Kälteschutz, Schallschutz, Brandschutz,* Heft 38, 1–4, 1996. With permission.)

is $12.4\,\text{m}^2$. Thus, if we refer to the theoretical result (Equation 7.42) the area effect in this case might be up to 14 dB increased transmission loss.

15.4.4 Acoustical results

An overview of the measurement results is presented in Table 15.4.

The measured airborne sound insulation between two rooms in the vertical direction is shown in Figure 15.15, both for the experimental building and the traditional building. The difference is around 8 dB to 10 dB in the entire frequency range. The thickness of the slab is not the same in the experimental building and the traditional building, but this cannot account for more than 1 dB difference. So, the efficiency of the principle in the experimental building is clearly demonstrated. In the horizontal direction, the airborne sound insulation is also very high. At the low frequencies 50 Hz to 80 Hz, the average transmission loss is around 35 dB, which is quite good for a light-weight double wall.

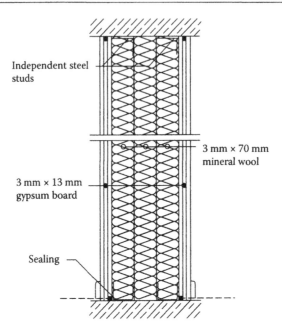

Independent steel studs

3 mm × 70 mm mineral wool

3 mm × 13 mm gypsum board

Sealing

Figure 15.13 Section of the double walls used between the apartments on 'BRF Kajplatsen'. (Reproduced from Ljunggren, S., *wksb, Zeitschrift für Wärmeschutz, Kälteschutz, Schallschutz, Brandschutz*, Heft 38, 1–4, 1996. With permission.)

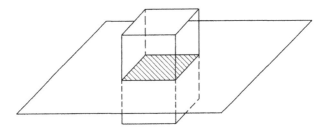

Figure 15.14 Principle of the area effect. The area of the slab is much larger than the area separating the source room and receiver room. (Reproduced from Ljunggren, S., *wksb, Zeitschrift für Wärmeschutz, Kälteschutz, Schallschutz, Brandschutz*, Heft 38, 1–4, 1996. With permission.)

Table 15.4 Single-number results of measured sound insulation on 'BRF Kajplatsen'

	Airborne, horizontal (dB)	Airborne, vertical (dB)	Impact, vertical (dB)
Experimental	$R'_w = 61$	$R'_w = 68$	$L'_{n,w} = 46$
Traditional	–	$R'_w = 59$	$L'_{n,w} = 51$

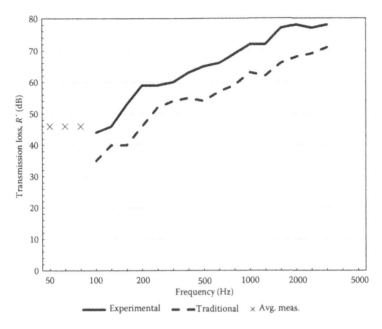

Figure 15.15 Measured airborne sound insulation in the vertical direction on 'BRF Kajplatsen'. The crosses are the average result at 50 Hz to 80 Hz in the experimental building.

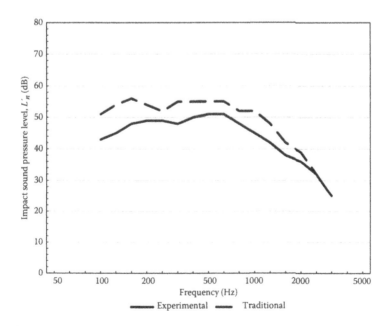

Figure 15.16 Measured impact sound insulation in the vertical direction on 'BRF Kajplatsen'.

The measured impact sound insulation in the vertical direction is shown in Figure 15.16. The advantage of the area effect is also obtained here, but it is less than for the airborne sound insulation, around 5 dB to 8 dB in most of the frequency range. At the high frequencies, the effect decreases, but fortunately it is the low frequencies that matters here. The explanation for this behaviour is that with impact excitation, the radiated sound from the slab consists of a near field radiation as well as resonant radiation (see Equation 10.14) The relative contribution of the near field increases with frequency and, thus, the area effect for impact sound decreases at high frequencies.

15.5 'YLÖJÄRVI' APARTMENTS, FINLAND

15.5.1 General

Three apartment houses were erected in 1996 on the occasion of a Finnish Housing Fair in Ylöjärvi. The houses have three stories and 20 apartments in total. The building concept was developed by Tampere University of Technology on the initiative of Skanska Oy and Finnforest Oy as part of the Nordic Wood Project (see next Section 15.6).

15.5.2 Purpose of experiment

The purpose was to find a method to improve the sound insulation in wooden apartment buildings by reduction of flanking transmission. The problem to be solved was the development of a resilient structural joint with which the non-bearing walls could be joined to the bearing beam-to-column frames.

15.5.3 Building concept

The building concept is a column-beam structure made from laminated veneer lumber (LVL) wood. This is type D in Figure 11.1. The prefabricated ribbed slab elements are made from 51 mm thick LVL with 299 mm high ribs having a spacing of 300 mm. The slabs are 1800 mm wide and can have a possible span of 7 m. Details of the floor and the junctions are shown in Figure 15.17. More information is found in the work of Keronen and Kylliäinen (1997).

15.5.4 Acoustical results

The main problem in the research and development project was finding a solution with sufficient impact sound insulation at low frequencies, i.e. below 100 Hz. Impact sound pressure levels were measured from 50 Hz to 3150 Hz, and in addition to the ISO 717-2 reference curve, Bodlund's

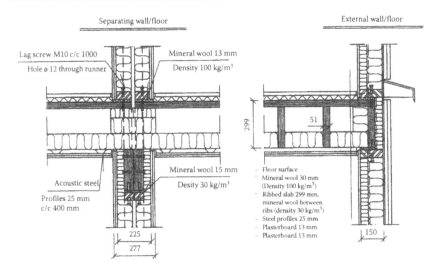

Figure 15.17 Construction details of the floor and the junctions to wall and façade. The building in Ylöjärvi has two kinds of floor surface: vinyl on 3 mm×13 mm plasterboard (30.8 kg/m²) or birch boarding on plastic foam layer on 2 mm×13 mm plasterboard (18.8 kg/m²). (Reproduced from Keronen, A., Kylliäinen, M., *Sound Insulation Structures of Beam-to-Column Framed Wooden Apartment Buildings*. Publication 77 Structural Engineering, Tampere University of Technology, Finland, 1997. With permission.)

Table 15.5 Single-number results of measured sound insulation in 'Ylöjärvi' apartments

	Airborne, horizontal (dB)	Airborne, vertical (dB)	Impact, vertical (dB)
Vinyl/birch floor	$R'_w = 64$	$R'_w = 67/62$	$L'_{n,w} = 44/39$
	$R'_{w,50} = 59$	$R'_{w,50} = 63/58$	$L'_{n,w,50} = 47/47$

Source: Keronen, A., Kylliäinen, M., *Sound Insulation Structures of Beam-to-Column Framed Wooden Apartment Buildings*. Publication 77 Structural Engineering, Tampere University of Technology, Finland, 1997.

Note: Results in the vertical direction both for vinyl floor and birch floor are indicated by a slash: */*.

reference curve was applied (see Section 5.8.3). An overview of the measurement results is presented in Table 15.5. Examples of the detailed measurement results are presented in Figures 15.18 and 15.19. The results are very satisfactory.

The following abbreviations are used in Table 15.5: $R'_{w,50} = R'_w + C_{50-3150}$, $L'_{n,w,50} = L'_{n,w} + C_{I,50-2500}$.

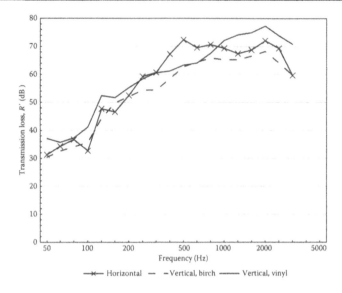

Figure 15.18 Measured airborne sound insulation in the horizontal direction and in the vertical direction both with vinyl and wooden flooring in 'Ylöjärvi' apartments. (Adapted from Keronen, A., Kylliäinen, M., *Sound Insulation Structures of Beam-to-Column Framed Wooden Apartment Buildings*. Publication 77 Structural Engineering, Tampere University of Technology, Finland, 1997.)

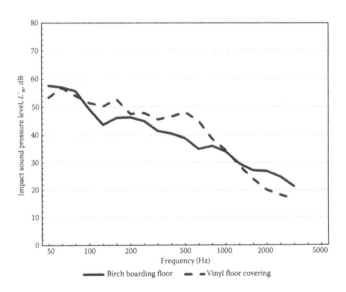

Figure 15.19 Measured impact sound insulation in the vertical direction with two different floorings, vinyl or wood in 'Ylöjärvi' apartments. (Adapted from Keronen, A., Kylliäinen, M., *Sound Insulation Structures of Beam-to-Column Framed Wooden Apartment Buildings*. Publication 77 Structural Engineering, Tampere University of Technology, Finland, 1997.)

15.6 NORDIC WOOD PROJECT

15.6.1 General

In the years 1993–1997, the so-called Nordic Wood project has lead to a series of experimental wooden buildings being developed and built in the Nordic countries (Nordic Wood, 1997; Hveem et al., 2000).

15.6.2 Purpose of experiment

The emphasis was on wood as an attractive building material. However, the two obvious challenges in wooden multi-storey houses were fire protection and sound insulation. Thus, the purpose was to initiate research and give support to experimental buildings that could provide possible solutions to these problems.

15.6.3 Acoustical results

The greatest acoustical challenge in multi-storey wooden houses is the impact sound at low frequencies. An overview of measurement results of impact and airborne sound insulation from some of the experimental buildings is found in Table 15.6. All impact sound insulation measurements include frequencies down to 50 Hz, and in addition to the ISO 717-2 reference curve, Bodlund's reference curve was applied (see the explanation in Section 5.8.3).

Table 15.6 Single-number results of measured sound insulation of floor constructions in the Nordic Wood project

Place	Thickness (mm)	$L'_{n,w}$ (dB)	$L'_{n,w,50}$ (dB)	L'_B (dB)	R'_w (dB)	$R'_{w,50}$ (dB)
Hørsholm, DK	375	44–47	52–54	58–62	59–63	–
Ylöjärvi, FI	487	36–44	41–47	49–56	62–67	–
Vik/Vikki, FI	400	48–53	54–58	62–67	58–62	–
Uleåborg/Oulo, FI (RL-slab)	407	44–49	49–54	55–62	61–65	–
Uleåborg/Oulo, FI (I-beam)	505	51–54	54–57	62–66	61–65	–
Solbakken, N	555	46–48	58–60	65–66	62–65	58–59
Wälludden, S	439	50–52	52–54	60–62	56–58	55–56
Orgelbäken, S	530	48–51	50–52	57–59	60–63	56–59

Source: Hveem, S. et al., *Trehus i flere etasjer. Lydteknisk prosjektering.* (Wooden multi-story houses. Acoustical design, in Norwegian). Anvisning 37, Norges byggforskningsinstitutt, Oslo, 2000.

The following abbreviations are used in Table 15.6: $L'_{n,w,50}=L'_{n,w}+C_{I,50-2500}$, L'_B is impact sound evaluated with Bodlunds reference curve, $R'_{w,50}=R'_w+C_{50-5000}$.

The experience from the Nordic Wood project is that good sound insulation is possible for airborne sound and impact sound above 100 Hz, but impact sound insulation below 100 Hz is an issue. The best result in this aspect is the solution in Ylöjärvi, which is described in the previous Section 15.5.

REFERENCES

Byggforskningsrådet (1991). *Det tysta huset. Bätre ljudkomfort i nyproducerade bostäder.* (The quiet house. Improved acoustical comfort in new-built dwellings, in Swedish). Statens råd för byggforskning, T9:1991, Stockholm, Sweden.

J. Gath and N.O. Karstoft (1992). *ECO-HOUSE. Et nyt dansk byggesystem.* (ECO-HOUSE. A new Danish building system, in Danish). Dansk ståldag 1992, Birch & Krogboe A/S. (9 pages).

S. Hveem, P. Hammer, A. Homb, A. Keronen, J.H. Rindel (2000). *Trehus i flere etasjer. Lydteknisk prosjektering.* (Wooden multi-story houses. Acoustical design, in Noregian). Anvisning 37, Norges byggforskningsinstitutt, Oslo 2000.

A. Keronen and M. Kylliäinen (1997). *Sound Insulation Structures of Beam-to-Column Framed Wooden Apartment Buildings.* Publication 77 Structural Engineering, Tampere University of Technology, Finland.

KKS Enterprise (1992). *Scanhouse – et nyt dansk byggesystem.* (Scanhouse - a new Danish building system, in Danish). Eight pages leaflet, KKS Enterprise A/S, Hvidovre, Denmark.

S. Ljunggren (1993). Sound insulation in buildings of concrete with large span floors. *Proceedings of Noise 1993*, 31 May – 3 June 1993, St. Petersburg, Russia, 83–88.

S. Ljunggren (1995). A new quiet house in Stockholm. *Acta Acustica* 3, 283–286.

S. Ljunggren (1996). Eine neue kosteneffektive Lösung zur Verbesserung der Schalldämmung von Wohngebäude. (A new cost-efficient solution to improving the sound insulation of dwellings, in German). *wksb, Zeitschrift für Wärmeschutz, Kälteschutz, Schallschutz, Brandschutz*, Heft 38, 1–4.

Nordic Wood (1997). *Flervånings trähus.* (Multi-unit wooden houses, in Swedish). T. Hansson (Editor). Nordic Timber Council and Träinformation, Stockholm.

Index

Printed and bound by CPI Group (UK) Ltd, Croydon, CR0 4YY

01/11/2024

01782614-0015